International Conference on Materials Sciences and Nanomaterials (ICMSN 2017)

IOP Conference Series: Materials Science and Engineering
Volume 230

Barcelona, Spain
14 – 16 July 2017

ISBN: 978-1-5108-4997-6
ISSN: 1757-8981

Printed from e-media with permission by:

Curran Associates, Inc.
57 Morehouse Lane
Red Hook, NY 12571

CURRAN ASSOCIATES INC.
proceedings.com

Some format issues inherent in the e-media version may also appear in this print version.

Printed by Curran Associates, Inc. (2017)

For permission requests, please contact the Institute of Physics
at the address below.

Institute of Physics
Dirac House, Temple Back
Bristol BS1 6BE UK

Phone: 44 1 17 929 7481
Fax: 44 1 17 920 0979

techtracking@iop.org

Additional copies of this publication are available from:

Curran Associates, Inc.
57 Morehouse Lane
Red Hook, NY 12571 USA
Phone: 845-758-0400
Fax: 845-758-2633
Email: curran@proceedings.com
Web: www.proceedings.com

Table of contents

Volume 230

International Conference on Materials Sciences and Nanomaterials
14 July to 16 August 2017, Barcelona, Spain

Accepted papers received: 14 August 2017
Published online: 3 October 2017

Preface

Metals and Alloys

Metal Processingand Surface Treatment

Materials Chemistry and Engineering Materials

Materials Physics

Building Materials and Structural Engineering

Mechanical Engineering and Manufacturing Process

IOP Conf. Series: Materials Science and Engineering **230** (2017) 011001 doi:10.1088/1757-899X/230/1/011001

PREFACE

It is our great pleasure to welcome you to 2017 International Conference on Materials Sciences and Nanomaterials (ICMSN 2017) which has been held in July 14-16, 2017 Barcelona, Spain. ICMSN 2017 is dedicated to issues related to Materials Sciences and Nanomaterials.

The major goal and feature of the conference is to bring academic scientists, engineers, industry researchers together to exchange and share their experiences and research results, and discuss the practical challenges encountered and the solutions adopted. Professors from United Kingdom, Romania, France are invited to deliver keynote speeches regarding latest information in their respective expertise areas. It will be a golden opportunity for the students, researchers and engineers to interact with the experts and specialists to get their advice or consultation on technical matters, sales and marketing strategies.

This proceedings present a selection from papers submitted to the conference from universities, research institutes and industries. All of the papers were subjected to peer-review by conference committee members and international reviewers. The papers selected depended on their quality and their relevancy to the conference. The volume tends to present to the readers the recent advances in the field of Nanomaterials, Electronic Materials, Material Chemistry, Materials Processing Engineering, Structural Material, etc.

We would like to thank all the authors who have contributed to this volume and also to the organizing committee, reviewers, speakers, chairpersons, sponsors and all the conference participants for their support to ICMSN2017.

Prof. Maria Tomoaia-Cotisel
Babes-Bolyai University of Cluj-Napoca, Romania
July 27, 2017

IOP Conf. Series: Materials Science and Engineering **230** (2017) 011002 doi:10.1088/1757-899X/230/1/011002

Organizing Committee

Conference Chairs

Prof. Alexandre Maitre, Universite de Limoges, France

Prof. Maria Tomoaia-Cotisel, Babes-Bolyai University of Cluj-Napoca, Romania

Technical Program Committee Chairs

Prof. Yuyuan Zhao, University of Liverpool, UK

Prof. Angelo Maligno, University of Derby, UK

Organizing Chair

Ms. Yang Dan, Hong Kong Chemical, Biological & Environmental Engineering Society, Hong Kong

Publication Chair

Ms. Ivy Hou, Chemistry and Materials Society (CMS-HKCBEES)

Technical Program Committee

Prof. Anca Silvestru, Babes-Bolyai University, Romania

Prof. Shafique Ahmed Arain, Shah Abdul Latif University, Khairpur, Pakistan

Assoc. Prof. Kar Seng (Vincent) Teng, Swansea University, UK

Assoc. Prof. Doina Lutic, Alexandru Ioan Cuza University of din Iasi, Romania

Assoc. Prof. Claudia Valentina Cimpoiu, Babes-Bolyai University, Romania

Assoc. Prof. Aurora MOCANU, Babes-Bolyai University, Romania

Assoc. Prof. Claudia Cimpoiu, Babes-Bolyai University, Romania

Assoc. Prof. Keigo Kamata, Tokyo Institute of Technology, Japan

Assoc. Prof. Astuty bt. Amrin, UTM Razak School of Engineering and Advanced Technology, Malaysia

Assoc. Prof. Md. Maksudur Rahman Khan, Universiti Malaysia Pahang, Malaysia

Assoc. Prof. A L Roy Vellaisamy, City University of Hong Kong, Hong Kong

Assoc. Prof. Anshuman Srivastava, SIET, Allahabad India

Dr . Alexandru Bogdan Andreescu, TRANSILVANIA University of Brasov, Romania

Dr. Volkan Eskizeybek, Canakkale Onsekiz Mart Universiy, Turkey

Dr J Antonio Zapien, City University of Hong Kong, Hong Kong

Dr. Mohamad Zaki Hassan, Universiti Teknologi Malaysia, Malaysia

Dr Anshuman Srivastava, Shambhunath Institute of Engineering and Technology, Allahabad, India

IOP Conf. Series: Materials Science and Engineering **230** (2017) 011003 doi:10.1088/1757-899X/230/1/011003

Peer review statement

All papers published in this volume of *IOP Conference Series: Materials Science and Engineering* have been peer reviewed through processes administered by the proceedings Editors. Reviews were conducted by expert referees to the professional and scientific standards expected of a proceedings journal published by IOP Publishing.

IOP Conf. Series: Materials Science and Engineering **230** (2017) 012001 doi:10.1088/1757-899X/230/1/012001

Experimental Research on the Factors Affecting Fusion Zone Size and Mechanical Properties in Crack Arrest by Current Detour and Joule Heating

Jing Yu[1] , Yanchuan Liu[2] and Bo Song[1]

[1]Marine Engineering College, Dalian Maritime University, Dalian 116026, China
[2]Dalian Special Equipment Inspection Institute, Dalian 116031, China
E-mail: yj.0730.kb@163.com

Abstract. When pulsed current passes through the specimen containing a crack, since the cannot flow through the crack, the current has to detour around the crack tip, and the current density at the crack tip is excessively high. Owing to Joule heating, local melting occurs and a circle hole is formed at the crack tip. In this work, with 316 stainless steel as a specimen, the crack arrest experiment is performed on high pulsed current discharge device of type HCPD-I. The factors affecting the fusion zone size are studied. By making tensile test and fatigue test, relationships between the molten hole size and the crack arrest effect are derived. The experimental results indicate that the molten hole size is proportional to crack length and discharge parameter. The discharge parameter to the best effectiveness of crack arrest can be obtained.

1. Introduction

To remanufacturing blanks with remanufacturability (named after remanufacturing, it possesses the residual life which can complete a whole service cycle), advanced surface technology[1,2] is applied to recover the surface size and make the remanufactured components superior to the original ones or advanced manufacturing technique[3] is adopted to process the parts to satisfy the assembly requirements. In past, the remanufacturing blanks with cracks were considered as irreparable. The core parts of large mechanical equipment (such as engineering machinery, ship, airplane, large-scale compressor) are costly, highly value-added, complexly processed and possess high technological requirements. During processing of these parts with cracks, it is vital to arrest propagation of the cracks so as to guarantee effectiveness of the repair process to achieve the purpose of prolonging the life of products.

Retarding or arresting the crack growth can be achieved by reducing crack tip stress intensity[4], introducing residual compressive stress[5] and reducing crack tip stress concentrations[6]. Drilling stop-hole at the crack tip or in the direction of crack propagation is commonly used in practice. However, the effectiveness of crack arrest depends on the machining precision and the size of stop-hole. Yet, new stress concentration and fatigue cracks can initiate at edge of the stop-hole. In addition, drilling stop-hole is only suitable for surface cracks; it is not applicable to internal defects or embedded cracks.

During recent years, pulsed current is widely applied as a very promising non-equilibrium processing technology for practical solution, such as the modification of materials[7], refining the solidification structure[8], improving the fatigue of metals and healing damage [9].

As known, when a conductive metal containing a crack is subjected to pulsed current, since the current cannot flow through the crack, the current has to detour around the crack tip, and then the current density at crack tip increases significantly. The local temperature increases owing to Joule heating that initiates crack tip melting, a round molten hole is formed, and the curvature radius of the crack tip is increased. In this way, mechanical stress concentration can be reduced or even eliminated and large compressive stresses could be generated around the crack tip. This phenomenon causes prevention of crack growth. A brief review of the performed studies using Joule heating of pulse current for crack arrest is presented because this subject forms the basis of the presented work.
The effects of electromagnetic fields produced by current pulses on the kinetics of crack growth were examined in the work of GOLOVIN[10]. The authors found out that when the current pulses were passed at appropriate time, the crack growth was prevented. The papers[11,12] reported that the temperature field and the stress field of the specimens containing two and three dimensional cracks were derived by the methods of integral transformation and complex variables functions. The pulsed current discharge experiments and mechanical tests on embedded crack and welded joints were reported in the work of FU[13,14]. In papers[15,16], the phase change zone and microstructure near the crack tip were analyzed after performing discharge experiments. It was found that the grain size was refined and toughness, hardness, and corrosion resistance were increased. These increases were good enough for retarding crack growth and improving the resistance to stress corrosion. A plate containing a crack, under high electric current, was analyzed based on thermo-electro-structural coupled theory. The temperature-dependent characteristic parameters[17] and phase change[18] were also considered in the works.
However, the factors affecting the fusion zone size and the relationship between the fusion zone size and the effectiveness of crack arrest have not been evaluated quantitatively. The technology of crack arrest by pulsed current has not been applied in modern industry. In the present work, the 316 stainless steel is put under investigation. The crack arrest experiments are performed on a high pulsed current discharge device of type HCPD-I. The size of fusion zone under various discharge parameters and geometries of specimens are quantified. By making tensile test and fatigue test, the effectiveness of crack arrest is analyzed, and the discharge parameter to the optimal effective of crack arrest is determined.

2. Experimental methods

2.1. Specimen Preparation

The material investigated is austenite 316 stainless steel with 1.8 mm in thickness. The chemical compositions of the test material determined using EDS (energy dispersive spectrometer) are listed in table 1, and the mechanical properties of it are plotted in table 2. In this paper, the main purpose is to investigate the effect of current detour and Joule heating at the crack tip. In this regard, the slit can be substituted for a crack. The configuration of specimen is shown in figure 1. A unilateral slit is prepared by wire-electrode cutting in the center of longer side that penetrates throughout the thickness. The root radius of the crack tip is r=0.1mm. The crack lengths are a=10,12.5 and 20mm, respectively, for the three specimens.

Table 1. Chemical composition of the 316 stainless steel (wt.%).

C	Si	Mn	Fe	Ni	Cr	Mo
7.03	0.53	1.10	63.14	9.62	16.40	2.18

Table 2. Mechanical properties of the 316 stainless steel.

Tensile strength/ MPa	Yield stress/ MPa	Elongation/ %	Reduction of area/%
620	310	30	40

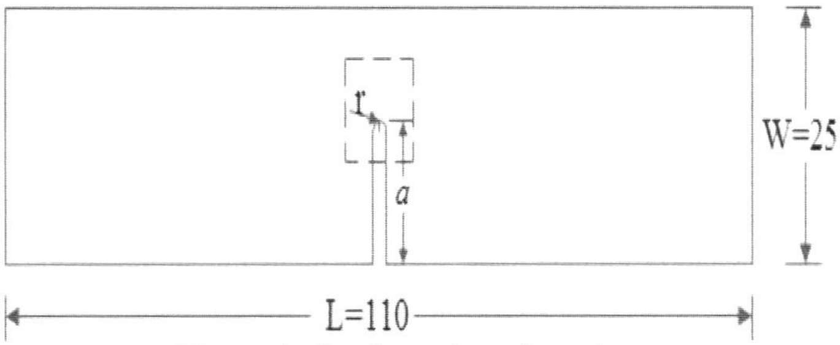

Figure 1. Configuration of specimen.

2.2. Discharge experiment

The crack arrest experiments were performed under ambient on a self-made high pulsed current discharge device of type HCPD-I, as shown in figure 2a. It is composed of electric governor, high-tension transformer, rectifying silicon stack, current limiting resistor, high-tension switch, capacitors and load. The basis parameters are as follows:

(1) There are ten capacitors (MWF12kV-20μf) and the energy storage capacity of each is 1 kJ;
(2) The discharge voltage is 4kV-10kV and continuously adjustable;
(3) The peak current of discharge is 250 kA;
(4) The pulse periodic time is 40μs-100μs.

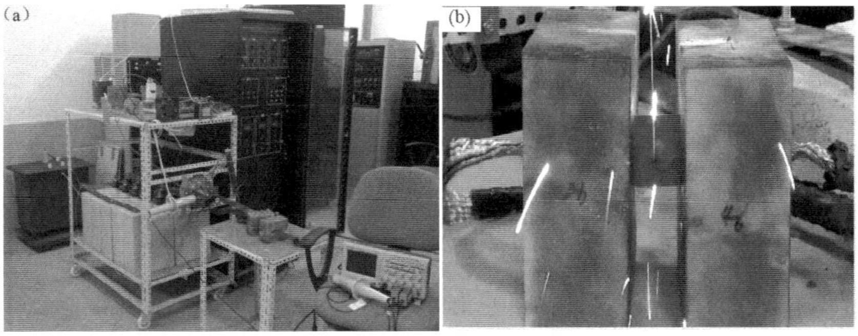

Figure 2. The discharge device and process (a) discharge device (b) discharge process.

In this study, two factors affecting the fusion zone size: discharge parameters (voltage U and capacitance C) and geometry of specimen a/W (the ratio is the slit length versus the specimen width) were investigated. The various discharge parameters were respectively fed to the specimens. Each specimen was only treated once by one pulse. The pulsed current was applied into the specimen through two copper electrodes, as shown in figure 2b. In the discharge process, local fusion and eruption occurred at the crack tip because of Joule heating accompanied by sparks and intense sound. As a result, a small hole was formed and the crack tip became blunt. With increase of discharge parameter and crack length, the current and heat concentration ahead of crack tip was larger and the melting process gets more intense.

2.3. Morphology and microstructure observation

After the discharge treatment, the metallographic sample was cut from the position corresponding to the fusion zone of specimen and the geometry of the sample was 10mm×10mm as illustrated in figure 1. The surface was polished with emery papers of grain number from #200 to #1500 and was finished up to a mirror plane by buffing with diamond powder of 1 μ m grain diameter. The sample was etched by solution containing 17% nitric acid, 50% hydrochloric acid and 33% glycerin. The optical metalloscope was used to observe the morphology and microstructure around the crack tip, and the diameters of molten holes were measured. The attained data were worked upon to obtain the average value.

2.4. Mechanical property test

Owing to the larger range of fusion zone size, the tensile test and fatigue test were conducted on specimens with a=20mm (a/W=0.8) treated by various discharge parameters. When the specimen fractured, the tensile strength, displacement corresponding to the tensile strength and cycles were recorded to analyze the mechanical properties of specimen so as to assess the effectiveness of crack arrest.

To investigate the tensile strength and plasticity of the specimen untreated and treated by different discharge parameters, the tensile tests at ambient temperature were performed on AO LAN DESL-5T universal material test machine. The tensile speed was 1.0 mm / min. The tests stopped when the specimen fractured. The tensile strength σb and the displacement d corresponding to tensile strength were recorded.

The tensile fatigue tests were performed in an atmosphere under sine load conditions using electro-hydraulic servo testing machine SHIMADZU EHF-U. All of tests were run at a stress ratio of R=10, a maximum force of F_{max}=200N and the frequency of f=20Hz. The load cycle number N was recorded when the specimen fractured.

3. Results and discussion

3.1. Morphology and microstructure around crack tip after discharge

In this section, the specimen with a=20mm（a/W=0.8）treated by discharge parameter 180μf-7,000V is investigated. Figure 3 shows the microstructure around crack tip after discharge. As shown in figure 3(a), after performing crack arrest, the shape of the crack tip becomes obtuse and smooth. There are no serrated convexes or concaves, neither are there any secondary cracks. As the current cannot flow through the crack, the current has to detour around the crack tip, which causes the current density to increase excessively because of Joule heating. This results in local melting at the crack tip. The partition around the crack tip is distinct. As shown in figure 3(b), it is the edge of the molten hole covered by fusion metal (white-bright layer). The microstructure of area A is very fine. That is because pulsed current increases the degree of supercooling of fusion metal, reduces crystallization and refines solidification structure[19]. Figure 3(c) shows area B located outside the white-bright layer. Color change is visible, which means the local temperature is higher in this region. It does not reach the fusion point but does reach the austenitic phase change temperature. Area B is actually the heat affected zone (HAZ). The rapid heating increases the nucleation rate. After discharge, the growth of nucleated grains is restrained by the surrounding cool matrix. This, consequentially, results in formation of fine acicular austenite. The original microstructure in area C remains unchanged during discharge as shown in figure 3(d). The reason is that the effect of detour and Joule heating is weak over there and, thus, the temperature of matrix stays low. It is the characteristic that the non-defected part of component remains unchanged. The boundaries around areas A, B, and C are clearly detectable and they show a huge temperature gradient around the crack tip. At the instant of pulse current passing through the specimen, the temperature around the crack tip increases sharply and it initiates melting. On the other hand, the matrix remains almost at the room temperature. The expansion of the fusion zone and the HAZ with high temperature is restricted by the surrounding matrix which is at low temperature. Therefore, a huge compressive stress, which is a superposition of phase transition stress and thermal stress, is formed around the crack tip. This is beneficial for suppressing generation of new microcracks and preventing crack growth.

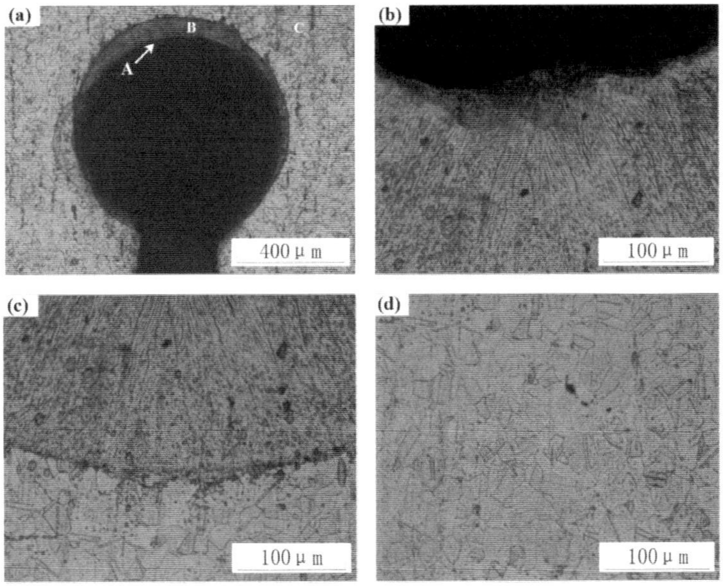

Figure 3. The microstructure around crack tip after discharge (a) the morphology of the crack tip after discharge (b) the microstructure of the wihit-bright layer (c) the microstructure of the HAZ (d) the microstructure of the matrix.

3.2. Fusion zone size versus discharge parameter

The specimen geometry in this section is the same as that described in figure 1 and the crack lengths are a=10, 12.5 and 20 mm, respectively, for the three specimens. The various discharge parameters are fed to the shorter sides. Figure 4 shows the diameter of molten hole versus discharge parameter on specimen with a/W=0.5. When the discharge parameter is smaller, no obvious melting occurs at the crack tip and the radius of curvature of crack tip is 0.1mm as before performing discharge. But a snuff colored HAZ is observed around crack tip. When the discharge parameter is large enough, the crack tip begins to melting and a circular hole is formed. With increase in discharge parameter, input energy increases so that the melting process gets more severe and diameter of molten hole becomes larger.

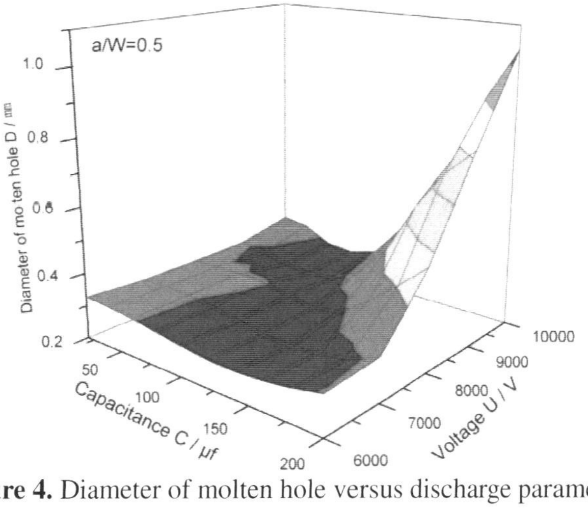

Figure 4. Diameter of molten hole versus discharge parameter.

When the voltage U increases, the size of molten hole becomes larger. Figure 5 shows the average diameter of molten hole under different discharge voltage at C=180μf. With the increase of capacitance, the diameter of molten hole also becomes larger. The average diameter of molten hole under different capacitance at U=10000V is shown in figure 6.

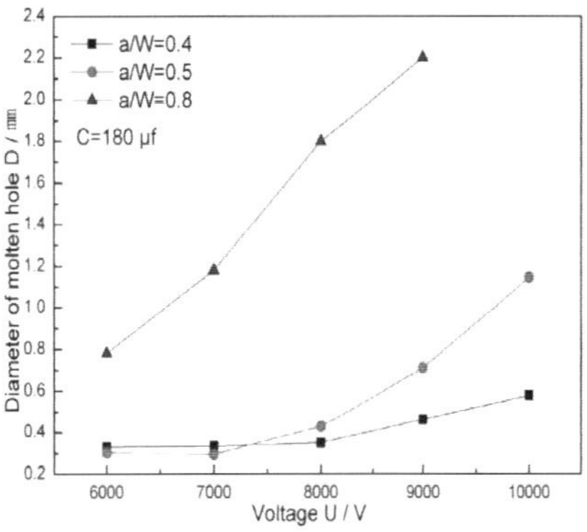

Figure 5. Diameter of molten hole versus voltage.

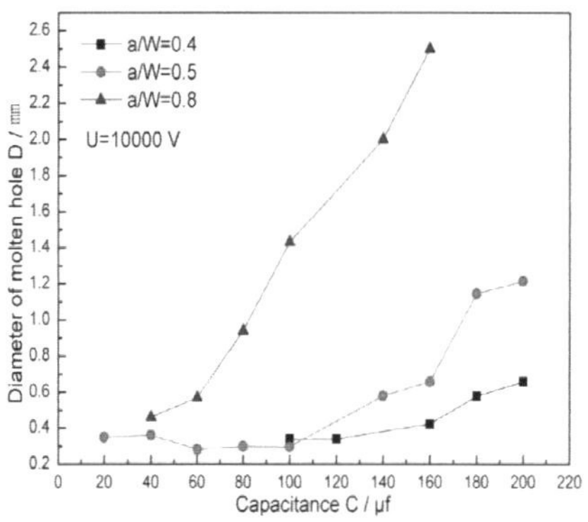

Figure 6. Diameter of molten hole versus capacitance.

With the increase of voltage and capacitance, the current flowed through the specimen increases. Furthermore, owing to high current density around crack tip, Joule heating increases significantly. Therefore, the diameter of molten hole can be adjusted by the discharge parameter.

3.3. Fusion zone size versus a/W

In this section, the dimensions L, W and r is kept constant. The slit length *a* is used as a variable. Figure 5 and 6 also shows the diameter of molten hole *D* versus *a/W* under different voltage and capacitance, respectively. Larger diameter of molten hole is obtained when *a/W* is larger. This phenomenon is due to the higher concentration of the electric current at the crack tip. On the contrary, the specimen with smaller *a/W* can get the smaller size of molten hole. When the discharge parameter is the same, the molten hole appears first on the specimen with larger *a/W*. It can be concluded that the crack tip cannot melt when the crack length relative to specimen width was too small unless higher current is subjected to specimen. The results are in agreement with the numerical simulation results [15].

3.4. Melting area size versus mechanical properties

The specimen with *a*=20mm（*a/W*=0.8）was used in this section. The tensile strength σ_b displacement corresponding d to tensile strength and fatigue life N were used to analyze the relationship between diameter of molten hole and mechanical properties.

When the diameter of molten hole is *D*=1.18mm, the maximum tensile strength is σ_b=582.84 MPa, increased by 1.12% in comparison with original specimen. Figure 7 shows the diameter of molten hole is less than 2mm, the tensile strength of specimen changes slightly; the diameter of molten hole is larger than 2mm, the strength of specimen is significantly weakened, and the tensile strength decreases sharply when the diameter of molten hole increases. On the other hand, after treated by pulsed current, the strength of the specimen could be increased due to the refined microstructure and the high-density dislocation around the crack tip[20].

Figure 8 exhibits when the diameter of molten hole is *D*=1.43mm, the displacement corresponding to tensile strength is *d*=12.436mm, increased by 36.86% compared with that of non-discharge specimen. The result indicates that the appropriate size of molten hole can improve the elongation and plasticity of the specimen. In terms of the room temperature tensile behavior, coarse phase provides low elongation to failure. After treated by pulsed current, the microstructure around crack tip becomes very fine and more homogeneous (as shown in figure 3c) and has the best plasticity[21]. In addition, the directionally drift electrons can generate an electron wind when pulsed current is passing through the specimen. It can enhance the mobility of dislocation and annihilate the point defects, thus can prevent the formation of microcracks[22]. On macroscopic aspect, the specimen deforms in the tensile direction, which results in the significant improvement in plasticity[23].

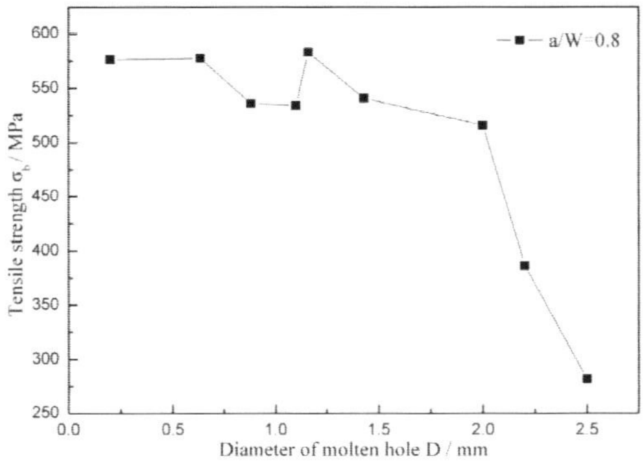

Figure 7. Diameter of molten versus tensile strength.

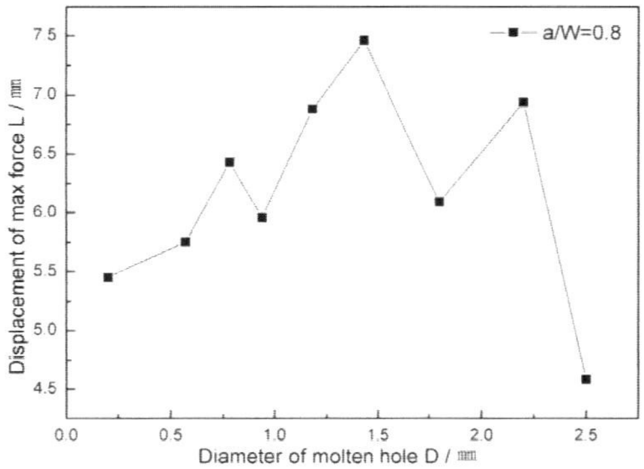

Figure 8. Diameter of molten hole versus displacement of max force.

Figure 9 illustrates when the diameter of molten hole is D=1.18mm, the fatigue life of specimen is N=21424, prolonged by 41.6% compared with that before discharge. After treating by pulsed current, the crack tip gets blunt, the singularity of crack tip is removed and the crack initiation is barricaded. The results of tensile test show the improvement of tensile strength and plasticity. The formation of acicular austenite (as shown in figure 3c) results in the decentralization of the deformation and reduction in stress concentration. It suppresses the grain-boundary cracks and crack propagation so as to prolong the fatigue life[24]. Moreover, a large compressive stress which is a superposition of phase transition stress and thermal compressive stress is generated in discharge process. It would cause diversion of the crack propagation, and is also good for improving the service life of components[25].
Regarding the results of mechanical properties tests above, when the diameter of molten hole is D=1.18mm, the mechanical property is improved remarkably than original specimen. In other words, the crack arrest effect under the discharge parameter 180μf-7000V is the best.

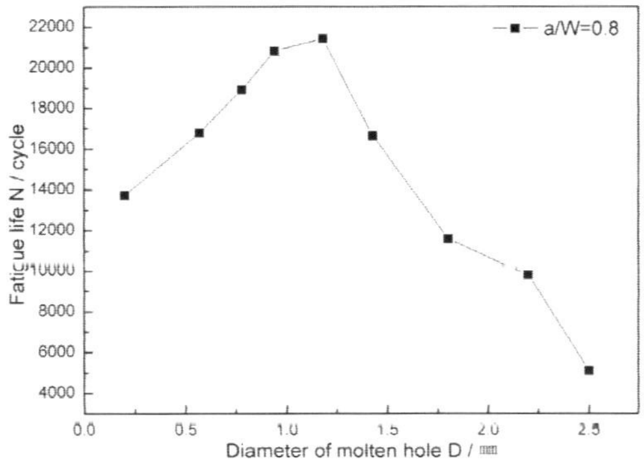

Figure 9. Diameter of molten hole versus fatigue life.

Thus, it can be concluded that when the size of molten hole is too small or too large, it cannot prevent the crack propagation, but also weakens the strength of specimen. The better effectiveness of crack arrest can be obtained by choosing the optimal discharge parameter. The proper discharge parameter can be determined by experiments combined with numerical simulation.

4. Conclusion

In this study, the factors affecting the fusion zone size and the relationship between diameter of molten hole and the effectiveness of crack arrest by current detour and Joule heating effect are investigated. The main results are summarized as follows:

(1) The previous experimental and analysis results show that current detour and Joule heat release by pulsed current can effectively prevent the crack propagation and prolong the service life of component.
(2) The main factors affecting the fusion zone size are discharge parameter (voltage and capacitance) and geometry of specimen (the length of crack). The diameter of molten hole is proportional to the discharge parameter and *a/W*.
(3) The specimen with *a/W*=0.8 is as example. The mechanical properties comprehensively considered, under the discharge parameter 180μf-7,000V, the diameter of molten hole is *D*=1.18mm and the crack arrest effect is the best, the tensile strength increased by 1.12%, the displacement corresponding to tensile strength increased by 26.19% and the fatigue life prolonged by 41.6%.
(4) The diameter of molten hole with better crack arrest effect can be obtained by adjusting the discharge parameter. The tensile strength, plasticity and fatigue life would be improved. If the model of the relationship between the fusion zone size and discharge parameter and geometry of component can be established, the technology of crack arrest using pulsed current can be applied in remanufacturing.

5. Reference

[1] Lin H C, Wang P C, Wu T F, Lin M C, Yeh M T and Lin K M. 2012 Surface coatings of super magnesium-lithium alloys. *Mater. Sci. Technol.* 28 799-803.
[2] Tian Y S and Ge R. 2013 Laser fabrication on nickel aluminide coatings on Al2024 alloy. *Mater. Sci. Technol.* 29 314-318.
[3] Wu X. 2007 A review of laser fabrication of metallic engineering components and materials. *Mater. Sci. Technol.* 23 631-640.
[4] Tu J F and Paleocrassas A G. 2011 Fatigue crack fusion in thin-sheet aluminum alloys AA7075-T6 using low-speed fiber laser welding. *J Mater. Process. Technol.* 211 95-102.
[5] He B Y, Soady K A, Mellor B G, Morris A and Reed P A S. 2013 Effects of shot peening on short crack growth rate and resulting low cycle fatigue behavior in low pressure turbine blade material. *Mater. Sci. Technol.* 29 788-796.
[6] Song P S and Shieh Y L. 2004 Stop drilling procedure for fatigue life improvement. *Int. J Fatigue*. 26 1333-39.
[7] Sosnis O V, Gromova A V, Suchkova E Y, Kozlov E V, Inanov Y F and Gromov V E. 2005 The structural-phase state changes under the pulse current influence on the fatigue loaded steel. *Int. J Fatigue.* 27 1221-26.
[8] Li X B, Lu F G, Cui H C and Tang X H. 2013 Effect of electric current pulse on flow behavior of Al melt in parallel electrode process. *Mater. Sci. Technol.* 29 226-233.
[9] Zhou Y Z, Guo J D and Gao M. 2004 Crack healing in a steel by using electropulsing technique. *Mater. Lett.* 58 1732-36.
[10] Golvin Y I, Finkel V M and Sletkov A.A. 1975 Effects of Current Pulse on Crack Propagation Kinetics in Silicon Iron. *Problemy Prochnosti* 86-91.
[11] Fu Y M, Tian Z G and Zheng L J. 2006 Analysis on the thermal stress field when crack arrest in an axial symmetry metal die using electromagnetic heating, *Appl. Math. Mech.* 8 53-55.
[12] Zhang H C, Yu J, Hao S Z and Peng Y F. 2013 Application of Electro-magnetic heat effect on arresting the crack in remanufacturing blank. *J. Mech. Eng.* 49 21-28. (in Chinese)
[13] Fu Y M, Zhou H M, Wang J L and Zheng L J. 2012 Analysis of Crack Arrest by Electromagnetic Heating in Metal with Oblique-Elliptical Embedding Crack. *Key. Eng. Mater.* 525-526 404-408.
[14] Fu Y M, Chai X, Zheng L J and Li H. 2011 Pulse discharge strengthening of 16Mn welded joint and mechanical performance. *Adv, Mater, Res*. 197-198 1460-1463.
[15] Wang P and Bai X Z. 2011 Phase transformation stress and its influence for arresting crack propagation using electro-heating effect. *China Mech. Eng.* 22 980-984. (in Chinese)
[16] Gao D K, Li H, Fu Y M and Bai X Z. 2001 Acted a pulse current to prevent the thermal fatigue crack subcritical extending. *J Mech. Eng.* 37 28-31. (in Chinese)

[17] Liu T J C. 2008 Thermo-electro-structural coupled analyses of crack arrest by Joule heating. *Theo. Appl. Fract. Mech.* 49 171-184.
[18] Liu T J C. 2011 Finite element modeling of melting crack tip under thermo-electric Joule Heating. *Eng. Fract. Mech.* 78 666-684.
[19] Barnak J P, Sprecher A F and Conrad H. 1995 Colony (grain) size reduction in eutectic Pb-Sn casting by electropulsing. *Scripta Mater.* 32 879-884.
[20] Zhao Y G, Ma B D, Guo H C, Ma J, Yang Q and Song J S. 2013 Electropulsing strengthened 2GPa boron steel with good ductility. *Mater. Design* 43 195-199.
[21] Song H, Wang Z J and He X D. 2013 Improving in plasticity of orthorhombic Ti2AlNb-based alloys sheet by high density electropulsing. *Trans. Nonferrous Met. Soc. China* 23 32-37.
[22] Lin H Q, Zhao Y G, Ma J and Jiang Q C. 2008 Effects of pulse electric current simulation on the microstructure, mechanical properties and thermal fatigue behavior of cast-hot-working die steel. *ISIJ Int.* 48 212-217.
[23] Guo X N, Shen Y F, Zhou Y Z, He G H and Zhou B L. 1999 Effect of a single high current density electropulsing on the mechanical properties of H62. *Chinese J Mater. Res.* 13 73-75.
[24] Sosnin O V, Gromova A V, Ivanov Y F, Konovalov S V, Gromov V E and Kozlov E V. 2005 Control of austenite steel fatigue strength. *In.t J Fatigue* 27 1186-91.
[25] Qin R S and Su S X. 2002 Thermodynamics of crack healing under electropulsing. *J Mater. Res.*17 2048-52.

Acknowledgements
The authors appreciate the financial support from Liaoning Science & Technique Project (20161064), Fundamental Research Funds for the Central University (Grant number 3132016012 and 3132017014) and Key Laboratory of Ship-Machinery Maintenance and manufacture for Ministry of Transportation (Grant number KF-2016-05).

Variability of Mechanical Properties and Weight for Reinforcing Bar Produced in Saudi Arabia

F. Djavanroodi[1a], A. Salman[2]

[1]Department of Mechanical Engineering, Prince Mohammad Bin Fahd University, Al Khobar, KSA
[2]Department of Civil Engineering, Prince Mohammad Bin Fahd University, Al Khobar, KSA
Email: fdjavanroodi@pmu.edu.sa

Abstract: Under the category of material property variations, the variability of the physical and mechanical properties of reinforcing steel affects the performance of reinforced concrete structures. In Saudi Arabia, these properties have minimum requirements, as detailed by ASTM International Standards A 615. In this study, the variability of the weight and mechanical properties of reinforcing steel produced throughout the Saudi Arabia is evaluated experimentally. The results were analyzed to evaluate which manufacturers satisfy the minimum requirements established by ASTM International. 130 ASTM 615 grade 60 samples from different manufacturers were collected and tested to obtain yield strength, tensile strength and elongation. The numbers of samples tested for percent of nominal weight were 96. A statistical analysis of steel rebar's properties is conducted. EasyFit (5.6) software is utilized to determine the distribution type and to perform the statistical analysis. The analysis showed that yield, tensile, and elongation follows different types of continuous distributions. Finally, control charts are generated for the three tests in order to identify values above and below the 3 sigma. The results show that less than 1.5%, 3% and 7.3 %of the steel failed to meet minimum ASTM standards for yield strength, tensile strength, and weight respectively. All the samples satisfied the ASTM 615-15 condition for elongation.

1. Introduction

Steel is one of the basic building blocks of the modern world. Automobiles, appliances, bridges, oil pipelines, and buildings, all are made with steel. While steel manufacturing has existed for centuries, the process for making steel continues to evolve. Establishments in this industry produce steel by melting iron ore, scrap metal, and other additives in furnaces. Conventional reinforced concrete is a composite material of steel bars embedded in a hardened concrete matrix [1–8].

ASTM standards establish minimum criteria for the mechanical properties of reinforcing steel. With the exception of ASTM A 706, however, they do not set maximum limits nor do they address instances where properties may significantly exceed the standards. This may be of concern because member behavior can differ from the planned response if material properties are significantly higher than those used in the design. For instance, if the reinforcement is too strong in a reinforced concrete flexural member, it is over reinforced. If the member is overloaded, this can result in brittle failure, with the concrete crushing before the steel yields. For members subjected to severe lateral loads, an increase in flexure strength can increase the induced shear forces on the member, also resulting in a brittle failure.

An understanding of the variability of steel properties is also useful in the development of statistically based expressions for member strength, which are used in the development of reliability-based strength-reduction factors in design codes. For these reasons, it is worth examining the actual values of the mechanical properties of the reinforcing steel as compared to those used in design.

The mechanical properties of reinforcing steel are controllable in the manufacturing process, but variations between manufacturers and between heats for the same manufacturer do exist. Results may be influenced by several factors including, but not limited to the rate of loading, bar cross-sectional area, and variations in the composition of the steel. All reinforcing steel manufacturers must perform tests on their products to verify that they meet the ASTM standards. These tests measure the yield strength, tensile strength, elongation, and weight per foot (or percent of nominal weight) of the reinforcing steel.

There has been number of statistical studies dealing specifically with the variability of the mechanical properties of reinforcing steel [9-14]. In these studies, variations in yield and tensile strengths were examined. These variations is believed to be caused by variation in the rolling practices and quality control measures used by different manufacturers, as well as possible variations in cross-sectional area, steel strength, and rate of loading. To develop statistical descriptors for the mechanical properties of reinforcing steel, Mirza and MacGregor [9] studied the results of about 4000 tensile tests. The sample included rebar's having wide range of diameters (9.5-57.3mm) and two grades of steel (yield strength = 276 MPa or 414 MPa). The means and standard deviations of the mill test yield strengths were found to be 337 MPa and 36.1 MPa for Grade 276 steel, and 490 MPa and 45.6 MPa for Grade 414 steel, respectively. Joshi and Ranganathan [15] analyzed statistical data on yield strength and modulus of elasticity of reinforcing steel bars from rolling mills and building sites. In a study on steel reinforcing bars used in Turkey, Akyz and Uyan [16] agreed with the requirements of Turkish Steel Rebar Specification Standard TS-708. In Saudi Arabia, Arafah [17] used an experimental program to develop probabilistic models for compressive strength of concrete and yield strength of reinforcing steel produced in the country. A total of 955 concrete samples and 434 samples of steel bars were randomly collected from construction sites. Galasso et al. [18] carried out statistical analysis of reinforcing steel properties based on about 200 test data. The data included a wide range of reinforcing steel bars with diameter between 12 and 26 mm made in Italy. Jibrin and Ejeh [19] studied the Chemical composition of reinforcing steel in the Nigerian Construction Industry. A total of 14 companies supplied nineteen samples were tests. Most of the bars showed absence of some element such as Molybdenum, Vanadium, etc. which is a strength and coefficient of weldability determinants BS4449. Also, it is shown that the high percentage of elements such as Silicon and Phosphorus impacted negatively on the strength and deformation characteristics of the bars.

Saudi Arabia's steel demand has made the country one of the largest consumer in the GCC region. The country also accounts for significant number of construction activities in the Middle-East region. Over the past decade, steel consumption in the Kingdom has increased considerably buoyed by construction boom, growing investment in the real estate sector and rapid infrastructure developments. In addition, the steel industry has witnessed tremendous growth in terms of production, as various players are expanding their production capacities to meet the soaring steel demand [20].

The aim of this work is to study the variability of the mechanical and weights of reinforcing steel produced throughout the Saudi Arabia by experimental work and evaluate manufacturer's performance in satisfying the minimum requirements established by ASTM International A615. The variability of the mechanical and weight of steel reinforcing bars is evaluated and expressions are developed to represent the probability distribution functions for yield, tensile strength, and elongation. 130 samples from different manufacturers were collected and tested to obtain yield strength, tensile strength and elongation. 96 samples were tested to obtain the percent of nominal weight. A statistical analysis of the rebar properties is conducted for acceptably large numbers of data to identify the variability of the mechanical properties, weight of steel reinforcing bars produced in the KSA under ASTM A 615. Expressions are developed to represent the probability distribution functions for yield, tensile strengths and elongation. The results are very important for the long term financial stability and mechanical viability and structural safety for this sector in KSA.

2. Experimental Method

Samples of steel bars ASTM A615 Grade 60 were collected and labeled from eight local steel manufacturing companies in Saudi Arabia namely Usaimi (US), SABIC (HA), Al Rajhi (RA), Al Ittefaq (IT), Jazira (JA), Watania (WA), Taybah (TA) and Al Yamamah (YA). A total of 130 specimens were randomly selected from the manufactures stockpiled. All the samples were prepared and tested for the weight and mechanical properties as per ASTM 615-15. These tests were performed at five different laboratories namely Saudi Arabia standard organization, SABIC, Al-Hoty Stanger, Al Ittifaq and Imperial College London. Instron universal testing machine were used to perform the mechanical tests.

3. Results and Discussion

The tests result of mechanical properties and percent of nominal weight of the ASTM 615 grade 60 samples from different manufacturers indicated the followings. The steel bars exhibited variability in yield and tensile strength with minimum and maximum of 403, 620.3MPa and 611.1, 733.2MPa, respectively. The average yield strengths and tensile strength for bars were 535 and 664.7 MPa, respectively. On the other hand, the percentage elongation showed much less variability. The maximum, minimum, and average were 19, 9.7 and 14.7 respectively.

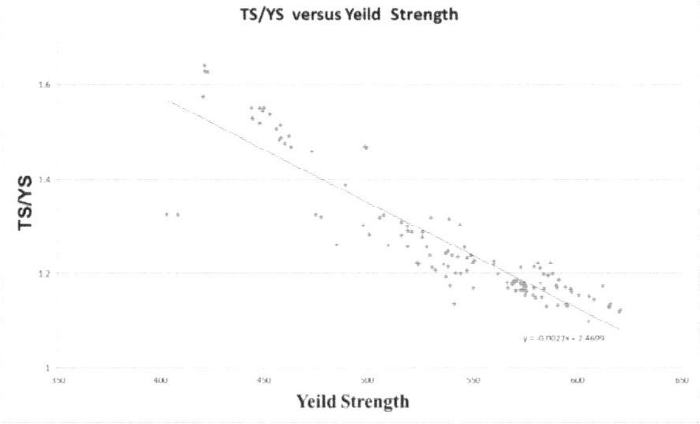

Figure 1. The influence of Yield Strength (YS) on the ratio (TS)/YS

Fig. 1 shows the influence of yield strength (YS) on the TS/YS ratio, as expected the ratio decreases with increasing the yield strength. This will reduce the ductility of the structure. Note of Caution: There have been many occasions in the past where over strength reinforcing bars (Higher Yield and/or Higher Tensile Strength) are innocently accepted by the Design Engineer or passed on by the supplier as the specified grade in the mistaken belief that stronger is necessarily better. There are number of rebar's in the list which have high values of yield strength. There could be number of reasons for high strength values, this can be due to the poor chemical composition control manufacturing process. More recently, structures have been designed using plastic design concepts whereby the ability of the structure to yield and redistribute load without catastrophic failure is required. In such cases the post-yield behavior of the steel assumes an increasing importance. In engineering terms, the parameter selected to idealize the ability to withstand plastic loading is the tensile/yield ratio. The unique material strength properties TS & YS are individually important to consider and control as they influence the behavior of structure. Taken together as the Ratio TS/YS (known as the "Strain Hardening Value" in European practice), it indicates the ductility capacity of the structural member or component where it was used. The ASTM A615/A 615/M do not specify or requires any value for this ratio. A large TS/YS ratio means a greater energy absorption capability before failure. In addition, larger deformations are experienced which could serve as visible warning to building occupants prior to total failure or collapse. Less Brittle behavior therefore is experienced. Insufficient value of TS/YS value or the Strain Hardening Value leads to high concentrations of strain and initiates earlier steel fracture at ultimate loads.

- The high ratio of TS/YS helps to impart ductility to structures by:
- Assuring that significant energy absorption and dissipation occur during inelastic deformation.

- Preventing the premature failure of reinforcing bars due to brittle behavior.
- Guaranteeing that plastic hinging develops at intended locations.
- Avoiding premature failure due to strain concentrations.

The statistical analysis including the values of maximum, minimum, average, median and standard deviation for yield strength, tensile strength and elongation are calculated. Tables 1 and 2 depict the summary result of the three tests; yield strength, tensile strength, and elongation. EasyFit (5.6) software was performed in order to obtain required results. The three tests follow different types of distributions as follows:

a

Descriptive Statistics

Statistic	Value	Percentile	Value
Sample Size	130	Min	420
Range	200.31	5%	444.11
Mean	540.87	10%	453.01
Variance	2697.6	25% (Q1)	516
Std. Deviation	51.938	50% (Median)	549.29
Coef. of Variation	0.09603	75% (Q3)	579.25
Std. Error	4.5553	90%	595.07
Skewness	-0.75279	95%	609.71
Excess Kurtosis	-0.34122	Max	620.31

b

Descriptive Statistics

Statistic	Value	Percentile	Value
Sample Size	130	Min	611.15
Range	122.04	5%	638.55
Mean	675.99	10%	642.1
Variance	480.81	25% (Q1)	666.41
Std. Deviation	21.927	50% (Median)	675
Coef. of Variation	0.03244	75% (Q3)	693.49
Std. Error	1.9232	90%	701.9
Skewness	-0.2883	95%	708.98
Excess Kurtosis	0.52514	Max	733.19

c

Descriptive Statistics

Statistic	Value	Percentile	Value
Sample Size	130	Min	9.5
Range	9.5	5%	11
Mean	14.676	10%	12
Variance	4.4747	25% (Q1)	13.45
Std. Deviation	2.1153	50% (Median)	14.5
Coef. of Variation	0.14414	75% (Q3)	16.625
Std. Error	0.18553	90%	17.5
Skewness	-0.03274	95%	18
Excess Kurtosis	-0.49587	Max	19

Table 1. Descriptive Statistics (a) Yield Strength; (b) Tensile Strength (c) Elongation

Yield strength follows "Johnson SB distribution", it is unimodal distribution and is parameterized by four numbers (γ: shape parameter; μ: location parameter; δ: shape parameter; and σ: scale parameter), as shown in Fig. (2.a). Accordingly; the mean and standard deviation are 540.87Mpa and 51.938Mpa based on Johnson SB distribution, as shown in Table 4. The skewness is to the left; it indicates that the tail on the left side of the probability density function is longer or fatter than the right side as shown in Fig. (2.a). Hence, the mean is less than the median value. However, this result is considered acceptable according to ASTM standards requirements.

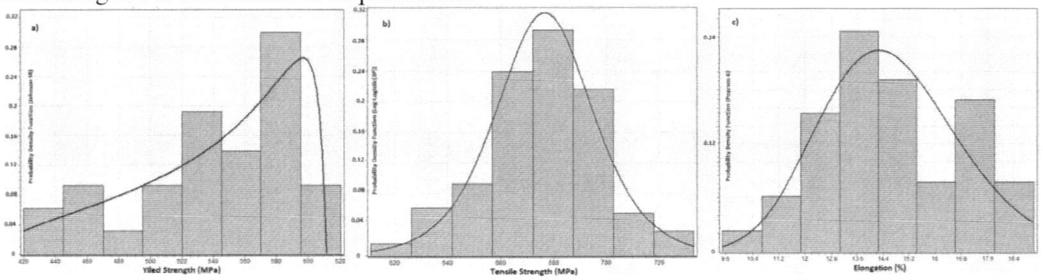

Figure 2. Distribution Type: (a) Yield Strength (Johnson SB); (b) Tensile Strength (Log-Logistic (3P)) (c) Elongation (Pearson 6)

Tensile strength follows "Log-Logistic distribution", which is similar to log-normal distribution but has heavier tails. It has several different parameters such as a scale parameter (α) and shape parameter (β) as shown in Fig. (2.b). The result shows that the mean and standard deviation values of tensile strength are 675.99Mpa and 21.927Mpa, as shown in Table 4, which are considered acceptable according to ASTM standards requirements. The coefficient of variant (CV), which is a function of mean and standard deviation, is 3.2%, which is very low compared to yield strength and elongation. By another word, the tensile strength test has less variation than yield strength and elongation.

Elongation follows "Pearson 6 distribution" as shown in Fig. (2c), it is one member of Pearson family. This type of distribution is used in risk analysis. It has four parameters; shape parameters ($\alpha 1$, $\alpha 2$), scale parameter (β), and location parameter (γ). Table 4 shows that the range of elongation value is between "9.5%" to "19%". The coefficient of variation (CV) is equal to 14.4%, which is very high compared to yield and tensile strength. Hence, chemical tests are required to identify the elongation results and to get thoroughly about steel behaviors regarding elongation.

IOP Conf. Series: Materials Science and Engineering **230** (2017) 012002 doi:10.1088/1757-899X/230/1/012002

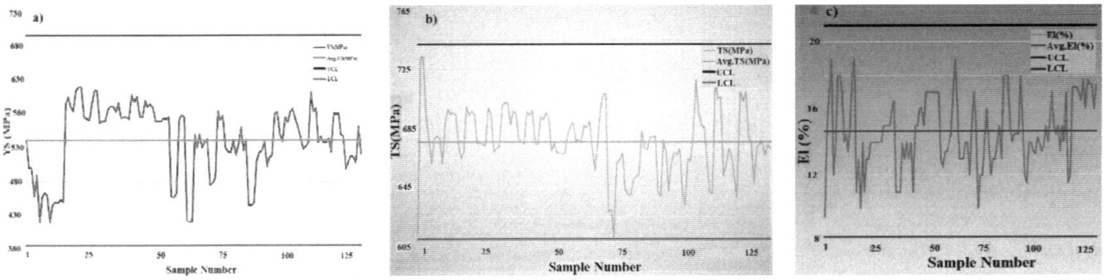

Figure 3. Control Charts (a) Yield Strength; (b) Tensile Strength; (c) Elongation

Fig. 3 depicts control charts of the three tests; yield strength, tensile strength, and elongation respectively. The control chart is utilized to identify values above and below 3 σ of mean value. However, the results show that there is no any value above or under the three limits of 3 σ. For yield strength test; the under, mean, and upper are 385.06Mpa, 540.87Mpa, and 696.86Mpa respectively. For tensile strength test; the under, mean, and upper are 610.21, 675.99, and 741.78 respectively. Finally, for elongation test; the under, average, and upper are 8.33%, 14.4%, and 21.02% respectively. In comparing the data with the minimum standards set forth by ASTM, the following are observed. The value of yield strength recorded for all samples were analyzed. Of those, 3 (1.5%) did not meet the minimum ASTM standard for grade 60. The same analysis for tensile strength on 130 samples showed that only 6 (3.0%) did not meet the minimum ASTM standard for grade 60. Finally, the elongation recorded for 130 samples showed that, all have met the minimum ASTM standard for grade 60. However, the analyses on 96 rebar's samples show that approximately 7 (7.3 %) failed to meet minimum ASTM standards weight.

Function	Yield Strength (MPa)	Tensile Strength (MPa)	Elongation %
Distribution Type	Johnson SB	Log-Logistic (3P)	Pearson 6
Skewness	Left	Left	Left
Maximum	620.31	733.19	19
Minimum	420	611.15	9.5
Mean	540.87	675.99	14.676
Standard Deviation	51.938	21.927	2.115
Coefficient. of Variation - CV	9.6 (%)	3.2 (%)	14.4 (%)
Validation of Control Chart (± 3 σ)	100 %	100 %	100 %

Table 2. Statistical analysis of mechanical properties of A 615 Grade 60 steel rebar's

4. Conclusion and recommendations

Mechanical tests conducted on 130 samples of locally manufactured steel reinforcing rebar's in Saudi Arabia. The results of the tests have shown that less than 1.5 and 3% of the steel failed to meet minimum ASTM standards for yield strength and tensile strength, respectively. All the samples satisfied the ASTM 615-15 condition for elongation. More ever, the analyses on 96 rebar's samples tested for weight showed that approximately 7 (7.3 %) failed to meet minimum ASTM standards requirements. It is also evident that, for most bars sizes of A 615 Grades 60, the mean for the mechanical properties are not situated at the midpoint of the data range, indicating non-normal distributions, therefore, the characteristic mechanical properties of locally produced steel bars is not consistent. The following recommendation and benefits can be deduced from this study.

1. The results are very important for the long term financial stability and mechanical viability and structural safety for this sector in KSA
2. Saudi Arabia Standard organization should make studies on mechanical characteristics of the rebar at least every 5 years to collect a large set of data from the manufacturers as to observe the consistency in the production process.
3. Saudi Arabia Standard Organization board should make sure, all the local producers follow the standard.
4. The expected immediate and long term benefits from this study include:

Assessment of current production of rebar's in Kingdom;
Identification of potential improvements to existing standard, for reduced risks of accepting lower quality materials;
Using variability analysis procedures for assessing product variability based on production data

References

[1] Adom-Asamoah M and Kankam C K 2008 *J Materials and Design 29 1125–1130*
[2] Topcu I B and Sengel S 2004 *Cement Concrete Res 34 1307–12*
[3] Gonzalez-Fonteboa B and Martinez-Abella F 2007 *Constr Build Mater 21(4) 887–93*
[4] Senthamarai R M and Devadas Manoharan P Cement 2005 *Concrete Compos 27 910–3*
[5] Xiao J, Li J and Zhang C 2005 *Cement Concrete Res 35 1187–94*
[6] Ebrahimi M and Djavanroodi F 2014 *Progress in Natural Science: Materials International 24 (1) 68-74*
[7] Kankam C K and Adom-Asamoah M 2006 *J Material Design 27 928–34*
[8] Djavanroodi F, Omranpour B and Sedighi M 2013 *Materials and Manuf. Proc. 28 (3) 276-281*
[9] Allen D E 1972 *Building Research Note No. 85, Ottawa, Canada*
[10] Mirza S A, and MacGregor J G 1979 *J of the Structural Division ASCE 105 921-937*
[11] Nowak A S and Szerszen M M 2003 *ACI Structural Journal, Vol. 100 377-382*
[12] Shaeri M H, Djavanroodi F, Sedighi M, Ahmadi S and Salehi M T 2013 *J Strain Analy 48 (8) 512-521*
[13] Lim W T 1991 *Master's Thesis, University of Canterbury, New Zealand*
[14] Djavanroodi F, Pirgholi A and Derakhshani E 2013 *Materials and Manuf. Processes 25 (8) 864-872*
[15] Joshi C P and Ranganathan R 1988 *J of the Inst. of Eng., Civil Eng. Division, 68 6 309-312*
[16] Akyz S and Uyan M 1992 *Technical J of Turkish Chamber of Civil Engineers, 3, 1, 136-138*
[17] Arafah A M *Magazine of Concrete Research 1997 Vol. 49, No. 180, 185-193*
[18] Galasso C, Cosenza E and Maddaloni G 2010 *Proc. of the 14th Euro. Conf. on Ear. Eng. 3715-3722*
[19] Ibrin M U J and Ejeh S P 2016 *Inter. J. of Civil & Envi. Eng. IJCEE-IJENS 13*
[20] http://www.metalsteelsaudi.com/?page_id=17

Acknowledgement
This research was supported by Prince Mohammad Bin Fahad University. The support is greatly acknowledged.
We are also thankful to Dr. M. M. Al Motari at Saudi Arabia Standard Organization (SASO) for providing their laboratories and assistance for the project. The support is greatly acknowledged.
We would like to show our gratitude to Dr K. F. Al-Hajeri, Mr M. Yaghoub at SABIC, Saudi Basic Industries Corporation who provided experimental assistance for the project. The support is greatly acknowledged.
We would like to show our gratitude to Mr. H. M. Dowla at Al Ittifaq steel that provided experimental assistance for the project. The support is greatly acknowledged.
Finally, we thank I. Babbili1; I. Ansari1; R. Al-Roqaiti1; O. Qassar2; N. Qahtani1; S. Aladhyani1; A. Alunaiz1; M. Almatrook1; M. Najar3, A. Asiz1 for their valuable input and assistance.

Application and Prospects of High-strength Lightweight Materials used in Coal mine

Pan He
Jiangsu Province Xuzhou Technician Institute, Xuzhou, Jiangsu Province, China
E-mail: sideno311@163.com

Abstract. This paper describes some high-strength lightweight materials used in coal mine, and if their performance can meet the requirements of underground safety for explosion-proof, anti-static, friction sparks mine; and reviewed the species, characteristic, preparation process of high-strength lightweight materials for having inspired lightweight high-strength performance by modifying or changing the synthesis mode used in coal mine equipment.

1. Introduction

The mechanical strength of the material is an important indicator on the project, it is an important measure of material carrying capacity. The traditional structural materials mainly iron and steel, which can be improved metallurgy, heat, cold, etc in strength and hardness. However, with the development of technology, portable energy device is widely used, such as automotive, aircraft, its shell required light weight, low energy consumption and a certain intensity. As a result of its low energy consumption and green modern philosophy, lightweight, high strength material has been widely used and developed. Down hole coal industry has high risk, in order to ensure coal mining enterprises develop sustainable, safety and inhibition of major accidents[1].So the use of automated equipment to replace the manual operation to go down is an inevitable trend of development, while to extend underground work time lightweight device is one of the important factors, portable energy device and mine rescue robot used underground are all urgent need for the development and application of high-strength lightweight materials. The complexity of the underground environment asked higher demands on equipment and materials, miners working underground space which is small, complex geology, easy fire, gas explosion, flood and other extremely hazardous accidents[2] while mine rescue robots are applied mine disaster scene catastrophic, sudden, secondary and destructive characteristics[3], and under such an environment vulnerable to encounter a variety of secondary explosions and other dangerous conditions[4], in order to protect mine safety, equipment and materials selection strict control standards.

2. Requirements for coal-mine materials

2.1 Explosion-proof, flame retardant

For housing material used underground, nation has clear limits, GB38361-2000 provide the metal shell material containing light metal, the total content of aluminum, titanium and magnesium are not allowed more than 15% (mass percentage) for class I electrical equipment manufacturing, moreover titanium and magnesium total contenting are not allowed more than 6%. Steel, tensile strength of not less than 120 MPa can be used for class I drill hand-held or bracket (supplied with the unit and plug), portable instrumentation, the lamp housing. Manufacture Class II-electrical-equipment housing Material, amount of magnesium must less than 6%. However the flameproof enclosure material of underground Robot was not stipulated clearly[5]. Actually, exposed on the robot must be flame retardant material, such as a housing material and rail.

2.2 Antistatic, non-sparking requirements

Equipment for coal mine, the housing material must be prevented explosion, also should meet the requirements of anti-static and low friction generating sparks. GB 38361-2000 provides class I electrical equipment should be designed to prevent ignition of dangerous electrostatic charge structure when the shell surface is plastic which surface area more than $100cm^2$ and surface insulation resistance measured less than 1GΩ at a temperature of (23 ± 2) ° C, relative humidity of (50 ± 5)% than perform plastic flame test, aging test light plastic, light-alloy friction sparks safety test, after these can be used.Friction sparks safety testing methods and determination of metal for coal mine is stipulated by GB/T13813, light metal material which should through free drop hammer impact test, after 32 times test gas do not be ignited, considered acceptable conversely ineligible. In 1985 the State promulgated the standard MT113-85 "Coal mine safety inspection of non-metallic polymer products specifications", which specifies two standards for the use of the polymer down hole, namely ① antistatic properties, surface resistance of less than 1×108Ω, ② flame retardant properties, oxygen index greater than 27.

2.3 Strength requirements

Mechanical tests should be applied in the case include impact and drop tests, impact test testing standards are hemispherical impact head, 25mm diameter made by hardened steel and hammer weight 1kg drop vertically from h on housing surface (derived from the impact energy E, h = E / 10; h, m; E, J). The level of mechanical risk is judged by the impact energy E. Drop test is samples are dropped from the height of 1m in using state to flat concrete surface 4 times (Drop position of the sample is determined by the test unit). Non-plastic housing device at temperature of (20 ± 5) ° C, when the material data shows impact resistance reducing at low temperature the test shall be performed at the lowest temperature specified in temperature range. Housing should not cause damage to the equipment, not affect the proof requirements, minor damage and small recessed surface which can not be considered after mechanical tests. In addition, Class I electrical equipment should meet the plastic shell heat and cold tests. According to the standard GB3836.1-2000 Explosion specified in 8.3, Class I drill hand-held or bracket (supplied with the unit and plug), portable instrumentation, the lamp housing, the ensile strength must less than 120 MPa. Light alloy housing should pass the examination of friction spark test can be used according to the provisions of GB13813.

3. Current study on lightweight materials used in coal mine

3.1 Metal materials

3.1.1 Aluminum

Explosion and friction sparks security concerns must be considered for light metals used in coal mine. Currently aluminum alloy is main mining material, as ZL401, ZL104, ZL102 cast aluminum alloy, etc. Flameproof housing of coal drill motor minted only by cast aluminum alloy grades for ZL401[6]. Zhou Dan[7]etc. manufactured flameproof compartment body of small flameproof mobile robot platform by ZL401, the compartment body has a cylindrical and rectangular housing with the volume of 8L, its internal explosion pressure were 0.55MPa and 0.51MPa, a full set of equipment has external dimensions of 560 mm × 590 mm × 250 mm, the quality is 21kg. If the equipment made by steel, the weight is more than 40kg. However the sparks and static friction problems were not mentioned in this paper. For friction sparks safety problems of high-strength aluminum, Su changxue[8] thought, could be solved by two technical solutions regulating the content of the recipe by chemical elements and preparing protective coatings. However this method exists durability, then the security concerns of friction sparks for light alloy lift vessel was raised friction sparks safety indicators depending on heat treatment after aluminum alloy composition determined, the key is prescription system, high-strength aluminum alloy LC15 not only can be enhanced the mechanical strength to meet the requirements of hoisting container but also used in down hole with reliable safety friction sparks after the bipolar overage (CGSI) Processing.

3.1.2 Steel

Needless to say steel material has high strength but its quality is great, it is difficult to carry, install and separate, besides poor mobility. All of these are determined by the properties of the material itself, if we want to reduce the quality restructuring is the key way. Minimize the usage amount of steel with meeting the strength requirements. Steel material is often used as proof system of coal mine rescue robots because lightweight aluminum is prohibited due to containing combustible magnesium. Above a certain grade of steel is used for explosion-proof robot system in order to ensure a sufficient proof strength, its thickness is 4-6mm[9] has larger mass. Explosion-proof system is a generally closed metal housing, charged elements are all inner, if explosion occur flame is prevented out by high strength sealed housing thus protecting combustion of coal mine gas explosion. While housing steel is too heavy has high energy consumption and poor mobility as consumption of mobile systems. Therefore, to optimize the weight and strength of the explosion-proof enclosure effective, stiffness become a huge challenge[10]. Wang changlong[9] proposed a static positive pressure explosion-proof technology as protection measures for mine rescue robots, the whole sealed enclosure using a thin steel plate Q235A chassis and cover using angle iron welded frame only. Hermetic housing was filled with protective gas to reduce the oxygen content below 1% (Volume ratio). This way can prevent the generation of internal explosion fundamentally, thus greatly reduced the demand for flameproof enclosure and only meet the mechanical strength requirements of robot enclosure. The weight of robot was reduced greatly and the mobile performance of it was promoted extremely with the structure.

3.2 Organic composite materials

With the research and development of high-strength material, polymer materials gradually entered the era of the stage it is not limited to metallic material able to meet the strength requirements. In order to sophisticate material lightweight, Wang tongyuan[11] proposed a "plastic steel" concept used in the coal industry, because of polymers having a light weight, corrosion resistance, high strength, easy processing, excellent performance. Li Heng[12] made climbing robot using polyoxymethylene (POM) as the body shell in line with principles of lightweight and compact.

Plastic products used for coal mine equipment should be the flame retardant and antistatic properties. The methods for polymer materials antistatic contain external coating and internal mixing. The internal mixing method is much wide, it is assimilated with antistatic agents. Its mechanism is mainly manifested in two aspects, the one is continuous film which has a certain moisture absorption and ionic formed on the surface of the polymer material could reduce surface resistivity and leaked generated electrostatic charge. The other is suppressing and reducing the generate electrostatic charge by enhancing the lubricating property of the surface and reducing coefficient friction. Sheng shouzhai[13] add carbon black as a conductive filler in PVC to reduce the surface resistivity, improve flame retardancy, mechanical properties threw adding flame retardants and modifiers, while experiments show that flame retardant anti-static PVC sheet can replace wooden plates used for coal mine.

4. Prospect of lightweight, high strength material

Currently investigate Lightweight, high strength materials can be divided into two categories: one is light metal material, one is fiber-reinforced polymer composite materials. High-strength aluminum alloy is widely used in coal mine but its strength is little poor as robot flameproof enclosure material. Titanium alloy are widely used in industry with its low mass, high strength, high toughness, high corrosion resistance, and other excellent performance, but the price is too expensive and used in coal mine need to modified some defect.

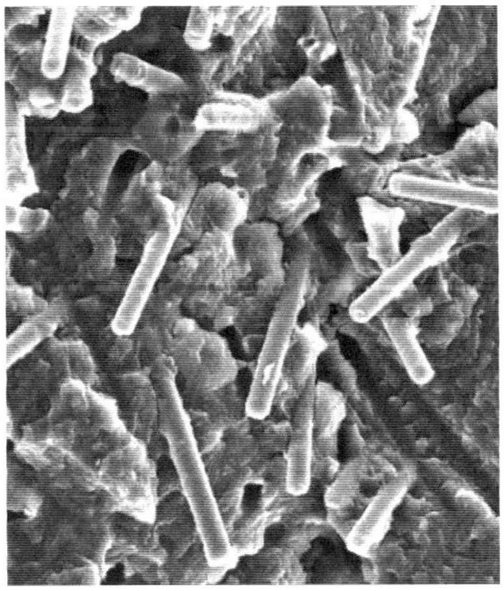

Figure 1. SEM for carbon fiber

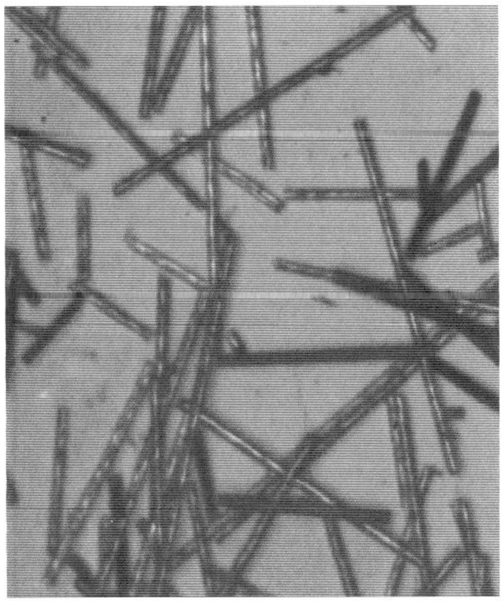

Figure 2. Micro-examination for carbon fiber

IOP Conf. Series: Materials Science and Engineering **230** (2017) 012003 doi:10.1088/1757-899X/230/1/012003

Table 1. Performance index of carbon-fiber cloth.

	High strength grade I	High strength grade II
Weight(g/m²)	300	200
Tensile strength (MPa)	⩾4900	⩾4100
Elastic modulus (GPa)	⩾240	⩾210
Elongation(%)	⩾2.0	⩾1.8

Fiber-reinforced polymer composite materials (figure 1and figure 2) is another focus. Carbon fiber as a high-performance fiber, which has a high specific strength, high modulus ratio, creep fatigue, expansion coefficient and the friction coefficient of performance, has been one of the most important composites reinforced material (table1). In addition Kevlar as reinforcing fibers also have very strong mechanical properties, it is used for military protection, under the same conditions its weight is half of the glass protective steel, made into laminated sheet its toughness is 3 times of steel and can withstand repeated impact. These are the directions for future development.

5. Reference

[1] J. Y. Gao. Coal Mine Detect and Rescue Robot Technique Research. International Conference on Information and Automation. June 22 -25, 2009

[2] W. X. Chen. Study on Risk Monitoring and Emergency Rescue Method of Key Accident in Coal. Shandong University of Science and Technology. 2005

[3] J. G. Liu , Y. C. Wang, B. Li. Research Status and Key Performance and Prospect of Disaster Rescue. Chinese Journal of Mechanical Engineering, 2006，42(12)：1-12.

[4] S. Qi, B. L. Li, Y. Chen. Present Situation of Research and Application of Rescue Robot in Coal Mine and Problems to be Solved. Mining machine. 2012.6(40)：7-10

[5] L. W. Jia, B. Zhu, R. Q. Qi. Explosion-proof Design of Light Wheel Type Abandoned Mine Detecting [J]. Mining machine. 35(4). Vol.35No.04 Apr. 2014

[6] Y. W. Li. Design and Machining of Flameproof Shell for Coal Mine Disaster Relief . Coal Mine Machinery.2009. 30(2).

[7] D. Zhou, Y. J. Zhang, R. Q. Jia. Discussion on the construction of higher occupation education training base of coal mine safety. Journal of North China Institute of Science and Technology .2012. 9(2).

[8] C. X. Su, J. X. Zhang. Experimental Study on the Safety of Friction Sparks in Light Alloy Lifting. Coal Science and Technology .1996. 24(2).

[9] C. L. Wang, X. W. Li, F. Xu, J. S. Dong, Y. L. Feng. Static positive pressure explosion-proof technology in the mine rescue robot application. Coal Engineering. 2009. 11

[10] W. d. Wang, Y. Y. Su, D. M. Wu, Z. J. Du. Development of Search-and-rescue Robotsfor Underground Coal Mine Applications[J] Journal of Field Robotics. 31(3), 386–407 (2014)

[11] T. Y. Wang, Q. Q. Wang. Application and Development of Polymer in Coal Mine Production.

Technology and Innovation Management. 2011.31(1)
[12] H. Li. Research on Dynamic Characteristics of Climbing Wall Robot and Simulation Analysis. Qingdao University of Science and Technology. 2010. 4.
[13] S. Z. Sheng, Y. X. Li， J. L. Hao, Z. W. Ren. Development of Flame Retardant and Anti - static PVC Sheet for Coal Mine. Coal Science and Technology. 2014. 32(10)

IOP Conf. Series: Materials Science and Engineering **230** (2017) 012004 doi:10.1088/1757-899X/230/1/012004

Study on Cracking Mechanism of Hardened Planetary frame

Xinghui Li

Department of Architecturl Engineering,Jiangsu Prvince Xuzhou Technician Institute Xuzhou, Jiangsu, China
E-mail: 2804809396@qq.com

Abstract. Planetary carrier made by 45 steel appear quenching crack, which is analyzed in chemical composition, hardness test and metallographic microscopic structure. The reasons of quenching crack of planetary gear include the unreasonable structure of the planetary carrier, thinner annular wall on the base of the upper part, and in dangerous area of the 45 steel in the process of quenching. The faster cooling rate of quenching results in a centripetal stress with the thick-wall part, which is greater than the ultimate bearing capacity of the material.

1. The macroscopic morphology of quenching crack of the planetary carrier

The crack of 45 steel planetary carrier material in the quenching process, its location and morphology is shown in figure 1.

Figure 1. Position and morphology of quenching crack of planetary carrier.

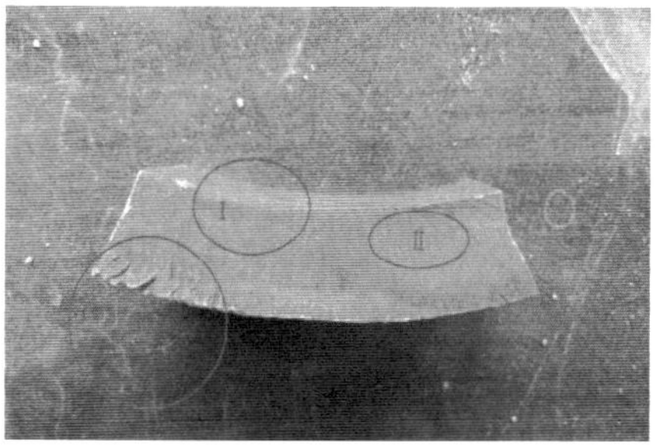

Figure 2 Macroscopic morphology of quenching crack.

Quenching crack appears on the inside of the annular upper part of the base of the planet carrier (thickness 10mm, depth 17mm, see figure 1). The crack extends radially to the inside of the annular portion at about 45 ° and along the circumferential direction, upwards and towards both sides, causing large crack of the annular portion. The macroscopic morphology of the quenching crack is shown in figure 2.

It can be seen in figure 2, the crack starts at position I, rapidly expands outward at position II, cracks at position III, which is a significant final rupture region.

2. Chemical composition of steel of planetary carrier

After analyzing spectrum of steel component of planetary carrier, the analytical results are shown in table 1.Table 1 shows that the chemical composition of the steel is in line with the 45 steel technical requirements in GB699-1988 "steel technical conditions of high-quality carbon structure"[1].

Table 1. Chemical composition of steel for planetary carrier (wt%).

Chemical composition	C	Si	Mn	P	S	Cr
Content	0.46	0.21	0.70	0.026	0.009	0.10
StandardGB 699－88	0.42～0.50	0.17～0.37	0.50～0.80	≤0.035	≤0.035	≤0.25

3. The hardness test of the planetary carrier

With the hardness test of the cracked parts, the average hardness of six parts is HRC31.2 (see table 2), in line with quenching and tempering requirements of 45 steel plate[2].

Table 2. Hardness test results.

Hardness /HRC	Average value /HRC
31，30，32，31，32，31	31.2

4. Metallographic microscopic structure of planetary carrier and its analysis

Choosing the crack site samples for metallographic microscopic structure analysis[3], as shown in figures 3 and 4, figure 3 shows the metallographic microscopic structure in the center of the annular site, and figure 4 shows the metallographic microscopic structure at the initial stage (Position I, figures 2).

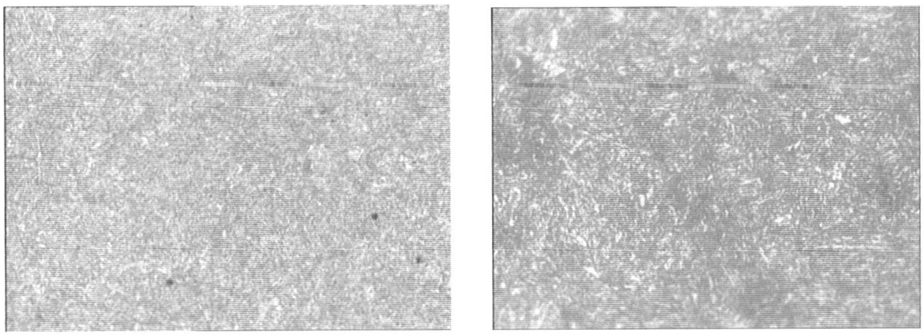

(a) ×200 (b) ×500

Figure 3. Metallographic microscopic structure in the center of the annular site.

The Metallographic microscopic structure of the crack at the initial stage is also the tempered sorbite structure with homogeneous distribution(see figure. 4), consistent with the theoretical results.

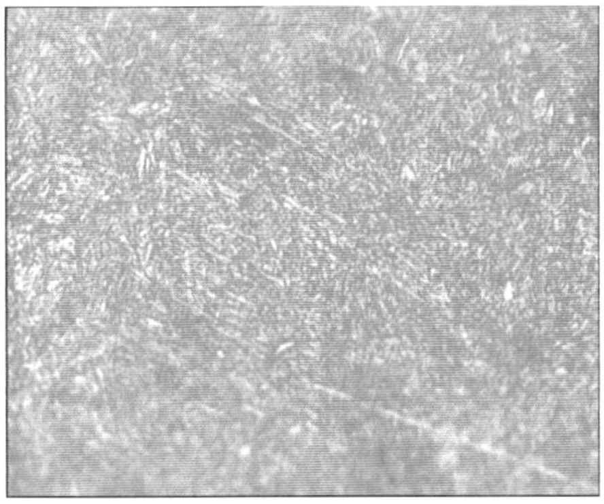

×500

Figure 4. Metallographic microscopic structure of crack at the initial stage.

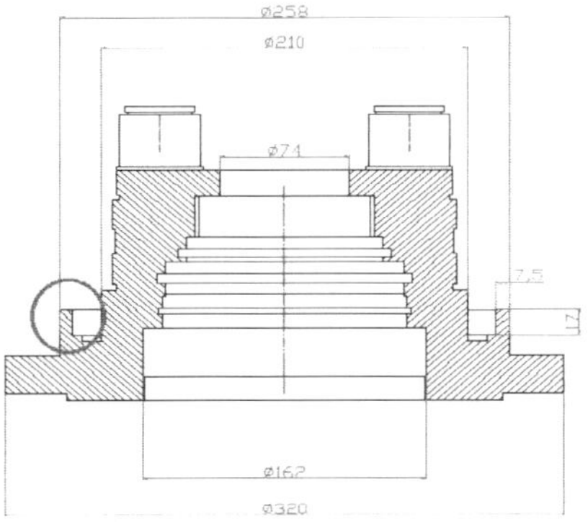

Figure 5 . Structure Dimension Drawing of the planet carrier.

Comparision of figure 3 and figure 4, the metallographic microscopic structure is completely the same in the crack at the initial stage and the center of the annular site, namely, the tempered sorbite structure with homogeneous distribution, without uneven organization and impurity that may cause cracks in the metallographic microscopic structure. Therefore, it is possible to eliminate the potential of quenching crack due to harmful tissue[4].

5. Analyzing the reason for hardening break of planet carrier

Under normal conditions, the position of NO.45 steel easy to cause hardening break, including the part with thick wall and mutation, sharp coner of stage(Radius of round angle is smaller)[5]. Figure 5 is the Structure Dimension Drawing of the planet carrier.

The position with red circle in figure 5 is actual position of quenching cracking, the thickness of wall, smaller than other position clearly, belong the position easy to cause hardening break. Due to the small thickness of the wall, the actual cooling rate will be faster than that with the thicker wall (Lampstand of Planet carrier).In the process of quenching cooling, the cooling contraction of the annular part will precede the lower part (Lampstand of the Planet carrier), resulting in the asynchronism of cooling shrinkage.

In a moment, when the annular part nearly finish the cooling shrinkage, due to slower cooling rate of the bottom(Lampstand of planet carrier), its cooling shrinkage behavior fall behind the annular parts, and then the bottom(Lampstand of planet carrier) will continue to cooling contraction; meanwhile, the cooling contraction of the bottom will be limited by the annular parts, there will be a centripetal (contraction) stress on the root, when the stress exceeds the ultimate strength of hardened and tempered NO.steel, which will cause quenching crack; The quenching crack will start from the weakest part of the annular part, and then rapidly expand, eventually complete crack. On the other hand, the NO.45 steel dangerous size of quenching is 8~15mm, the wall's thickness of annular part is just in the range, so that the possibility of quenching cracking greatly increased.Therefore, the size of the planet carrier is the direct reason to cause the quenching crack.

IOP Conf. Series: Materials Science and Engineering **230** (2017) 012004 doi:10.1088/1757-899X/230/1/012004

6. Conclusion

The chemical composition of the planet carrier's NO.45 steel meet the standard; The hardness after quenching and tempering meets the requirements; Through metallographic analysis, the metallurgical structure of the planetary wheel is tempered martensite structure distributed evenly, there is no proeutectoid ferrite, inhomogeneous microstructure, impurity distribution and other phenomena, it can be excluded the possibility that quenching crack caused by abnormal microstructure.

The reason for quenching crack of planet wheel is that: In the process of quenching, due to the thickness of the annular position on the Lampstand is flimsy and the cooling rate is faster, which leads to the asynchronism between the cooling contraction of the position with thick wall and the annular position. The cooling and contraction of annular part limit the cooling and contraction of the position with thick wall, producing a centripetal stress, it will crack when the stress is greater than the ultimate capacity of the material. The thickness of the wall is in the dangerous dimension of quenching, which may lead to it crack after quenching. Therefore, the structure size of the planetary carrier is the direct reason to cause the quenching crack.

7. Comments and suggestions

7.1. Optimize the size of planet carrier's structure

The annular groove on the upper part of the Lampstand should be processed after quenching and tempering.Or enlarge the radius of the fillet on the root of the annular groove.

7.2 .Reduce the cooling speed of the annular part on the Lampstand

Adding an annular flat steel in the annular groove on the Lampstand before quenching .Or improve the placement of the workpiece in the quenching basket.

8. Reference

[1] Xia Lifang 2008 Iron and Steel Material and its Technology for Heating Processing co (Shanghai: Harbin Institute and Technology Press) p 33–35

[2] An Yunzheng 1988 Technology for Heating Processing (Beijing: Machinery Industry Press) p 85–88

[3] Fan Dongli 2009 Heat Treatment Technical Manual (Beijing: Chemical Industry Press)

[4] Ye Weiping 2012 Practical metallographic examination of iron and steel materials (Beijing: Machinery Industry Press)

[5] Wang Guangsheng, Shi kangcai and Zhou jingen 2007 Analysis Defect of Metal Heat Treatment and Cases (Beijing: Machinery Industry Press) p 13–51

IOP Conf. Series: Materials Science and Engineering **230** (2017) 012005 doi:10.1088/1757-899X/230/1/012005

Effect of Pre-Strain on the Fatigue Behavior of Extruded AZ31 Alloys

Wu Yanjun
Department of Mechanical Engineering, Nanjing Communications Institute of Technology, Nanjing, PRC
yjwu_wendy@163.com

Abstract. An attempt is made to rationalize the effect of pre-strain history on fatigue behaviors of AZ31 magnesium alloy. Axial fatigue tests were conducted in the extruded and pre-compressioned AZ31 alloy under low cycle total strain control fatigue conditions. The pre-strain process influences the plastic deformation mechanism activated during fatigue deformation, especially during tensile loading, by enhancing the activity of detwinning mechanism. The low-cycle fatigue lifetime of extruded AZ31 alloy can be enhanced by the pre-compression process. And the hysteresis energy was successfully used to predict the low-cycle fatigue lifetime.

1. Introduction

The first systematic research works on the effects of pre-strain on cyclic behavior had been performed 40 years ago [1]. And there exists a threshold pre-strain for easy-cross-slip metals. Below this threshold, the cyclic stress-strain response doesn't depend on the pre-strain. Above it, the cyclic plastic response strongly depends on the pre-strain. While for planar-slip metals, the cyclic behavior is always sensitive to the pre-strain[2]. Contrary to other industrial materials, most magnesium alloys are with hexagonal close packed (hcp) crystal structure and they have very limited number of slip systems at room temperature. Besides slip, twinning is another important plastic deformation mode in magnesium alloys.

As such, the fatigue properties of magnesium alloys differ markedly from those deformed predominantly by slip, such as pure copper and aluminum alloys. Yin et al. [3] reported that the hysteresis loops of extruded AZ31 alloy were asymmetric between the compression and tension cycle. This indicated that twinning and detwinning alternated with the cycle loading. Twinning-detwinning led to complicated cyclic deformation and fatigue behavior of magnesium alloys [3-8]. These investigations suggested that twinning-detwinning played an important role in the cyclic response of magnesium alloys. Then pre-strain of magnesium alloy induced change in the twinning-detwinning characteristics, and influenced the fatigue deformation behavior. The main objective of this investigation was to rationalize the effect of pre-stain on fatigue behavior of an extruded AZ31 alloy.

2. Experimental

The starting material for this study was an extruded rod of commercial AZ31 (Mg-3wt.%Al-1wt.%Zn) alloy. Specimens for pre-compression process with a gauge size of 95mm in length and 20mm in diameter were prepared from the extruded rod, where their cylindrical axes were parallel to the extrusion direction. After pre-compressioning these specimens to the strain values of 3% and 5%, they were remachined to tension (2×3×10 mm3), and fatigue (14×6×5 mm3) specimens.

All tests were conducted with an Instron-8801 testing machine. A strain rate of 5×10-4/s was used for tensile tests. The cyclic frequency was 1Hz. Also, the fatigue experiments were carried out at room temperature under the strain ratio S= -1. Before testing, the samples were polished with 2000 grit silicon carbide abrasive paper, followed by electropolishing using a solution comprising 15 ml HClO4, 50 ml glycol and 180 ml ethanol to eliminate the residual stress of the surface layer. The microstructure and morphology of the specimens were characterized by optical microscope and a Quanta-200 scanning electron microscope. X-ray diffraction (XRD) was performed using a D/Max-1200X diffractometry with Cu Kα radiation operated at 20kV and 100mA, in order to identify the crystallographic orientation.

3. Results and discussion

The microestructures of the as-extruded and pre-strained AZ31 alloy are shown in Figure 1. The average grain size in initial state is approximately 30μm. The pre-compression process introduces plenty of twins into the microstructure. And with the progress of deformation, the twins initiated, gradually grew, and intersected with each other.

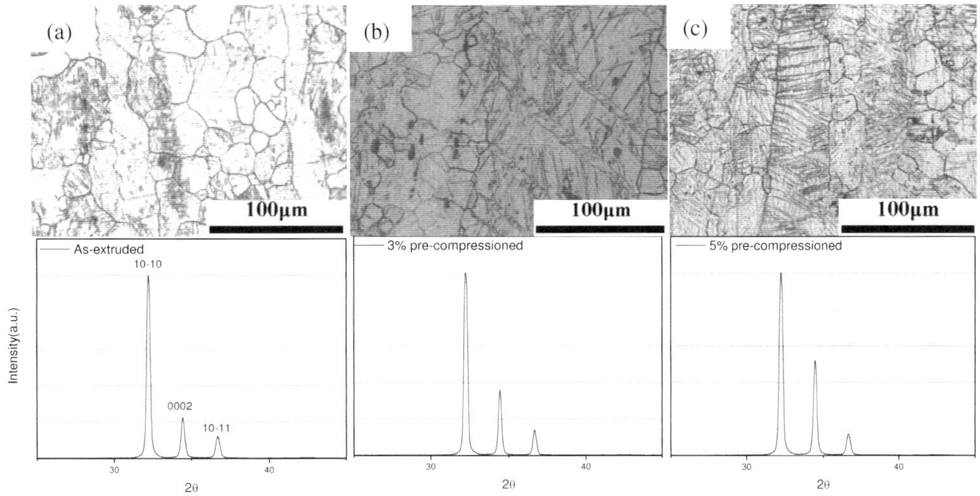

Figure1. Optical microstructures and XRD scans from the longitudinal planes of (a) the as-extruded (b) 3% pre-compression, and (c) 5% pre-compression materials.

To gain insight into the texture evolution of the material, the X-ray diffraction (XRD) is employed. Figure 1(a) shows the crystallographic orientation of as-extruded sample. The intensity of the (0002) peak, (denoted as I(0002)/I(1010) ratio [9]), is about 0.22, so the initial texture can be roughly considered as having a ring fiber texture with the basal planes parallel to the extrusion direction. This initial texture does not allow any twinning under tensile loading but twinning under compressive loading. As the $\{10\bar{1}2\}$twinning induces a crystallographic lattice reorientation of 86.3° in the twinned region, the intensity of basal poles aligned perpendicular to the ED (i.e., initial texture) gradually became weak with increasing pre-compression. On the contrary, the intensity of basal poles aligned parallel to the ED (i.e., twin texture) gradually increases because of the formation and growth of twins. As shown in Figure 1(b) and (c), the I(0002)/I(1010) ratio in 3% pre-compression specimen is about 0.35, while the I(0002)/I(1010) ratio in 5% pre-compression specimen is about 0.52. Hence, it is reasonable to conclude that the pre-compression specimen has a higher total volume fraction of $\{10\bar{1}2\}$ twins than the extruded specimen. The twin texture will be favored for detwinning during subsequent tensile loading because it undergoes tension parallel to the c-axis. This causes a significant decrease in flow stress.

Figure 2 presents the tensile and compressive stress-strain curves of pre-strained AZ31 alloys and Table 1 summarizes the mechanical properties. It is interesting to note that the as-extruded

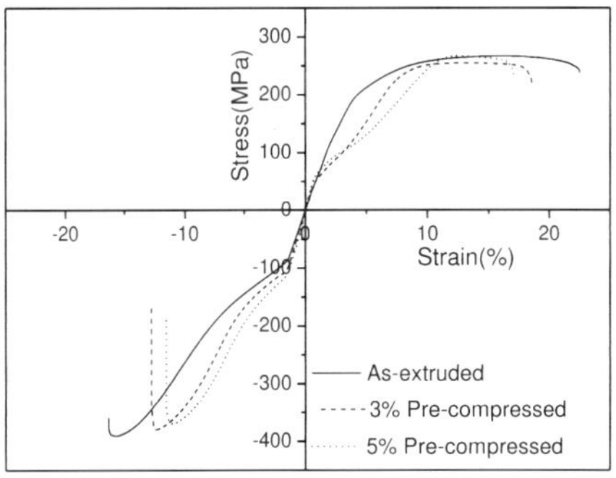

Figure2. Tensile and compressive stress-strain curves of pre-compression AZ31 magnesium alloys.

specimen yield at much higher strength as compared to the pre-strained specimen during the tensile deformation, and vice versa in the compression test. This asymmetric feature has been found to be due to the different deformation mechanisms which vary with the orientation of specimen and the direction of loading [5].

For HCP magnesium with a c/a ratio of approximately 1.622, tensile twinning on the $\{10\bar{1}2\}$ plane is activated by a compressive stress parallel to the basal plane or a tensile stress perpendicular to the basal plane [4]. Accordingly, specimen deformed by twinning reveals much lower yield strength than that deformed by slip. Consequently, the lower tensile yield stress, shown in Figure 2, indicates that twinning is the dominant deformation mechanism which arises from twins produced by pre-compression. Also, the as-extruded specimen showed the parabolic hardening of slip-dominated deformation, whereas the pre-strained specimen showed the sigmoidal (S-shaped) hardening of twin-dominated deformation [5]. The yield strength of the as-extruded is about 155MPa, decreasing to 66 and 54MPa for 3% pre-compression and 5% pre-compression, respectively.

Table 1. Mechanical properties of the extruded AZ31 Mg alloy with and without pre-compression.

Sample	TYS(MPa)	UTS(MPa)	CYS(MPa)	Elongation(%)
As-extruded	155	267	100	18
3% compression	66	267	109	12.5
5% compression	54	253	102	13.5

Representative hysteresis curves are shown in Figure 3(a) for the as-extruded sample at total strain amplitude of 0.3%. In the first cycle, the stress strain curve is asymmetric between tension and compression cycle. Upon the initial compressive yielding at ~90MPa, the sample showed little hardening and the maximum compressive stress at 0.3% was ~95MPa. This type of strain hardening plateau is typical of magnesium alloys which are deformed by twinning. Upon unloading, a significant pseudoelasticity was observed which was caused by detwinning [3,4].

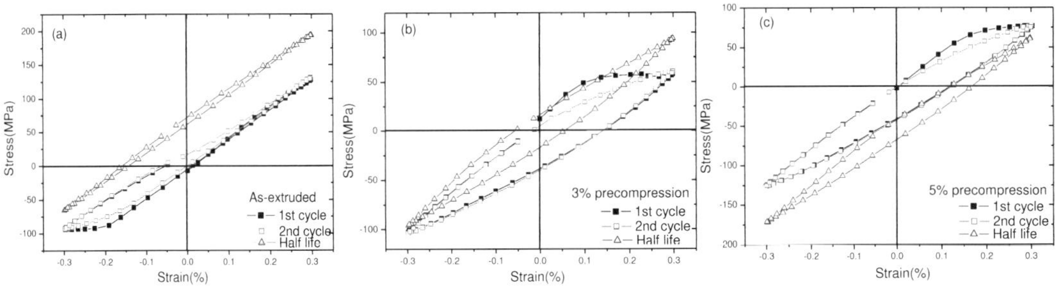

Figure3.Evolution of the stress-strain hysteresis loops with the number of cycles($\Delta\varepsilon_t/2$=0.3%): (a) the as-extruded, (b) 3% pre-compression, and (c) 5% pre-compression materials.

For pre-compression materials, as described earlier, initial twins formed by pre-compression process serve as a source for the detwinning mechanism during tensile loading. With the aid of initial twins, the first tensile loading would be accommodated by the detwinning mechanism alone and lead to tensile strain hardening plateau, as shown in Figure 3(b) and (c). It is commonly recognized that the pre-compression process decreases tensile flow stress, as shown in Figure 4(a). The mean stress at half-life is taken as the representing one. However, the effect of pre-compression on the hysteresis loops was dependent on the amount of pre-compression and the number of cycles. The residual twins gradually increase with increasing cycles so that the number of twins capable of detwinning decreases with increasing cycles [3,4]. As a result, after several cycles, it becomes insufficient to accommodate the imposed tensile strain by detwinning alone. This requires the activity of slip systems [10], which leads to the symmetric hysteresis loops. The total strain amplitude vs. number of cycles to failure for the pre-compression materials is depicted in Figure 5. It is apparent that the low-cycle fatigue lifetime can be enhanced by the initial $\{10\bar{1}2\}$ twins processed by pre-compression.

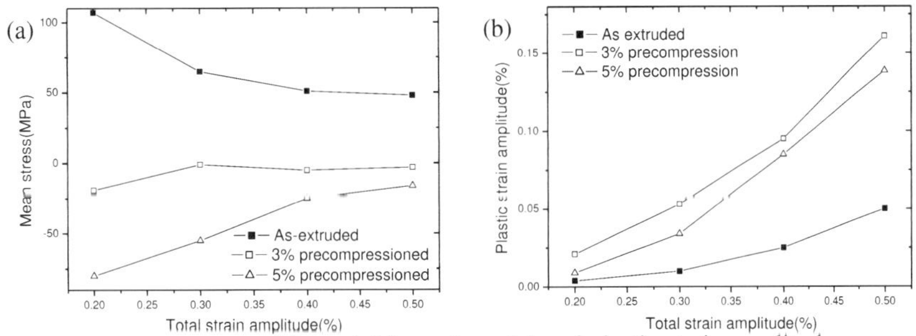

Figure4. Mean stress (a) and plastic strain amplitude (b) at 50% of fatigue life in dependence on total strain amplitude.

From the detailed observations of fracture surfaces, examples of fractographs are shown in Figure 6(a) and (b) for the as-extruded sample and 5% pre-compression sample at strain amplitude of 0.2% in the fatigue crack initiation and stable propagation zone respectively. In Figure 6(a) the as-extruded sample shows irregular surfaces going up and down in some random way. In contrast, the 5% pre-compression fracture mode is mostly facted and brittle-like.

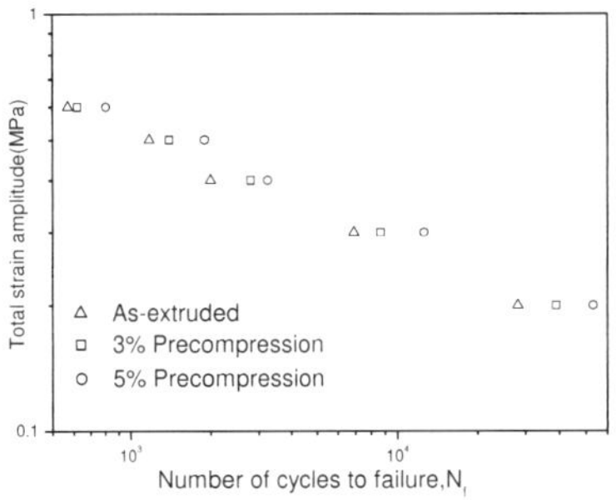

Figure5. Total strain amplitude as a function of the number of cycles to failure.

It is well known that tensile mean stress has a harmful effect on the fatigue resistance by accelerating crack initiation and propagation mechanisms, while the reverse is true for compressive means stress [5]. Based on this model, Park et al.[10] explained the effect of mean stress on fatigue lifetime under total strain amplitude. Actually, the assessment of the fatigue performance is not quite straightforward. The effect of mean stress on fatigue lifetime, proposed by Evans [11], is only valid for stress-controlled fatigue. While under strain-controlled conditions, except for mean stress, plastic strain amplitude, γpl, is another important factor influencing the fatigue lifetime. The amount of cyclic damage [12], Γ, can be expressed as $\Gamma = \sum_{i=1}^{n} 4\gamma_{pl,i}$. The variation of plastic strain amplitude does not follow the trend of mean stress. The extruded sample shows the highest mean stress, while its plastic strain amplitude is the lowest, shown in Figure 4(b).

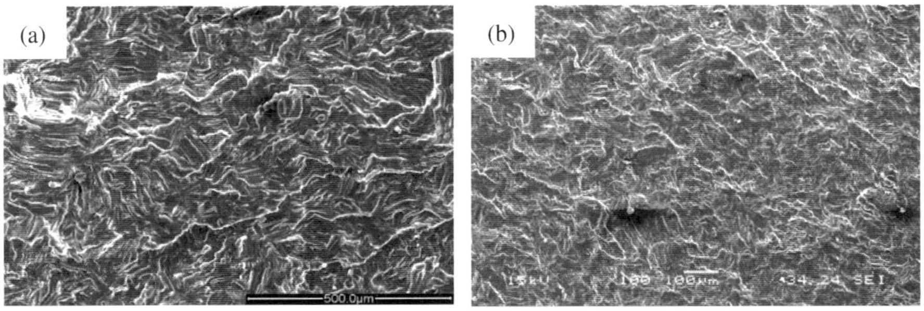

Figure6. (a and b) Typical fatigue fracture morphologies in crack stable propagation zone for the as-extruded and 5% pre-compression materials at total strain amplitude of 0.2%, respectively.

Actually, both the mean stress and plastic strain parameters can be included in the energy-related formulation. As such, the hysteresis energy can be regarded as a composite measure of an amount of fatigue damage per cycle. In this study, the tensile strain energy has been incorporated into hysteresis energy. Since there is certain variation on the hysteresis loops during cyclic deformation, the hysteresis energies were measured at half-life. As shown in Figure 7(a), it can be seen that the hysteresis energies of the extruded samples are the highest among those of pre-strained ones at all strain amplitudes. Accordingly, the low-cycle fatigue lifetime of the extruded samples is the lowest, as shown in Figure (5). Also, it is thought that the hysteresis energy is suitable for the prediction of low-cycle fatigue life of the pre-strained Mg alloy.

The devised Morrow model, proposed by Ellyin and Kujawski [13], can be expressed as follows:

$$\Delta W_t \cdot N_f^m = C \qquad (1)$$

where ΔWt, Nf, m and C are the hysteresis energy, fatigue life, and material constants, respectively. Fatigue life curve as a function of hysteresis energy is presented in Figure 7(b). It is clearly shown that the fatigue data for the pre-strained samples were well depicted with the total hysteresis energy. This implies that the total strain energy-based life prediction model provides a good prediction on the fatigue life behavior of the pre-strained Mg alloy.

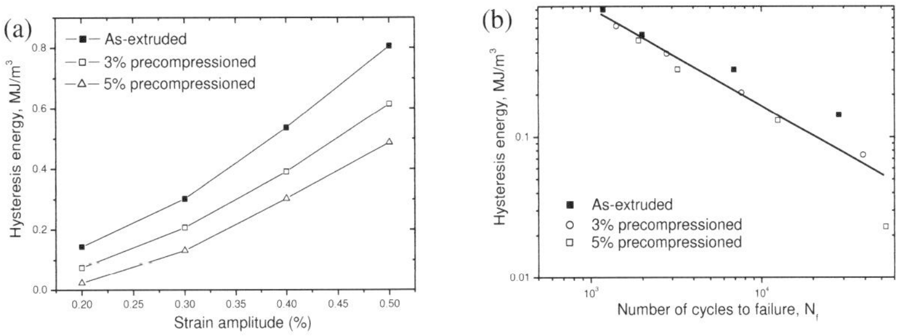

Figure7. The hysteresis energy (a) and life prediction using the energy-based model (b) of pre-compression materials.

4. Conclusions

From the above mentioned low-cycle fatigue study on pre-strained AZ31 alloy, the following conclusions can be drawn:

(1)The pre-compression process influences the plastic deformation mechanism activated during fatigue deformation, especially during tensile loading, by enhancing the activity of detwinning mechanism.

(2)The pre-strain process decreased the mean stress, but increased the plastic strain amplitude.

(3)The low-cycle fatigue lifetime can be enhanced by the pre-compression process. And hysteresis energy considering the mean stress can be used for the prediction of fatigue life.

Acknowledgements

This work was financially supported by Natural Science Foundation of Jiangsu Province (No. BK20161582), Jiangsu Overseas Research & Training Program for University Prominent Young & Middle-aged Teachers and Presidents, Scientific Research Foundation of the Nanjing Communications Institute of Technology(No.440105001) and Research Foundation of the Nanjing Communications Institute of Technology(No.JY1513),China.

References

[1] P.Lukas and M. Klesnil 1973 Cyclic stress-strain response and fatigue life of metals in low amplitude region *Mater. Sci. Eng.* **11** 345

[2] L.Kunz, P. Lukas, B. Weiss and D. 2001 Melisova Effect of loading history on cyclic stress-strain response *Mater. Sci. Eng. A* **314** 1

[3] S.M. Yin,H.J. Yang, S.X. Li, S.D. Wu and F. Yang 2008 Cyclic deformation behavior of as-extruded Mg-3%Al-1%Zn *Scr. Mater* **58** 751

[4] L. Wu, A. Jain, D.W. Brown, G.M. Stoica, S.R. Agnew, B. Clausen, D.E. Fielden and P.K. Liaw 2008 Twinning-detwinning behavior during the strain-controlled low-cycle fatigue testing of a wrought magnesium alloy *ZK60A, Acta Mater.* **56** 688

[5] Y.J. Wu, R. Zhu, J.T. Wang and W.Q. Ji 2010 Role of twinning and slip in cyclic deformation of extruded Mg-3%Al-1%Zn alloys *Scr. Mater* **63** 1077

[6]S. Hasegawa, Y. Tsuchida, H. Yano and M. Matsui 2007 Evaluation of low cycle fatigue life in AZ31 magnesium alloy *Int. J. Fatigue* **29** 1839

[7] M. Matsuzuke and S. Horibe 2009 Analysis of fatigue damage process in magnesium alloy AZ31 *Mater. Sci. Eng. A* **504** 169

[8]X.Y. Lou, M. Li, R.K. Boger, S.R. Agnew and R.H. Wagoner 2007 Hardening evolution of AZ31B Mg sheet *Int. J. Plast* **23** 44

[9] Y.N.Wang and J.C. Huang 2007 The role of twinning and untwining in yielding behavior in hot-extruded Mg-Al-Zn alloy *Acta Mater* **55** 897

[10] S.H. Park, S. Hong and C.S. Lee 2010 Role of initial {10-12} twin in the fatigue behavior of rolled

Mg-3Al-1Zn alloy *Scr. Mater* **62** 666

[11] P.R.V. Evans, N.B. Owen and L.N. McCartney 1974 Mean stress effects on fatigue crack growth and failure in a rail steel *Eng. Fract. Mech* **6** 187

[12] S. Suresh 1998 *Fatigue of Materials* (Cambridge :Cambridge University Press) p85

[13]F. Ellyin and D. Kujawski 1993 A multiaxial fatigue criterion including mean-stress effect *Advances in Multiaxial Fatigue* **1191** 55

IOP Conf. Series: Materials Science and Engineering **230** (2017) 012006 doi:10.1088/1757-899X/230/1/012006

The Experimental Research on Seismic Capacity of the Envelope Systems with Steel Frame

Jiuyang Li [1] ,Bingbing Wang [1] and Hengxu Li[1]
1 School of Civil Engineering, Changchun Institute of Technology, Changchun,China

E-mail: 429838648@qq.com, 504921917@qq.com, 634692900@qq.com

Abstract: In this paper, according to the present application situation of the external envelope systems steel frame in the severe cold region, the stuffed composite wall panels are improved, the flexible connection with the steel frame is designed, the reduced scale specimens are made, the seismic capacity test is made and some indexes of the envelope systems such as bearing capacity, energy consumption and ductility, etc. are compared, which provide reference for the development and application of the steel frame envelope systems.

1. Introduction

The application of steel frames in the civil buildings has become more and more mature. The research on light composite wall panels in the severe cold region has begun to take a shape, but the welding is mostly taken by the connection of composite wall panel with steel frame at present, so the welding connection has some advantages such as simple construction process and high speed, etc. and the bearing capacity is ensured in the case of no earthquake. In recent years, with the frequency happening of earthquakes, people have paid more attention to the seismic capacity of the building structures. The welding connection of the light composite wall panel with the steel frame is often become the weak link of the seismic capacity. Therefore, the external envelope wall panel is chosen and improved in the severe cold region, the flexible connection with the steel frame is designed and the experimental research on the seismic capacity of the envelope system (envelope wall panel and connection) is carried out in this paper.

2. Test conditions of the envelope wall and connection

According to the climatic conditions and the local energy conservation requirements in the severe cold region B, the stuffed composite wall panel is improved and the new flexible connection mode (bolt connection) is designed to establish the building envelope systems W1, W2 and W3, respectively and the 1:2 reduced scale specimen (1.5m×1.3m) is made to carry out the seismic capacity research in the test. See Table 1 for the types of W1, W2 and W3 wall panels and connection modes.

The structure of the stuffed composite wall panel and the detailed structure and process of the relevant connection are as follows:

Wall panel Y1: The stuffed composite wall panel Y1's structure: C20 volcanic ash light concrete panels are on both the external and internal sides with average thickness: 50mm, built-in Φ3@50HRB400 steel wire gauze; the thickness of the middle stuffed heat preservation layer are 200mm, EPS; the three-layer materials are wrapped withΦ8HRB400 steel grids around the wall panel to establish the framework with total thickness of the wall panel 300mm[1].

Wall panel (G1, G2): the new stuffed composite wall panel has been improved by the project group on the basis of the stuffed composite wall panel (Y1). Its structure: C20 volcanic ash light concrete panels are on both the external and internal sides with thickness of 50mm, built-inΦ5@200 CRB500 steel wire gauze; XPS for the middle stuffed heat preservation layer, with thickness of 150mm; the three-layer materials are stringed and connected with theΦ5CRB500 steel bar truss; the total thickness of the wall panel is 250mm[2].

Table 1 Operating Conditions of the building envelope systems W1, W2 and W3

Envelope System	Type of Wall Panel		Connection Mode	
W1	Y1	Existing	YZ1	New design
W2	G1	Improved	GZ1	New design
W3	G2	Improved	GZ2	New design

The connecting mode of the three wall panels and the steel frame in the experiment is the semi-outer hanging, that is, the inner concrete panel is put on the steel frame, the heat preservation layer and the external concrete panel are wrapped on the external steel frame to establish the semi-outer hanging connection[3]. Such a connection mode prevents the cold bridge of the principal structure. The connecting positions of the three wall panels with the steel frame are on the steel frame beam, that is, 1/4 of the span of the wall panel, with the detailed structure as follows:

See Figure 1 for the YZ1 connection mode, with the detailed structure as follows: in the wall panel processing, a slot steel frame (10×10×10cm, thickness of steel sheet: 5mm) or the same specification of U-bar is embedded in the designed connecting position; the lower flange of the steel frame is reserved with the bolt hole (12+1.5mm); the same as the upper flange of the design position of the steel frame. In order to reduce the impairing of the steel frame beam by the flange opening, a steel sheet can be welded and the same bolt hole can be opened on the steel sheet; after the wall panel is hoisted in place, the M12 bolt penetrates the flange, welded steel sheet and the bolt hole of the lower flange of the steel frame from the lower side of the upper flange of the steel frame, the bolt cap is screwed and finally the foam concrete or polyurethane are used to fill and block the steel frame.

See Figure 2 for the GZ1 connecting mode and structure: in the making of the wall panel, a I-shaped embedment is embedded in the designed connecting position (consisting of 2 steel sheets and 2 sleeves, dimensions of the steel sheet: 150×120×6mm, embedded in the internal and external concrete of the wall panel, it is required that the steel sheet embedded in the internal side of the wall panel and the internal side surface of the wall panel should be on the same level; the internal diameter of the steel sleeve isφ14, 210mm long, the sleeve and the internal side surface of the wall panel are on the same level, the built-in threaded sleeve is in the range of 1/2 of the length close to the internal side of the wall panel and the solid structure is in the residual part; the surface of the steel sleeve should be blocked so as to prevent the concrete falling into it.); the upper surface of the designed connecting position of the steel frame beam is welded with the angle steel of L90×10, the vertical limb is opened with 2 bolt holes (14+1.5mm); after the wall panel is hoisted in place, the bolts are used to connect the vertical limb of the angle steel and the sleeve in the embedment of the wall panel.

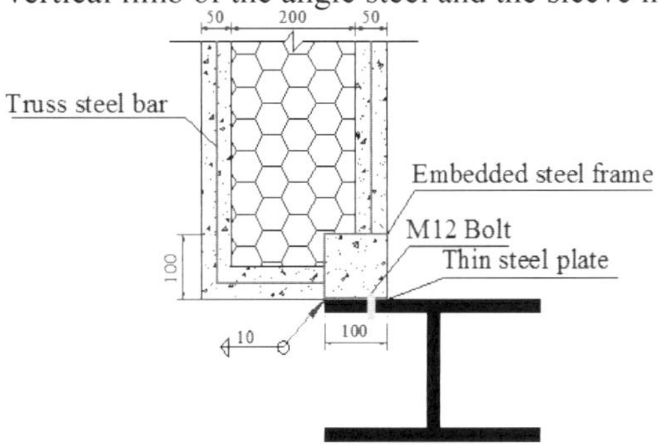

Fig.1 YZ1 Connecting Node

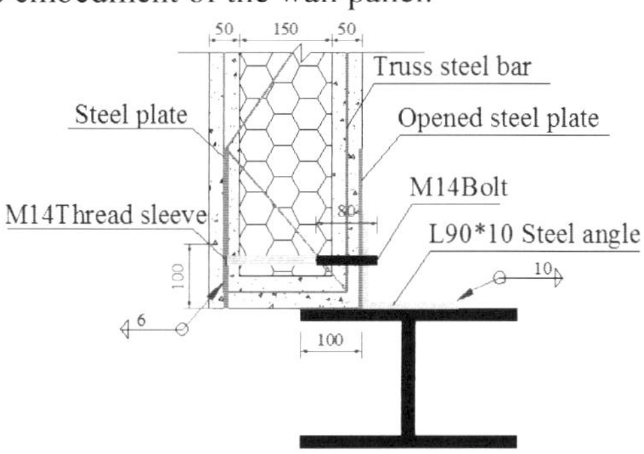

Fig.2 GZ1Connecting Node

Specimen GZ2: the connecting mode and process of GZ2 are completely the same with those of GZ1 and the only difference is that there is 1 bolt for each connecting node.

It is considered that there are the multiple specimens in the project which should be connected and loaded to the same frame. With the design and calculation; the loading frame is made. The Q345 steel is taken for the loading frame and support, which includes the support beam, two frame columns, two frame beams (upper beam and bottom beam), corbel and diagonal stay. The hot-roll H steel is taken for the steel beam and steel column; the type of the beam section is HW350×350×12×19; the type of the column section is HW350×350×12×19; the type of the support beam is HW350×450×12×19; the U bar is taken for the steel diagonal stay and the type of the section is [250×82×11×12. See Figure 3 for the vertical view of the loading frame and support and see Figure 4 after the wall panel is installed.

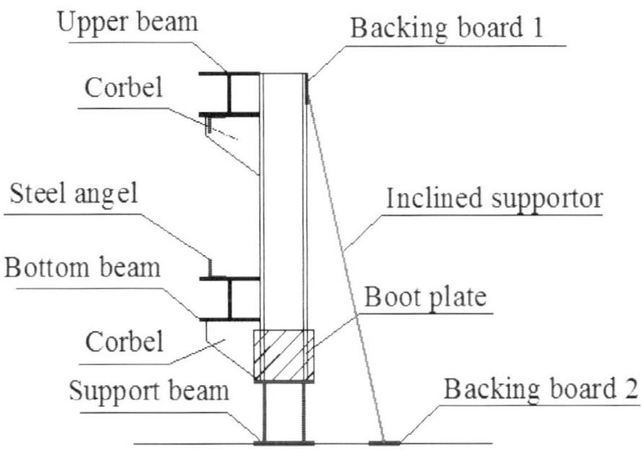

Fig. 3 Loading Frame and Support

Fig. 4 Loading Frame and Connecting Position

3. Loading test

The test adopts the pseudo-static low-cycle repeated loading to directly impose the simulated earthquake load on the composite wall panel and transmit the load to the steel frame via the connection to research the force mechanism and destruction form of the envelope system on the seismic effect. The loading plan is the mixed loading method of force control and displacement control. The destruction principle of the envelope system is that the envelope system fails when the load obviously declines (to less than 85% of the peak load) or one of the following phenomena occurs: ① the bolt is sheared; ② the local concrete of the wall panel is crushed; ③ the serious crack occurs on the surface of the wall panel and penetrates the wall panel.

In order to assure the accurate test data and reliable result analysis, before the formal loading in the test, pre-loading to the specimen are first carried out to test the working performances of the equipment and the loading system. First pre-impose 2kN, second to 5kN and then return to 0 and this repeats twice. In the pre-loading process, whether the strains or displacements of all the test points tend to increase in the linear way has been observed; when the unloading is 0, if the reading numbers of all the test points return to the initial readings. In the formal loading, the force control method has been first take to impose the horizontal repeated centralized force on the middle of the specimen via the hydraulic servo actuator, taking the horizontal pushing force of the actuator as the positive direction and the pulling force as the negative direction. First the pushing force is imposes and then the pulling force. The initial value is 0kN, taking $\pm 5kN$ as the loading step to gradually increase the load and keep each level of load for 2mins and then impose the next-level load. After the envelope system yields and the displacement control method is taken according to the 1 time yielding displacement and twice yielding displacement, etc. until the envelope system fails.

Measurement of displacements: In the test, the displacements of the wall panel on the effect of the load are collected by the displacement sensor of the actuator system. It is considered that the relative sliding may happen or is produced between the loading frame bracket and the floor on the effect of the

IOP Conf. Series: Materials Science and Engineering **230** (2017) 012006 doi:10.1088/1757-899X/230/1/012006

repeated load, so the displacement sensor is also laid on the bottom of the frame bracket so as to assure the accuracy of the displacement of the wall panel measured. The layout of the displacement gauges for the test loading is shown in Figure 5.

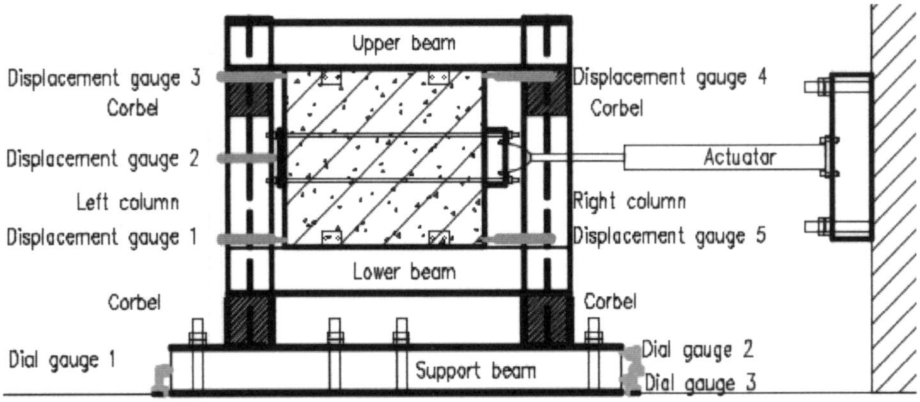

Fig. 5 Layout Figure of Displacement Gauges for the Test Loading

4. Test phenomena and analysis

For the W1 envelope system, in the loading process, before the repeated load is less than 20kN, no any crack is observed on the surface of the wall panel. The loading path and the unloading path nearly completely are coincide with each other and the wall panel is in the elastic phase; when it is loaded to 30kN, a few fine inclined cracks occur around the embedment in the lower left corner of the wall panel and on the back of the wall panel. The obvious displacement occurs to the specimen; when it is loaded to 40kN, the bolt at the connecting node of the wall panel and the connecting steel sheet are displaced to produce the clear friction noise; when it is loaded to 50kN, the connecting node produces the clear "bang bang". The deformation of the connecting node is expanded. The loading frame slides and the friction displacement occurs between the support beam and the anti-pull anchor arm; when it is continued to the 70kN cycle, the inclined cracks around the lower left corner and lower right corner of the wall panel in the pre-embedded specimens are obviously widened and deepened with the occurrence of the new cracks; when it is loaded to 80kN cycle, the inclined cracks are around the embedment in the lower left corner and lower right corner of the wall panel penetrates the wall panel. The bolt of the connecting node yields. The shearing deformation visible to human eyes occurs to the bolt; when it is loaded to 90kN, the bolt of the connecting node in the lower part of the wall panel is sheared, and at this time, the wall panel loses the bearing capacity and the loading is stopped.

For the W2 envelope system, before the repeated load is less than 70kN, the representations of the specimen and the node are basically the same with those before the Y1 wall panel cracks and later the crack occurs to the wall panel G1; when it is loaded to 80kN, the inclined crack in the upper part of the embedment in the lower left corner and lower right corner of the wall panel is continued to be widened and deepened. It is going to penetrate it and the new crack occurs. The accumulative injury becomes more and more obvious with the repeated load; when it is loaded to 100kN, a small part of the concrete in the lower right corner peels off. The inclined crack of the embedment in the lower left corner and the lower right corner of the wall panel penetrates the wall. A lot of cracks occur on the side of the wall panel; when it is loaded to 130kN cycle, the concrete at the embedment continues to be separated, and cracks occur to the concrete on the side of the wall panel. The "bang bang" noises are produced from inside the wall panel in the pushing and pulling conversion each time; when the load is increased to 165kN, the reinforced bar at the embedment of the wall panel is broken. The concrete at the node area is completely torn. The rigidity of the wall panel is greatly decreased and the degradation is serious. At this time, the wall panel loses the bearing capacity.

IOP Conf. Series: Materials Science and Engineering **230** (2017) 012006 doi:10.1088/1757-899X/230/1/012006

For the W3 envelope system, before the repeated load is less than 70kN, the representations of the wall panel and connection are basically the same with those of W2. When the G2 wall panel and the GZ2 connecting node are loaded to the 70kN cycle, the inclined crack in the upper part of the lower left corner and the lower right corner of the wall panel continues to be widened and deepened and the new cracks continuously occur; when it is cyclically loaded to 90kN, lots of inclined cracks occur around the embedment of the wall panel and the inclined cracks of the embedment in the lower left corner and the lower right corner. The cracks in the direction of the middle line on the back penetrate the wall; when it is loaded to 120kN cycle, the embedment concrete is seriously peeled off. The bolt of the connecting node of the wall panel is sheared. The structure loses the bearing capacity.

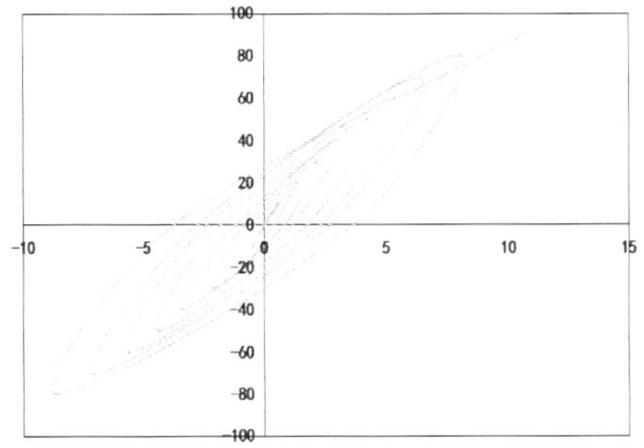

Fig. 6 (p-Δ) Hysteresis Curve of W1

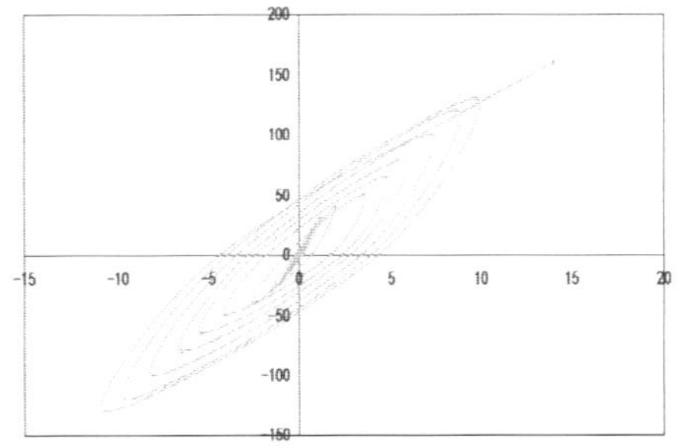

Fig. 7 (p-Δ) Hysteresis Curve of W2

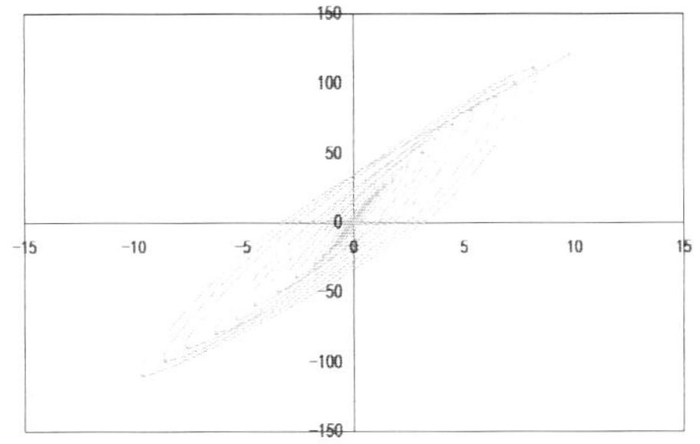

Fig. 8 (p-Δ) Hysteresis Curve of W3

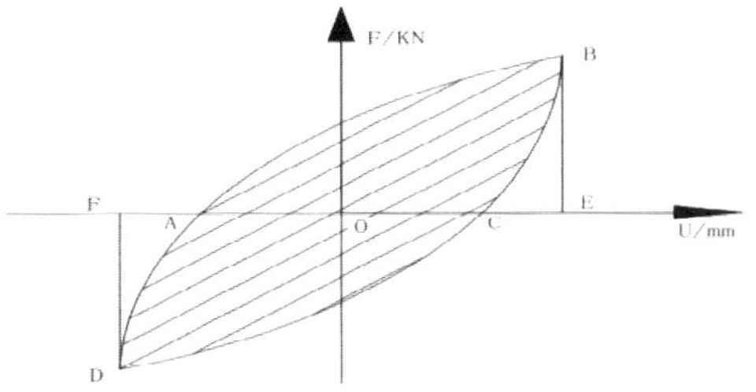

Fig. 9 Equivalent Viscous Damping Coefficient Calculation

In the test, the actuator loading system records the horizontal load-displacement (p-Δ) relational hysteresis curves of the models. The p-Δ hysteresis curves of all the models are shown in Figure 6, Figure 7 and Figure 8. Obviously, the bearing capacity of the envelope system W2 is the highest, W3 has one bolt less than W2 in each connection. Its bearing capacity is reduced, and the bearing capacity of the W1 envelope system is the lowest.

5. Energy consumption and ductility analysis

When an earthquake happens to the building structure, for the energy in the earthquake input structure, the structure has a process of absorbing and consuming the energy. The energy consumption capacity of the structural member is measured with the area enclosed by the (p-Δ) hysteresis curves. The area enclosed by the hysteresis curves in the loading phase reflects the magnitude of the energy absorbed by the structure. The area enclosed in the unloading phase reflects the energy consumed by the structure. The energy absorbed by the structure is converted to the energy consumption via the internal friction resistance or local injuries. The more the energy is consumed, the less possibility that the structure is destroyed. Therefore, the more saturated the hysteresis circle, the bigger the enclosed area, and the better the energy consumption performance of the structure. In the seismic test of the structure, the energy consumption capacity is one of the important indexes for investigating the seismic capacity of the structure[4]. The energy consumption coefficient E and equivalent viscous damping coefficient "he" are usually adopted to appraise the energy consumption performance of the structure, shown in

Figure 9. See Formula (1) and Formula (2) for calculation of the energy consumption coefficient E and equivalent viscous damping coefficient "he":

$$E = \frac{S_{(ABC+CDA)}}{S_{(\Delta OBE+\Delta ODF)}} \tag{1}$$

$$h_e = \frac{S_{(ABC+CDA)}}{2\pi \times S_{(\Delta OBE+\Delta ODF)}} \tag{2}$$

Whereas: $S_{(ABC+CDA)}$-----area enclosed by the hysteresis circle;

$S_{(\Delta OBE+\Delta ODF)}$ -----area of the triangle enclosed by the hysteresis circle peak value, coordinate origin and horizontal coordinate axis;

According to Formula (1) and Formula (2), the area of the hysteresis circle of the envelope system at the yielding load and limit load are calculated to obtain the energy consumption coefficient E.

It is concluded from Table 1: 1) In the yielding load, the equivalent viscous damping coefficient "he" and the energy consumption coefficient E of the envelope system W2 are the maximum; 2) In the limit load, the total energy consumption and equivalent viscous damping coefficient of the envelope system W3 are the maximum; 3) W2 among the three envelope systems has the maximum total energy consumption.

Table 2 Energy Consumption Coefficient E and Equivalent Viscous Damping Coefficient he of the Specimens

Envelope System	Wall Panel Name	Connection Name	Total Energy Consumption	Yielding Load		Limit Load	
			W(N.m)	*E*	*he*	*E*	*he*
W1	Y1	YZ1	1164.639	0.992	0.158	1.382	0.220
W2	G1	GZ1	2402.133	1.0	0.159	1.10	0.175
W3	G2	GZ2	1750.688	0.851	0.136	1.759	0.280

Ductility is an index reflecting the deformation capacity of the structure and member. It refers to the performance that the structure or member will not be destroyed when the external load is continuously acted on the structure or member and the deformation continues increasing in the plastic phase. As to the seismic capacity of the structure is concerned, ductility is more important than the bearing capacity. The ductility of the structure or member is usually expressed by the ductility coefficient. In this paper, the displacement ductility coefficient μ and displacement angle θ is taken to reflect the deformation capacity and the seismic capacity of the structure. It is calculated by taking Formula (3) and Formula (4) according to the Specification of Test Methods for Earthquake Resistant Buildings (JGJ101) [5]:

$$\mu = \frac{x_u}{x_y} \tag{3}$$

$$\theta = \arctan\frac{X}{H} \tag{4}$$

Whereas, x_y, x_u----displacement value of the structure or member in the yielding load and limit load, respectively;

H----Height of the structure or member

According to the test data of the loading process of each envelope system, the displacement ductility coefficients and displacement angles in all the phases are available in accordance with (3) and (4).

It can be seen from Table 2: The three envelope systems of cracking load are about 20 kN, roughly equivalent to load of non-structural members under 8 degree earthquake action. At this time, the crack displacement angles are in 1/600~1/1000, less than the limit value of 1/300 of the displacement angle between the elastic layers on the effect of small earthquakes on the multi-layer and high steel structures in the specification; Under the seismic action of the maximum deformation of the main structure equivalent to 8 degree earthquake, the envelope system has not yet yielded; When the envelope system is destroyed, the ultimate displacement angle in 1/90~1/150, is less than the standard of many high-rise steel structure elastic-plastic displacement angle limit of 1/50 requirements under the big earthquake. It will not affect the main structure. In addition, it can be seen that the three envelope systems have good ductility, of which the envelope system W1 has the best ductility.

Table 3 Displacement Angles θ and Ductility Coefficients of the building envelope systems

Envelope System	Crack Displace-men（mm）	Yielding Displace-ment(mm)	Yielding Displacement (mm)	Crack Displacement Angle	Yielding Displaceme nt Angle	Limit Displacement Angle	Ductility Coefficient
	x_c	x_y	x_u	θ_c	θ_y	θ_u	μ
W1	2.06	5.36	10.672	1/632	1/243	1/122	1.990
W2	1.21	7.19	14.032	1/1078	1/181	1/93	1.952
W3	1.36	6.47	9.784	1/960	1/201	1/133	1.512

6. Conclusion

With the experimental research on the seismic capacity of the three envelope systems consisting of 3 stuffed composite wall panels and 2 connection modes in the paper, the conclusion can be obtained as follows: 1) no serious destruction or separation occurs to the stuffed composite wall panels after the connections are destroyed. The wall panel is good as a whole and the test purpose of the project is achieved---the node is destroyed earlier than the whole; 2) the three envelope systems designed in the test meet the seismic capacity requirements of the structure in the bearing capacity, energy consumption and ductility in the specification. The seismic capacity is high and the differences between the comprehensive evaluation of the three systems in the capacity are small, all of which can be applied in the actual engineering; 3) 2 kinds of connection modes are the flexible connections, which prevent the common welding connection mode at present and achieve the structural requirement that no effect of the light wall panel on the principal rigidity of the frame structure; 4) in the case of meeting the local energy conservation requirements, the thickness of the G wall panel is smaller than that of the Y wall panel, which increases the using area of the building.

References

[1] Weidong Ji. Retaining wall plate in steel structure residence system technology research. [D]. *Tianjin University*, 2013: 4-5
[2] Hanhua Du. Tectonic Connection Technology and Economical Efficiency Study of Outer Enclosure System of Light Steel Residence in Jilin Province. [D]. Jilin. *Changchun Institute of Technology*. 2016: 26-32
[3] WeiGuo. Research on Structure Technology and Seismic Performance of Light Steel Structure Housing Sandwich Composite Panels. [D]. Jili. *Changchun Institute of Technology*. 2016: 31-54
[4] Lijiun Dou. *Precast Reinforced Concrete Composite Insulation of External Wall Plate Design and Construction*. [M]. Beijing: Higher education press. 2013：
[5] JGJ/T101-2015. *Procedures of the Aseismic Test Method*. [S]. Beijing: China building industry press, 2015.

Acknowledgements

This work was financially supported by the Jilin Science and Technology Development Planned Topics 《Tectonic Connection Technology and Economical Efficiency Study of Outer Enclosure

System of Light Steel Residence in Jilin Province》 [20140203014SF] and Jilin Education Development Planned Topics《Research on the Tectonic Connection Technology of the Envelope Systems with the Steel Frame in Cold Region》 [2015296]

IOP Conf. Series: Materials Science and Engineering **230** (2017) 012007 doi:10.1088/1757-899X/230/1/012007

An Experimental Study on Shrinkage Strains of Normal-and High-Strength Concrete-Filled Frp Tubes

Thomas Vincent[1] and Togay Ozbakkaloglu[2]

[1]School of Computer Science, Engineering and Mathematics, Flinders University, Adelaide, South Australia, Australia 5042
[2]School of Civil, Environmental and Mining Engineering, the University of Adelaide, Adelaide, South Australia, Australia 5005

Abstract. It is now well established that concrete-filled fiber reinforced polymer (FRP) tubes (CFFTs) are an attractive construction technique for new columns, however studies examining concrete shrinkage in CFFTs remain limited. Concrete shrinkage may pose a concern for CFFTs, as in these members the curing of concrete takes place inside the FRP tube. This paper reports the findings from an experimental study on concrete shrinkage strain measurements for CFFTs manufactured with normal- and high-strength concrete (NSC and HSC). A total of 6 aramid FRP (AFRP)-confined concrete specimens with circular cross-sections were manufactured, with 3 specimens each manufactured using NSC and HSC. The specimens were instrumented with surface and embedded strain gauges to monitor shrinkage development of exposed concrete and concrete sealed inside the CFFTs, respectively. All specimens were cylinders with a 152 mm diameter and 305 mm height, and their unconfined concrete strengths were 44.8 or 83.2 MPa. Analysis of the shrinkage measurements from concrete sealed inside the CFFTs revealed that embedment depth and concrete compressive strength only had minor influences on recorded shrinkage strains. However, an analysis of shrinkage measurements from the exposed concrete surface revealed that higher amounts of shrinkage can occur in HSC. Finally, it was observed that shrinkage strains are significantly higher for concrete exposed at the surface compared to concrete sealed inside the CFFTs.

1. Introduction

External confinement of concrete with fiber reinforced polymer (FRP) composites has become widely accepted as a technique for enhancing strength and ductility of normal- (e.g.,[1-3]) and high-strength (e.g., [4-9]) concretes (NSC and HSC). A large number of experimental studies into axial compressive behavior have been performed over the last two decades producing over 3000 test results, as discussed and assessed in the recent comprehensive review studies [10, 11] . The majority of early experimental studies focused on FRP-wrapped concrete columns (e.g., [12, 13]), with fewer studies reported on concrete-filled FRP tubes (CFFTs). Initial experimental studies on CFFTs (e.g., [14, 15]) along with more recent extensive studies on the axial compressive (e.g., [5, 16-20]) and seismic behavior of CFFT columns (e.g., [21-26]) have demonstrated the ability of CFFTs to develop very high inelastic deformation capacities, making them an attractive option for construction of new high-performance columns.

The use of higher strength concretes in the construction industry has been on a steady incline due to the superior performance and economy offered by HSC over NSC. These beneficial characteristics result in more cost-effective construction of bridges and multistory buildings. Recently, a significant amount of research has been performed on CFFTs manufactured with HSC (e.g., [6-9]) that has demonstrated the ability of these structural elements to display substantial strength and strain enhancements. However, the concrete shrinkage strain development for HSCFFTs is yet to be investigated. Concrete shrinkage may pose a concern for CFFT columns, as in these columns the curing of concrete takes place inside the FRP tube. This concrete shrinkage may influence the performance of the CFFT columns in the long term, and hence it is important to quantify and understand the shrinkage behavior of these members. This is of particular importance for CFFTs manufactured with HSC due to the potential of higher levels of shrinkage associated with higher strength concretes [27, 28].

Only two studies have been reported to date on concrete shrinkage in FRP-confined concrete, with both studies examining NSC specimens [29, 30]. Naguib and Mirmiran [29] monitored the concrete shrinkage development of a single CFFT specimen, with these strains monitored over a 12 month period. Karimi et al. [30] studied the compressive behavior of CFFT specimens with internal steel I-beams manufactured with and without a shrinkage reducing agent. However, to date, no experimental study has systematically investigated concrete shrinkage strain development within CFFTs.

To address this gap in the research, this paper presents an experimental investigation into concrete shrinkage measurements for normal- and high-strength CFFTs (NSCFFTs and HSCFFTs). The influence of strain measurement location and concrete compressive strength are examined. The results of the test program are first presented, followed by a detailed discussion on the observed influences.

2. Test database

2.1. Test specimens and materials

Six aramid FRP (AFRP)-confined cylindrical specimens, all with a 152 mm diameter and a 305 mm height, were prepared to monitor shrinkage development of the concrete in the CFFTs. Two different concrete mixes were used, with average compressive strengths of 44.8 and 83.2 MPa which are referred to as NSC and HSC, respectively. Three of the CFFTs were manufactured as NSC specimens, with the remaining three manufactured as HSC. The tubes of the CFFTs were made using two layers of AFRP, and the bottom and top surfaces of the specimens remained covered and uncovered respectively, for the entire duration of shrinkage strain measurements. Both the NSC and HSC were batched and mixed in the laboratory and consisted of crushed bluestone gravel as the coarse aggregate, with a 10 mm nominal maximum diameter. Superplasticiser was added to the HSC mix to ensure a workable concrete, which resulted in a measured slump of over 220 mm. The mix designs for the NSC and HSC mixes are reported in table 1.

Table 1. Concrete mix proportions

Concrete constituent	NSC	HSC
Cement (kg/m^3)	380	506
Silica fume (kg/m^3)	0	44
Sand (kg/m^3)	700	700
Coarse Aggregate (kg/m^3)	1050	1050
Water (kg/m^3)	220	172
Superplasticiser (kg/m^3)	0	12
Water to cement ratio	0.58	0.33

The specimens were manufactured using a wet lay-up technique, which involved wrapping epoxy resin saturated aramid fiber sheets around Styrofoam templates. The FRP tubes were prepared using two layers of 0.2 mm thick fiber sheet with fibers oriented in the hoop direction, and they had a single overlap region of 150 mm length. The fibers used in the manufacture of the AFRP tubes had an elastic modulus of 128.5 GPa and ultimate tensile strength and strain of 2390 MPa and 1.86% respectively, determined by coupon tests.

2.2. Instrumentation

The six CFFTs were instrumented with strain gauges and placed in a temperature controlled room for 6 months to monitor the shrinkage strain development of concrete. Each specimen was instrumented to measure concrete shrinkage in the radial direction for the CFFTs at 4 locations. A single 20-mm strain gauge was bonded to the top concrete surface and an additional 3 embedded strain gauges with a gauge length of 20 mm were located within the concrete at depths (*d*) of 15, 75 and 150 mm from the top concrete surface. All four strain gauges were aligned horizontally along the diameter of CFFTs to record shrinkage of concrete only in the radial direction. Specimens were placed in a controlled environment, with the temperature maintained at 22 ± 2 °C and relative humidity at 55 ± 10%. An overview of this instrumentation set-up is presented in figure 1.

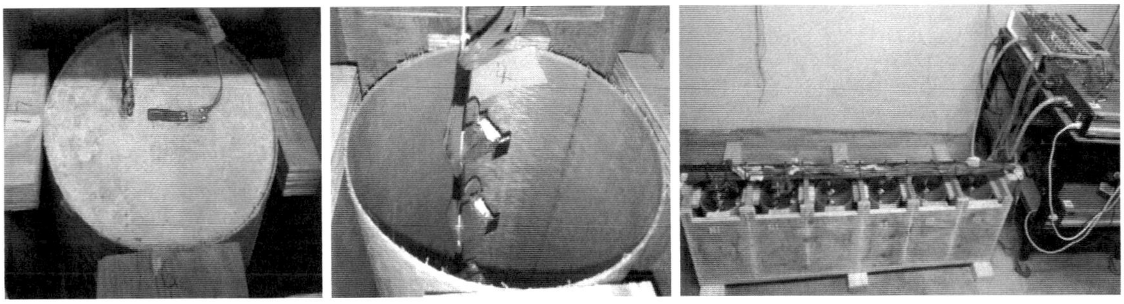

a) *b*) *c*)

Figure 1. Instrumentation of CFFTs: a) Typical surface strain gauge; b) Typical embedded strain gauges; c) AFRP tubes and data acquisition system prior to concrete pour

3. Test Results

The concrete shrinkage strain measurements of the NSC and HSC specimens are presented in figures 2 and 3, respectively. Measurements of strains began immediately after strain gauges were mounted to the top surface of the hardened concrete, approximately 24 h after the concrete was cast into the FRP tubes.

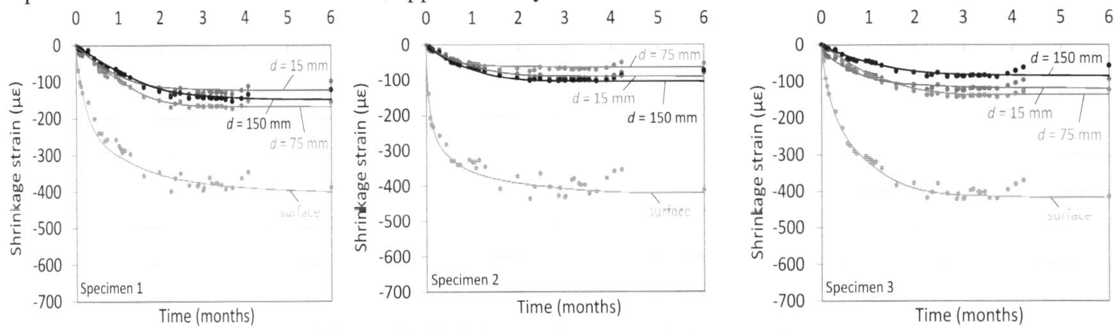

Figure 2. Shrinkage of concrete in NSCFFTs

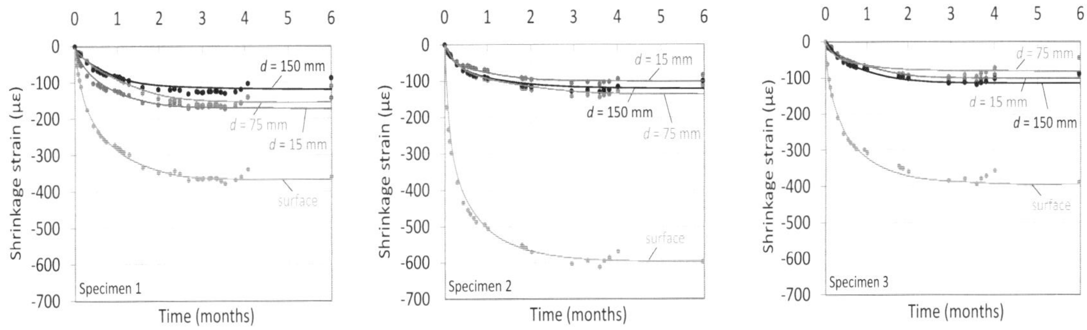

Figure 3. Shrinkage of concrete in HSCFFTs

It can be seen from the shrinkage strains of the NSCFFTs shown in figure 2 that NSC specimens recorded up to approximately 400 με of radial shrinkage after six months, with over 85% of this occurring during the first two months. A similar shrinkage behavior was evident for the HSC specimens shown in figure 3, with a maximum of around 600 με of shrinkage recorded at the end of the six-month period. As can be seen in figures 2 and 3, significant variations in shrinkage strain development can be seen when comparing values at the same locations from identically prepared specimens. A noticeable example of this is a comparison of surface measurements for HSCFFTs, where specimen 2 deviates from the general trend. These differences in measured shrinkage strains can be attributed to localized effects such as aggregate distribution.

3.1. Influence of strain measurement location

A comparison of the shrinkage strain development shown in figures 2 and 3 reveals that shrinkage strains recorded at the surface of the concrete are significantly higher than those recorded from embedment depths within the concrete of CFFTs. This experimental observation can be seen to be independent of concrete strength as it's equally applicable to NSCFFTs and HSCFFTs. This outcome can be attributed to less water loss experienced by the concrete within the CFFT, resulting in lower shrinkage strains. In addition to this observation, a comparison of shrinkage strains monitored from embedded strain gauges reveals that embedment depth has little to no influence on amount of shrinkage with the majority of embedded gauges recording approximately 100 με of shrinkage. Once again, this experimental observation can be seen to be equally applicable to NSCFFTs and HSCFFTs, and is in agreement with those presented in Naguib and Mirmiran [29] for the single NSCFFT examined.

3.2. Influence of concrete compressive strength

The influence of concrete compressive strength was examined through the comparison of shrinkage strain development at the same locations for NSCFFTs and HSCFFTs in figures 2 and 3, respectively. A comparison of the strains recorded at the top surface of the CFFTs reveals that shrinkage strain values are similar between NSC and HSC, except for one measurement that indicates that exposed concrete in HSCFFTs can develop significantly higher levels of shrinkage strains. A comparison of the shrinkage strains monitored from embedded strain gauges reveals that concrete compressive strength has little to no influence on development of shrinkage strains of concrete sealed inside CFFTs.

4. Conclusion

This paper has presented the results of an experimental study into the influence of strain measurement location and concrete compressive strength on concrete shrinkage strain measurements for CFFTs manufactured with normal- and high-strength concrete (NSC and HSC). Based on the results and discussion presented in this paper, the following conclusions can be drawn:

1. Shrinkage strains are significantly higher for concrete exposed at the surface compared to concrete sealed inside the CFFTs. It was found that these observations are equally applicable to NSCFFTs and HSCFFTs.
2. NSCFFTs and HSCFFTs typically develop comparable levels of concrete shrinkage strains, however HSCFFTs can develop significantly higher amounts. This observation of higher values was found to be limited to shrinkage strains measured on exposed concrete.

5. References

[1] Karabinis, A.I. and T.C. Rousakis, *Concrete confined by FRP material: a plasticity approach.* Eng. Struct., 2002. 24(7): p. 923 - 932.

[2] Jiang, T. and J.G. Teng, *Analysis-oriented stress-strain models for FRP-confined concrete.* Eng. Struct., 2007. 29(11): p. 2968 - 2986.

[3] Wu, Y.F. and Y.Y. Wei, *Effect of cross-sectional aspect ratio on the strength of CFRP-confined rectangular concrete columns.* Eng. Struct., 2010. 32: p. 32-45.

[4] Lim, J.C. and T. Ozbakkaloglu, *Confinement model for FRP-confined high-strength concrete.* ASCE, J. Compos. Constr., 2014. 18(4): p. 04013058.

[5] Vincent, T. and T. Ozbakkaloglu, *Influence of fiber orientation and specimen end condition on axial compressive behavior of FRP-confined concrete.* Constr. Build. Mater., 2013. 47: p. 814-826.
[6] Vincent, T. and T. Ozbakkaloglu, *Influence of overlap configuration on compressive behavior of CFRP-confined normal- and high-strength concrete.* Mater. Struct., 2016. 49(4): p. 1245-1268.
[7] Vincent, T. and T. Ozbakkaloglu, *Influence of concrete strength and confinement method on axial compressive behavior of FRP-confined high- and ultra high-strength concrete.* Compos. Part B, 2013. 50: p. 413-428.
[8] Ozbakkaloglu, T. and T. Vincent, *Axial compressive behavior of circular high-strength concrete-filled FRP tubes.* ASCE, J. Compos. Constr., 2014. 18(2): p. 04013037.
[9] Vincent, T. and T. Ozbakkaloglu, *Compressive behavior of prestressed high-strength concrete-filled aramid FRP tube columns: Experimental observations.* ASCE, J. Compos. Constr., 2015. 19(6): p. 04015003.
[10] Ozbakkaloglu, T., J.C. Lim, and T. Vincent, *FRP-confined concrete in circular sections: Review and assessment of the stress-strain models.* Eng. Struct., 2013. 49: p. 1068-1088.
[11] Lim, J.C. and T. Ozbakkaloglu, *Design model for FRP-confined normal- and high-strength concrete square and rectangular columns.* Mag. Conc. Res., 2014. 66(20): p. 1020-1035.
[12] Rousakis, T. and A. Karabinis, *Substandard reinforced concrete members subjected to compression: FRP confining effects.* Mater. Struct., 2008. 41(9): p. 1595 - 1611.
[13] Berthet, J.F., Ferrier. E., and H. P., *Compressive behavior of concrete externally confined by composite jackets. Part A: experimental study.* Constr. Build. Mater., 2005. 19(3): p. 223 - 232.
[14] Mirmiran, A. and M. Shahawy, *Behavior of concrete columns confined by fiber composites.* ASCE J. Struct. Eng., 1997. 123(5): p. 583 - 590.
[15] Fam, A. and S. Rizkalla, *Behavior of axially loaded concrete-filled circular fiber-reinforced polymer tubes.* ACI Struct. J., 2001. 98(3).
[16] Ozbakkaloglu, T. and T. Xie, *Geopolymer Concrete-Filled FRP Tubes: Behavior of Circular and Square Columns under Axial Compression.* Compos. Part B, 2016. 96: p. 215-230.
[17] Xie, T. and T. Ozbakkaloglu, *Behavior of Recycled Aggregate Concrete-Filled Basalt and Carbon FRP Tubes.* Constr. Build. Mater., 2016. 105: p. 132-143.
[18] Vincent, T. and T. Ozbakkaloglu, *Influence of slenderness on stress-strain behavior of concrete-filled FRP tubes: an experimental study.* ASCE, J. Compos. Constr., 2014. 19(1): p. 04014029.
[19] Lim, J.C. and T. Ozbakkaloglu, *Hoop strains in FRP-confined concrete columns: experimental observations.* Mater. Struct., 2015: p. 10.1617/s11527-014-0358-8.
[20] Chen, L. and T. Ozbakkaloglu, *Corner strengthening of square and rectangular concrete-filled FRP tubes.* Eng. Struct., 2016. 117: p. 486 - 495.
[21] Idris, Y. and T. Ozbakkaloglu, *Behavior of square fiber reinforced polymer-high-strength concrete-steel double-skin tubular columns under combined axial compression and reversed-cyclic lateral loading.* Eng. Struct., 2016. 118: p. 307 - 319.
[22] Seible, F., et al., *Development of advanced composite carbon shell systems for concrete columns in seismic zones.* 11th World Conference on Earthquake Engineering, 1996: p. Elsevier Science, Paper No. 1375.
[23] Yamakawa, T., P. Zhong, and A. Ohama, *Seismic performance of aramid fiber square tubed concrete columns with metallic and/or non-metallic reinforcement.* Reinf. Plas. Compos., 2003. 22(13): p. 1221-1237.
[24] Idris, Y. and T. Ozbakkaloglu, *Seismic behavior of high-strength concrete-filled FRP tube columns.* ASCE, J. Compos. Constr., 2013. 17(6): p. 04013013.
[25] Ozbakkaloglu, T. and M. Saatcioglu, *Seismic performance of square high-strength concrete columns in FRP stay-in-place formwork.* ASCE, J. Struct. Eng, 2007. 133(1): p. 44-56.
[26] Ozbakkaloglu, T. and Y. Idris, *Seismic behavior of FRP-high-strength concrete-steel double skin tubular columns.* ASCE, J. Struct. Eng, 2014. 140(6): p. 04014019.

[27] Bissonnette, B., P. Pierre, and M. Pigeon, *Influence of key parameters on drying shrinkage of cementitious materials.* Cem. Concr. Res., 1999. 29(10): p. 1655-1662.
[28] Kwan, A., W. Fung, and H. Wong, *Shrinkage of high-strength concrete and high-flowability concrete.* HKIE Transactions, 2010. 17(3): p. 25-33.
[29] Naguib, W. and A. Mirmiran, *Time-dependent behavior of fiber-reinforced polymer-confined concrete columns under axial loads.* ACI Struct. J., 2002. 99(2): p. 142-148.
[30] Karimi, K., M. Tait, and W. El-Dakhakhni, *Testing and modeling of a novel FRP-encased steel-concrete composite column.* Compos. Struct., 2011. 93(5): p. 1463-1473.

IOP Conf. Series: Materials Science and Engineering **230** (2017) 012008 doi:10.1088/1757-899X/230/1/012008

Study on the Key Technology of High Purity Strontium Titanate Powdcr Synthesized from Oxalic Acid Co-sediment Precipitation

Xiaoguo Bi[1], Yingnan Dong[1], Yingjie Li[1], Wei Niu[1], Jian Tang[1], Shuang Ding[1] and Meiyang Li[1]

1 Shenyang Institute of Engineering, Shenyang, P. R. China

E-mail: dyn4870@qq.com

Abstract. Oxalate coprecipitation is applied in this paper, high purity titanium tetrachloride, and after the purification of strontium chloride, match with a certain concentration of solution, oxalate and strontium chloride and titanium tetrachloride in 1.005:1.000 make strontium titanium mixture ratio, slowly under 60℃ to join in oxalic acid solution, aging around 4 h, get oxygen titanium strontium oxalate ($SrTiO(C_2O_4)_2 \cdot 4H_2O$) precipitation, after washing, drying and other process made oxygen titanium strontium oxalate powder.

1. Introduction

Oxygen titanium strontium oxalate as the preparation of strontium titanate lead reactants, and strontium titanate is an important raw material, the electronics industry to automatically adjust the heating element and manufacturing have demagnetizing effect of components. In the field of ceramics for making ceramic capacitors, sensitive piezoelectric ceramic materials, ceramic components, microwave ceramic components, so the study of the preparation of high purity titanium strontium oxalate oxygen has the necessity.[1-3]

In this paper the research of high purity titanium strontium oxalate oxygen is used in the preparation of strontium titanate optical single crystal, due to the preparation of titanium optical grade strontium single crystal of the purity of titanium strontium oxalate oxygen, strontium titanium ratio and is very tall to the requirement of chemical homogeneity, the experiment research of chemical coprecipitation preparation powder of high purity, uniform particle size distribution, specific surface area is big, good for doping raw material components in the mixture, reaction is in highly fragmented state of atomic, molecular, uniform product reaction, thorough and consistent structure, which can obtain high quality of high stability good oxalic acid strontium titanium powder, oxygen in liquid phase reaction, easy to control the molar ratio of titanium strontium, simple equipment, process, compact construction, low cost and is beneficial to industrialized production, so choose oxalate coprecipitation preparation of strontium titanate optical single crystal get satisfy oxygen titanium strontium oxalate powder.

2. Experiment

The chemical reagents such as strontium chloride, four titanium chloride, oxalic acid, ethanol and so on were used in this experiment. The experimental process is shown in figure 1. SEM images using SUPRATM 55 (Carl Zeiss AG) uptake.

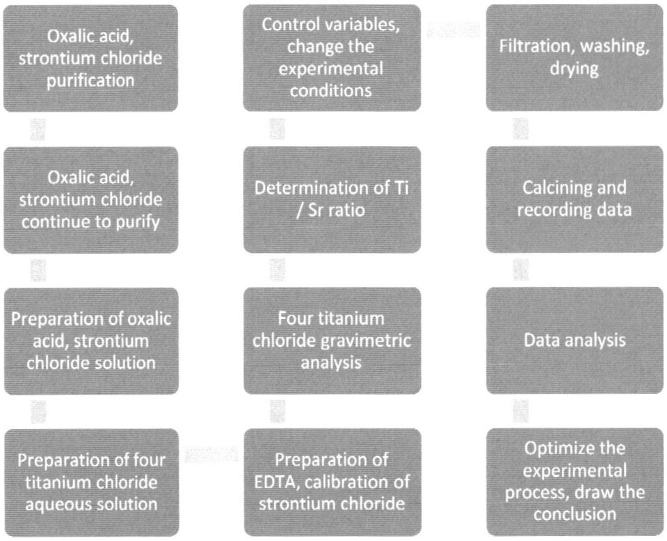

Figure 1. Flow chart of strontium titanate powder preparation.

3. Results and discussion

3.1. Preparation of four titanium chloride aqueous solution

Under normal temperature, $TiCl_4$ and water contact reaction, the formation of a large volume of yellow precipitate or white precipitate, and release a lot of reaction heat. The yellow precipitate is soluble in the hydrochloric acid solution generated by the reaction of $TiCl_4$ with water to form $TiCl_4$ hydrochloric acid aqueous solution ($TiCl_4$ solution). In fact, the reaction process between $TiCl_4$ and water is very complex, which is related to the concentration of titanium ion, solution temperature and solution acidity. When they begin to contact, they form hydrates. If the water is sufficient, the five ($TiCl_4 \cdot 5H_2O$) is preferentially formed, while the water content is low and the low temperature, the formation of a compound of water ($TiCl_4 \cdot 2H_2O$). Then the hydrate is hydrolyzed. In the process of hydrolysis, the chlorine atoms in $TiCl_4$ are gradually replaced by OH- groups, and HCl and $Ti(OH)_nCl_{4-n}$ (n =1,2,3,4) type compounds are formed. The reaction process is rapid at high temperature, the reaction is slow at low temperature, and the intermediate product can be precipitated; When the water is insufficient, the hydrolysis process can stay in an intermediate stage; When the water is excessive, the hydrolysis process of $TiCl_4$ can be carried out in the end. After a long period of time (aging) or heating, the titanic acid can be converted into a more stable H_2TiO_3[4,5] Four titanium chloride ($TiCl_4$) in water for $TiOCl_2$, directly into the ice $TiCl_4$ inside, it is best to fall on the ice cube, can also be slowly dropped into the ice water mixture, the water temperature should not exceed five degrees celsius. Four titanium chloride water solution preparation of the two key indicators, one is the feeding speed, the temperature is two feeding process, because four titanium chloride easily hydrolyzed into titanium dioxide (white precipitate), making the final preparation of strontium strontium titanate powder, titanium ratio deviation to obtain high purity strontium titanate powder.

In this study, three methods were used to inhibit the hydrolysis of titanium chloride and finally to obtain a colorless and transparent solution of titanium chloride. A: for four titanium chloride placed in the freezer to reduce its temperature; two: manufacturing low temperature electric oven, oven dry ice can be added to (or nitrogen) to reduce the temperature and humidity in the constant temperature box; three: four titanium chloride is slowly added into ice water mixture with deionized water or dry ice. And continue to join the ice in the process of dropping, in order to avoid the temperature to four titanium chloride hydrolysis degree becomes larger.

IOP Conf. Series: Materials Science and Engineering **230** (2017) 012008 doi:10.1088/1757-899X/230/1/012008

3.2. Effect of raw material purification

The recrystallization and purification of strontium chloride and oxalic acid were carried out. In order to study the influence of the purity of the reactants on strontium titanate powder, we set up two sets of control experiments to obtain the strontium titanate powder under the condition of flame fusion crystal growth. The results are listed in table 1.

Table 1. Effect of reactant purity on the product.

	Whether purification	T (℃)	Time (hours)	Productivity (g)	Theoretical strontium titanium ratio	Crystal case
1	No	60	4	83.541	1.005:1.000	Features are not clear
2	Yes	60	4	92.515	1.005:1.000	Cracks clear and transparent

The purification of oxalic acid and strontium chloride, strontium titanate powder particles are not uniform, poor mobility, and the calcined crystal appearance is not clear, the crack growth, the initial judgment is by the impurity in the reactants (iron or lead, barium ion) for purification after the normal production, powder uniform particle size, crystal clear appearance, improved transparency.

3.3. Effect of temperature and time on the coprecipitation of oxalic acid

We are at 55 degrees Celsius, 60 degrees Celsius, 70 degrees Celsius under the condition of keeping strontium and titanium than 1.000:1.000, precipitation time of 4 hours, postprocessing calcination temperature of 900 degrees Celsius conditions unchanged comparing experiment, crystal growth and characterization of SEM results in table 2. Under the condition of 55 degrees Celsius, the yield of strontium titanate is obviously low, which indicates that the precipitation is not complete. From the point of view of the crystal, 55 degrees Celsius and 70 degrees Celsius under the conditions of the powder containing different melting point substances (impurities), can not see the crystal when calcined. 60 degrees Celsius to get better powder growth.

Table 2. Effect of reaction temperature on products.

	T (℃)	Productivity (g)	Crystal growth	Average particle size (μm)
1	55	94.745	Material with different melting point	3.23
2	60	135.039	preferably	3.24
3	70	121.727	Material with different melting point	7.62

By scanning SEM on the morphology of strontium oxalate powders (Figure 2 and 3), we find that with the increase of temperature, strontium titanyl oxalate particles changes exponentially, at 40 degrees Celsius to 60 degrees Celsius, the crystal grain size changes slowly, when the temperature exceeds 60 degrees Celsius when the growth of crystal particle size quickly.Synthetic temperature on the crystal size effect is significant, the analysis result is: when the reaction temperature is low, the solubility of strontium titanyl oxalate crystal nucleus is very small, relatively easy to precipitate, but not conducive to the growth of the crystal nucleus, so the crystal size is smaller. With the increase of temperature rise, the solubility of strontium titanyl oxalate, nucleation precipitation is relatively difficult, but is conducive to the growth of the crystal nucleus, so the crystal size is larger, can be seen from the figure at 60 degrees Celsius temperature of the crystal size has increased significantly, from many aspects to consider to determine whether 60 degrees Celsius the best synthesis temperature centigrade.

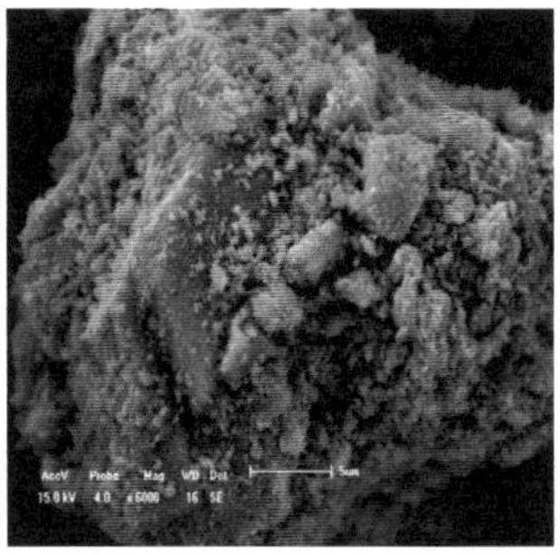

Figure 2. Scanning electron microscopy (SEM) of strontium oxalate powder

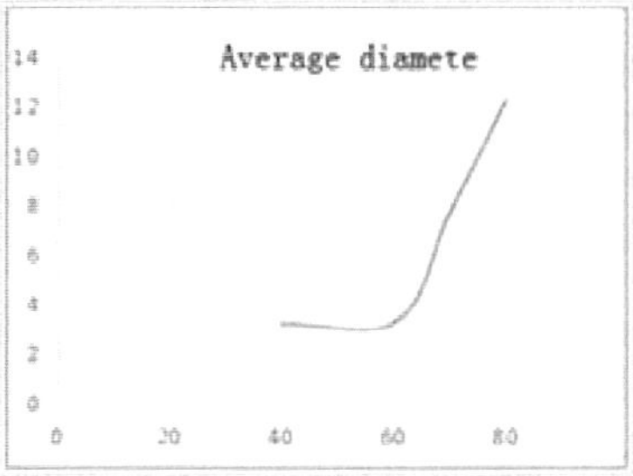

Figure 3. Relationship between temperature and particle size

At the same time, the results showed that the yield and properties of the obtained powders were not significantly different when the co precipitation reaction time was more than 4 hours, indicating that the precipitation was sufficient, and the total precipitation time was selected for 4 hours.

3.4. Effect of calcination temperature and time on the oxidation of strontium oxalate

The calcination process is the thermal decomposition process of strontium oxalate, and the calcination temperature is directly related to the thermal decomposition of the reactants. If the calcining temperature is too low or the time is too short, the decomposition will be incomplete, and the diffusion of strontium and titanium will be uneven. When the calcination temperature is too high, the particle size of strontium titanate powder will become larger, resulting in sintering. To get the optimum calcination temperature, strontium titanyl oxalate powder were obtained by orthogonal experiments at 800 degrees Celsius, 900 degrees Celsius, 1000 degrees Celsius and 3 hours, 4 hours and 5 hours under the condition of research shows that strontium titanate powder fluidity obtained 900 degrees Celsius and 4 hours under the condition of good easy, crystal growth.

4. Conclusion

Experimental results show that the main factors of influence on the purity of strontium titanate including reaction temperature and time, calcination temperature and time. By changing the principle of a single variable, the factors affecting each effect of strontium titanate were tested many times, and finally the optimum technological conditions of preparation of high purity strontium titanate powder by oxalate co precipitation method were finally determined.

(1) the reaction time 4 h, reaction temperature of 60 ℃ , Ti: Sr be equal to 1.000:1.005, the highest quality of strontium oxalate, after the calcination of strontium titanate powder particle size uniformity and high purity.

(2) When the calcination temperature is 900 ℃, the titanium oxalate is easy to decompose and crystallize into strontium titanate, the particle size is small, and the purity is high.

5. References

[1] S Kasap, S Piskin and H Tel 2012 *International journal for chemical aspects of nuclear science and technology* **100(12)** 925
[2] L Dai, S Wang and P Shu 2011 *CHINESE SCI BULL* **56(21)** 2267
[3] I Marozau, A Shkabko 2009 *Materials* **2(3)** 1388
[4] J XU, P Gong and X Qing. 2009 *Petrochemical Industry Application* **28(6)** 13
[5] Z Huang, K Li 2012 *Chenmical Intermediate* **10** 22

Acknowledgments
Thank you for the support of Foundation of Liaoning Educational Committee (ZXLGL-1614).

Influence of the β-$Mg_{17}Al_{12}$ phase morphology on the corrosion properties of AZ91HP magnesium alloy

Lingling Guo[1,2] and Jumei Zhang[1,3]

[1]School of Materials Science and Engineering, Xi'an University of Science & Technology, Xi'an 710054, China
E-mail:[2]781730142@qq.com, [3]feiyue-zjm@163.com

Abstract: The morphology of β-$Mg_{17}Al_{12}$ phase and corrosion behavior of AZ91HP magnesium alloy after spheroidizing treatment were investigated by optical microcope electrochemical and immersion tests in 3.5% NaCl at 25℃.The results show that the coarse divorced eutectic phase of AZ91HP cast magnesium alloy dissolve into Mg matrix during the isothermal process at 415℃, and the lameller β phase precipitated from magnesium solid solution as perlite-type precipitation during the slowly cooling. Next, the spheroidizing treatment at different temperatures for 20h was carried out, and the lameller β phase were spheroridizing by dissolved themselves. After spheroidizing treatment at 300℃ for 20h ,many small granular β phase are scattering within the magnesium matrix. The corrosion properties of AZ91HP magnesium alloy in 3.5% NaCl decreased obviously after spheroidizing treatment, the polarization measurement of the alloy can be up to -1.412V from -1.56V of the cast. The β-$Mg_{17}Al_{12}$ phase act as a corrosion barrier and hinder corrosion propagation ,if the second phase is in the form of a spherical morphology.

1. Introduction

Magnesium alloys being recognized as green materials in the 21st were widely applied in traffic tansport, electronic communications, aerospace, national defense and so on, due to their characteristics of lightness, high specific strength and specific, easily recycled, environmentally friendly. However, the chemical properties of magnesium lively and very easily oxidized in the air, the corrosion potential of magnesium and its alloys are very low ,it can't protect the matrix and easily prone to serious corrosion, has limited their further application[1-5];So it is meaningful to explore the way of improving the corrosion resistance of magnesium alloy. This study can be helpful to improve the property and application of magnesium alloy.

In the past few decades, there are many studies about the effects of phase composition on the corrosion behavior of magnesium alloys,β phase is the strengthening phase, it has stronger corrosion resistance than α phase, it plays two different roles in corrosion:(1) as cathode of corrosion cell, accelerating corrosion;(2) as a barrier of anode reaction, inhibiting corrosion; the different roles of β phase depends on the content of β phase [6].Therefore, in this paper, by studying the effect of spheroidizing temperature on AZ91HP magnesium alloy corrosion resistance, the corrosion resistance of magnesium alloys can be improved by getting small α phase and the appropriate content of β phase.

2. Experimental

This paper makes AZ91HP magnesium alloy as the research object, its chemical compositions are shown in table1.The as-cast magnesium alloy was cutted into 15×10×5mm samples; Homogenized treatment was performed at 415℃ for 24h with furnace cooling to room temperature, then through different spheroidizing treatment. The polished samples were immersed into the mixed solution of glycerin, hydrochloric acid, nitric acid and acetic acid about 8~10s for corrosion. Optical microstructure and corrosion microstructure were analyzed with OLYMPUS GX71 metallographic microscope(OM).The specimens were weighed for the original mass. In the process of Corrosion tests, the specimens were immersed in 3.5 wt.% NaCl solution, exposed to laboratory conditions at a temperature of 25℃.After the corrosion tests, the corroded specimens were photographed and then immersed into a ammonium solution at room temperature to remove the corrosion products. Afterwards the specimens were quickly washed with distilled water, dried using a hair dryer. At room temperature, polarization curve was tested by using cs double-cell electrochemical workstation in 3.5 wt.% NaCl solution.

Table1 Chemical compositions of AZ91HP samples (wt.%)

Al	Zn	Mn	Ni	Cu	Ca	Mg
8.005	0.593	0.199	0.001	0.001	0.001	Balance

3. Results and Discussion

The microstructure of the as-cast AZ91HP alloy and spheroidized at different temperature for 20h are shown in Fig.1.It is found that AZ91HP magnesium alloy had a two phase microstructure consisting of α-Mg matrix, divorced eutectic β-$Mg_{17}Al_{12}$ phases along the α grain boundaries in Fig.1(a).when the samples were spheroidized to various temperatures for 20h,the number and morphology of the second phase has changed greatly, the second phase were more and smaller. The layer second phase gradually melt with spheroidizing temperature increasing, Finally the second phase were diffusely distributed in the α-Mg matrix, and the structure morphology of the second phase has changed from the original layer to the granular. When the spheroidizing temperature was at 270℃,it can be seen from Fig.1(b) that the layer second phase has mostly remained ,Only few β phase fused into short rot.β phase showed short rod or granular in the α-Mg matrix with the increasing spheroidizing temperature, when the spheroidizing temperature was at 290℃ (Fig.1(d).Fig.1(e) shows that the layer second phase did not exist, and the granular second phase increased when the spheroidizing temperature was at 300℃.It can be observed from Fig.1(f) that the second phase was diffusely distributed in the α-Mg matrix and appeared granular structure after spheroidizing treatment at 310℃.

Fig.2 shows the corrosion morphology of AZ91HP alloy treated at different spheroidizing temperature immersing in 3.5wt.% NaCl solution for different time.The surface corrosion degree of the as-cast magnesium alloy was more serious than that of spheroidized samples, so the spheroidizing treatment improved markedly the corrosion resistance of AZ91HP magnesium alloy.It can be seen from Fig.2(a) that there was no obvious corrosion pit on the sample surface, but the overall sample surface was pitted corrosion,when the samples were immersed at 270 ℃for 2h; when the immersing

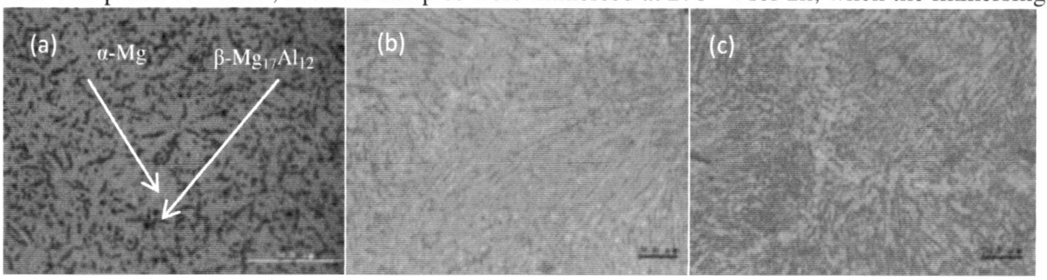

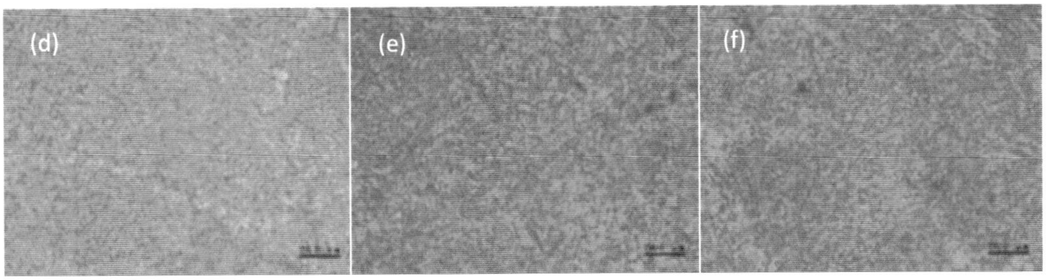

Figure 1.Microstructure of as-cast AZ91HP alloy and spheroidized at different temperature for 20h

(a)as-cast (b)270℃ (c)280℃ (d)290℃ (e)300 ℃ (f)310℃

time increased to 96h,the surface corrosion degree of AZ91HP magnesium alloy aggravated gradually, corrosion points got together and gradually got bigger ,and the corrosion went deeply from the samples surface to the internal, the corrosion morphology changed from the pitting corrosion to honeycomb(Fig.2(b)).Compared with the morphology of spheroidizing temperature at 270℃ and 310℃,the corrosion degree of magnesium alloy was less when the spheroidizing temperature was at 290℃ (Fig.2(c)(d)).Above pheomenon is related to the β phase content, the early granular β phase content is less,β phase acted as a cathode and accelerated the dissolution of magnesium matrix; granular β phase content increased at 290℃,the magnesium matrix was isolated, the corrosion product increased and the corrosion degree reduced with the increasing of corrosion time; when spheroidizing temperature added to 310℃(Fig.2(e)(f)),particle content was more, with the extension corrosion time, the magnesium substrate has been eroding, maked β phase lost support and fell off, accelerated the corrosion of magnesium alloys.

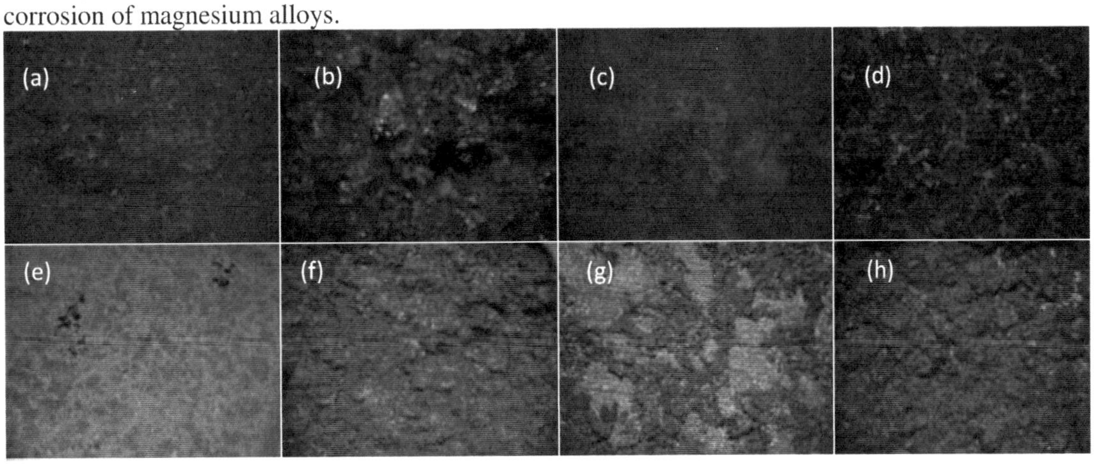

Figure 2.The corrosion morphology of AZ91HP alloy treated at different spheroidizing temperature immersing in 3.5wt.% NaCl solution for different time(a) 270℃×2h (b) 270℃×96h (c) 290℃×2h (d) 290℃×96h (e)310℃×2h (f))310℃×96h (g)as-cast×2h (h)as-cast×96h

The electrochemical polarization curve of the AZ91HP magnesium alloys in 3.5wt.% NaCl aqueous solution was shown in Fig.3.Thecorrosion potential: 300℃>290℃>310℃>270℃>280℃> the as-cast. It can be seen that the best corrosion resistance was at 300℃;the corrosion current: the as-cast>270℃>310℃>280℃>300℃>290℃.It can be seen that the best corrosion resistance was at 290℃.Although the results of the corrosion resisitance were different, the corrosion potential at 300℃ were greater than that at 290℃,The 300℃measured corrosion current was a little higher than the 290℃ measured corrosion current.comprehensive analysis shows that the spheroidizing temperature at 300℃ has good corrosion resistance.

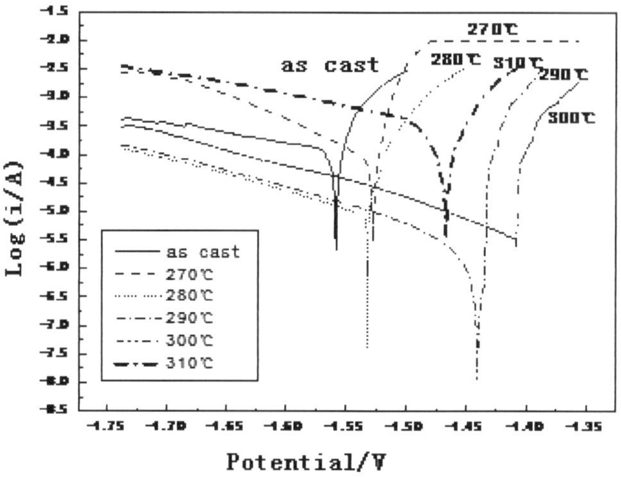

Figure 3.The electrochemical polarization curve of AZ91HP magnesium alloys of in 3.5wt.% NaCl aqueous solution

4. Conclusions

The as-cast AZ91 magnesium alloy was composed of α-Mg and the divorced black β-$Mg_{17}Al_{12}$ phases, and the latter were precipitated and irregularly distributed in α-Mg matrix with network during the non-equilibrium crystallizing processes. In the spheroidizing process, with the increasing of spheroidizing temperature, the layer β second phase continuously melted. the microstructure form of β phase changed from the original layer to the granular, after spheroidizing treatment at 300℃ for 20h ,many small granular β phase are scattering within the magnesium matrix. The corrosion resistance of AZ91HP magnesium alloy increased firstly and then decreased with the increasing of spheroidizing temperature; When the spheroidizing temperature was at 300℃,the content and form of the β phase was the most appropriate, and also showed the optimum corrosion resistance .

References

[1] Y J Li and A T Tang. *Materials Review*,2013,27(**9**):125-130.
[2] G Y Zhang, C J Li, J M Sun, P L Mao and Z Liu.*Foundry*,2015,64(**10**):1001-1004.
[3] W C Neil,M Forsyth, P C Howlett, C R Hutchinson and B R W Hinton.*Corrosion Science*,2011,**53:**3299–3308.
[4] H Y Xu and Z Y Li. *Journal of Chinese Society for Corrosion and Protection*, 2013,33(**4**): 298-305.
[5] J L Zhang, Y L Liu, J Zhou, S B Wang and B S Xu. *Transactions of materials and heat treatment*,2014, 35(**5**):64-68.
[6] T Zhang, Y Li and F H Wang. *Corrosion Science*,2006, **48**:1249–1264.

Acknowledgements

This work was financially supported by the Scientific Research Program Funded by Shaanxi Provincial Education Department (Program No.14JK1465) and State Scholarship Program(CSC NO. 201508615021) .

IOP Conf. Series: Materials Science and Engineering **230** (2017) 012010 doi:10.1088/1757-899X/230/1/012010

Influence of Laser Shock Texturing on W9 Steel Surface Friction Property

Yujie Fan[1,2], Pengfei Cui[2], Jianzhong Zhou[1], Yibin Dai[2], Erbin Guo[2], Deye Tang[2]

[1] School of Mechanical Engineering, Jiangsu University, Zhenjiang 212013, China
[2] School of Mechanical Engineering, Jiangsu University of Science and Technology, Zhenjiang 212000, China
E-mail:fanyujie1979@163.com

Abstract: To improve surface friction property of high speed steel, micro-dent arrays on W9Mo3Cr4V surface were produced by laser shock processing. Friction test was conducted on smooth surface and texturing surface and effect of surface texturing density on friction property was studied. The results show that, under the same condition, friction coefficient of textured surface is lower than smooth surface with dent area density less than 6%, wear mass loss, width and depth of wear scar are smaller; Wear resistance of the surface is the best and the friction coefficient is the smallest when dent area density is 2.2%; Friction coefficient, wear mass loss, width and depth of wear scar increase correspondingly as density of dent area increases when dent area density is more than 2.2%. Abrasive wear and adhesive wear, oxidative wear appear in the wear process. Reasonable control of geometric parameters of surface texturing induced by laser shock processing is helpful to improve friction performance.

1. Introduction

High speed steel has high strength, toughness and good heat resistance. It commonly used in manufacture of complex tools, such as lathe tools, milling cutters, drills, taps, etc. [1,2]. Wear is the main failure mode of high speed steel cutting tools in machining process. High speed steel tools should have good friction properties to ensure the stability of cutting force and the surface quality of workpiece as the key basic parts and it should also have good wear resistance to guarantee the dimensional accuracy of workpiece and tool life.

At present, there are several ways to improve wear resistance and service life of cutting tools. New green cutting technology was used to reduce tool wear by cooling and lubricating in the cutting process, such as Minimum Quantity Lubrication (MQL), spray cooling, liquid nitrogen cooling, low temperature air cooling cutting, water vapor cooling cutting technology [3,4]. Coating technology [5]combines the characteristics of high strength, toughness of matrix, high hardness and wear resistance of coating. It can improve the wear resistance of the tool without reducing its toughness. But coating technology has some shortcomings, including complex process, high requirements for equipment, long period, high production cost, etc. With higher demand of tool cutting performance, the requirements for coating material are increasingly hasher. Kawasegi N[6] designed micro- and nano-grooves textures on the tool rake by laser processing, and used the tool in turning experiments of aluminum alloys. Results show that surface textures produced beneficial lubricating film on the surface of tool, which effectively reduced the friction. However, laser thermal effect acted in the process they used. It can have a detrimental residual tensile stress which destroys the integrity of the material surface for the effect of ablation and vaporization on material. New idea to solve the problem appears with development of Laser Shock Processing (LSP). It avoids material surface ablation because surface deformation caused by laser shock is force effect. Furthermore, the microstructure refining, improvement of surface hardness and beneficial distribution of residual compressive stress induced by laser shock process can significantly improve the wear resistance of tool surfaces. Therefore, laser shock texturing on high speed steel tools for reshaping and modification can improve antifriction and antiwear properties, and increase the service life of the tools.

So far there has been little research on improvement of friction property and wear resistance of high speed steel material treated by LSP. In this paper, different micro-dent arrays were produced on W9Mo3Cr4V surface by LSP, and friction properties of textured surface was studied. The results provide good guidance for later research on improving wear resistance of cutting tools by LSP.

2. Experiments

2.1. Specimen preparation.

The material of specimens is high speed steel W9Mo3Cr4V. The specimens were cut into rectangular shape with dimensions 20 mm × 20 mm × 4 mm (length × width × height). Before laser shock experiment, the specimen surface was polished to a mirror finish with different SiC abrasive papers (from 600# to 1200#), and surface roughness was up to 0.05μm. Finally the treated surface was cleaned by anhydrous alcohol.

2.2. Experiment of laser shock texture.

The experiments were performed using a Spitlight2000 Nd: YAG pulse laser. Wavelength adopted in laser shock texture is 1064 nm. K9 glass with thickness of 4 mm acts as confinement layer. Aluminum foil with thickness of 100 μm produced by 3M Co., Ltd. was adopted as the absorbing material.

In order to study the influence of dent arrays distribution on friction and wear experiment, spot diameter were 1.2 mm, 1.4 mm, 1.6mm and 1.8 mm, and center distance of dent were 0.5 mm, 1 mm, 1.5mm under the same laser energy of 800 mJ. The treated area size was 10 mm×8 mm (length×width). The texture parameters and specimen number are presented in Table 1.

Table 1. Experiment parameters of laser shock.

Specimen no.	Spot diameter (mm)	Center distance (mm)	Specimen no.	Spot diameter (mm)	Center distance (mm)
1	1.2	1.5	4	1.4	1
2	1.2	1	5	1.6	1
3	1.2	0.5	6	1.8	1

2.3. Friction test.

The specimens were cleaned by anhydrous alcohol after laser processing to remove surface residual impurities. Dry friction tests were conducted using a ball-on-disc configuration on a UMT-2 type friction wear testing machine produced by CETR Company. The upper specimen was a GCr15 steel ball with diameter 9.5 mm, and the lower specimen was sample treated by laser shock texturing. The diameter of the wear track was 6 mm. The tests were carried out with a load of 20 N, rotation speed of 100 r/min and time 20 min.

3. Results and discussion

3.1. Morphology of the micro-dent.

LEXT OLS4000 3D laser scanning confocal microscope was used to measure the morphology of micro-dents after laser shock. Figure.1 is the morphology of the micro dents. The micro-dents of different locations on the specimen were measured. To obtain the average value of dent diameter and depth, based on the assumption that ***d*** is the diameter of the dent and ***l*** is the side length of the micro-dent unit (see figure. 1b), then the dent area density can be calculated as follow:

$$S_p = \frac{\pi d^2}{4l^2} \quad (1)$$

Table 2 shows the parameters of texturing in detail.

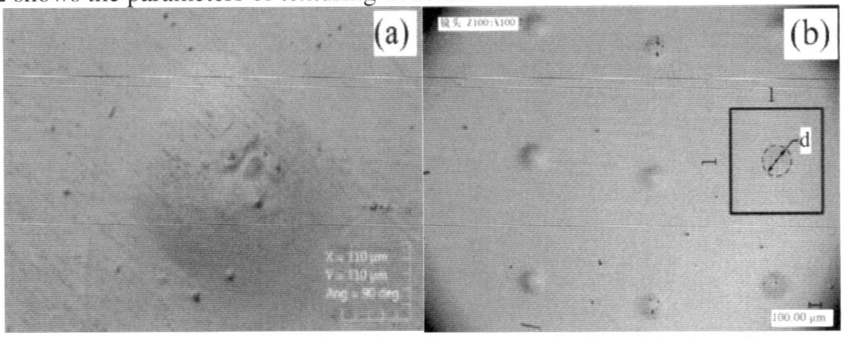

Figure. 1. The morphology of the micro dent (a) single dent (b) micro-dent arrays.

Table 2. Parameters of texturing.

Specimen no.	Diameter of dent (μm)	Depth of dent (μm)	Dent area density (%)	Specimen no.	Diameter of dent (μm)	Depth of dent (μm)	Dent area density (%)
1	247.875	4.216	2.20%	4	270.101	3.701	5.70%
2	247.875	4.216	4.90%	5	292.372	3.414	6.60%
3	247.875	4.216	19.60%	6	324.864	3.016	8.00%

3.2. Analysis of friction coefficient.

Figure. 2a shows the friction coefficient curves of smooth specimen and specimens with different distances of dent center, with laser energy of 800 mJ and spot diameter of 1.2 mm. It can be seen that the friction coefficient of textured specimen with dent center distance 0.5 mm is higher than the smooth surface. The friction coefficients of textured surfaces are less than smooth surface, and decrease with the increase of dent center distance from 0.5 mm to 1.5 mm. The friction coefficient is the minimum when the dents center distance is 1.5 mm. Figure. 2b shows the friction coefficient curves of textured specimens with different spot diameters with laser energy of 800 mJ and dent center distance of 1 mm. It can be seen that the friction coefficient gradually increases as the spot diameter from 1.2 mm to 1.8 mm, the friction coefficient does not increase significantly when the spot diameter increases from 1.4 mm to 1.8 mm.

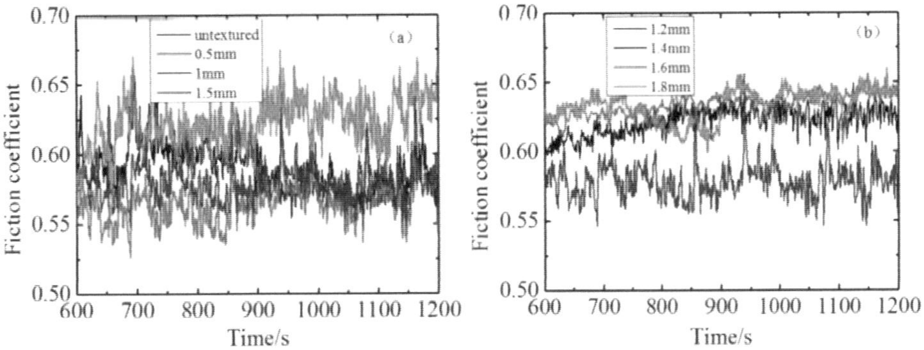

Figure. 2. Change of friction coefficient with time (a) different distances of dent center (b) different spot diameter.

Figure. 3 shows the friction coefficient curves with different dent area densities under laser energy of 800 mJ, and laser spot diameter is 1.2 mm, 1.4 mm, 1.6 mm, 1.8mm and dent center distance is 0.5 mm, 1 mm, 1.5 mm. It can be seen that the friction coefficient of textured surface with dent area density less than 6% is lower than smooth surface. Friction coefficient is the smallest when the dent area density is 2.2%. With the increase of dent area density from 2.2% to 6.6%, the friction coefficients increase rapidly, then an inflection point appears when the dent area density is close to 7.5%. When the dent area density increases continuously from 7.5% to 19.6%, the increments of friction coefficients is not obvious. Analysis suggests that the textured surface has the ability of capturing abrasive particles to a certain extent and play an important part of reducing furrow effect, so the friction coefficients are reduced. While the dent area density increases, the friction coefficients also increase. The reason is that the increase of dents number makes the surface roughness increased, and the effect of capturing abrasive particles and reducing furrow is less than the frictional resistance effect caused by micro dents, so friction coefficient increases. The distance between adjacent dents is large as the dent area density is small. The larger proportion of smooth area and dents area makes lower friction torque [7, 8], so the friction coefficients decrease.

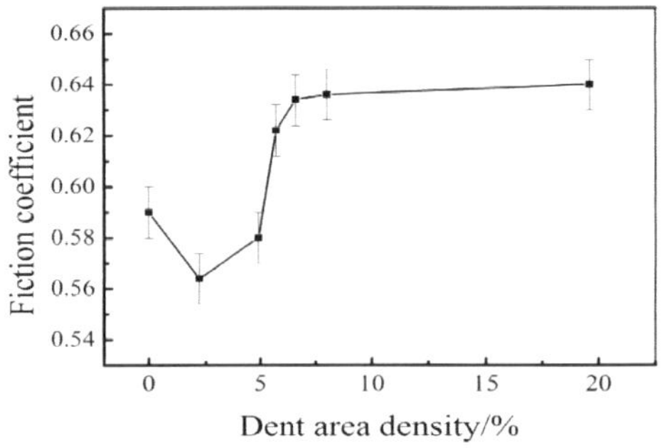

Figure. 3. Change of friction coefficient with dent area density.

3.3. Analyses of wear loss.

Evaluation indexes of wear loss are mainly wear mass loss, bulk worn loss and linear wear loss. Due to the uniform density of material, wear mass loss is chosen to evaluate the wear resistance of material in this experiment. Wear mass loss is the mass difference before and after wear test. In order to reduce errors caused by manual measurement, each specimen cleaned by ultrasonic and dried, was measured repeatedly for 3 times before and after weighing. Figure. 4 shows the wear mass loss of specimens with different dent densities measured by electronic balance (precision of 0.1 mg).

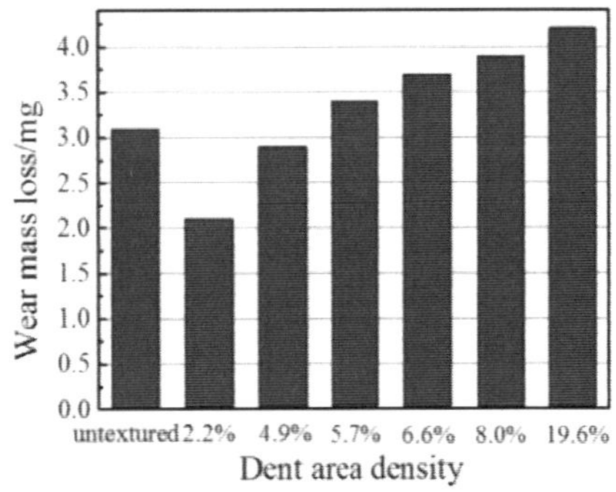

Figure. 4. Wear mass loss under different dent area density.

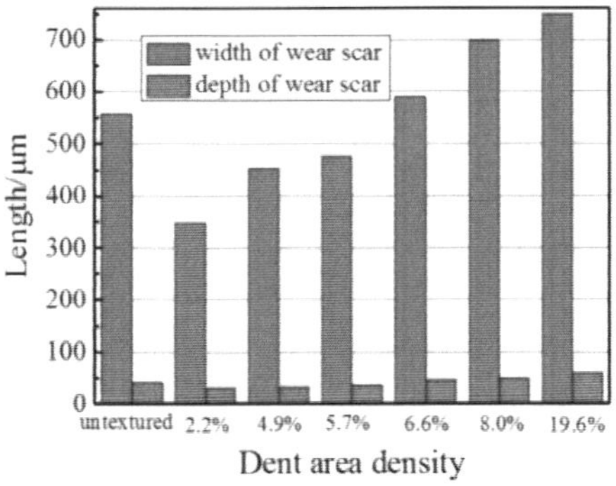

Figure. 5. Width and depth of wear scar under different dent area densities.

It can be seen from Figure. 4, the wear mass loss is consistent with friction coefficient under dry fiction. When the dent area density is less than 4.9%, the wear mass losses of textured specimens are less than smooth specimen's. The wear mass loss increases correspondingly as density of dent area increases.

By measuring the width and depth of wear scar to estimate the bulk worn loss, it can be also used as the evaluation index of wear loss. The width and depth of wear scar were measured by laser scanning confocal microscope. Figure. 5 shows the widths and depths of wear scars with different dent area densities.

The widths and depths of wear scars are less than smooth surface's when the dent area density is less than 5.7%, which is inconsistent with change of the friction coefficient and wear mass loss. By analysis that smooth specimens exist serious adhesive wear during the process of wear, and abrasive dust adhere to both sides of wear scar under the squeezing action of steel ball, resulting in too large measured values of smooth specimen's wear scar. The width and depth of wear scar decrease correspondingly as density of dent area decreases.

It is inferred that the dents capture abrasive particles during the wear process when the dent area density is low, then abrasive wear reduces. On the other hand, residual compressive stress is produced in the treated area by laser shock, leading to improvement of hardness and wear resistance. However, when the dent area density increases to some extend, the contact area of friction pair is reduced. The average contact stress also increases, which becomes the main factors affecting antiwear properties of material surface, so the wear rate increases.

3.4. Analyses of worn surfaces.

Figure. 6a shows the wear trace micrographs of textured specimen which was not cleaned after the dry friction experiment. It can be seen clearly that micro dent is covered with abrasive dust, indicating that the dents have ability of capturing abrasive dust in the process of friction. To a certain extent, it is good for reducing abrasive wear, thus reducing the surface wear. Brown rust appeared inside the wear trace. Analysis suggests that high temperature was produced in the process of friction, and the material is more susceptible to oxidation. Figure. 6b shows the serious adherent phenomenon on the both sides of wear trace. This is because steel ball squeeze material towards two sides during the motion, resulting in accumulation and adhesion of abrasive dust. From the both figures, we can see a number of parallel furrows were formed along the direction of sliding on the worn surface, which is caused by abrasive wear. Abrasive wear is that the material on the surface is pushed to the front of the steel ball, or the surface material is pushed to the both sides of motion direction, and therefore the hard particles which fell from surface produced furrows on the surface of material under the force [9].

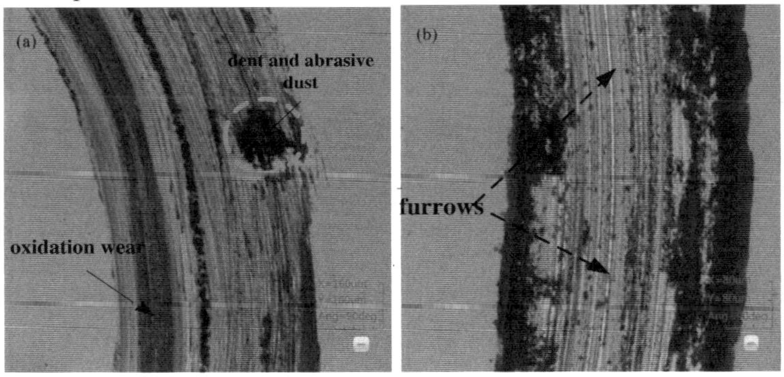

Figure. 6. Micrographs of wear trace (a) wear race of textured specimen (b) wear race of untextured specimen.

4. Conclusions

Influence of the dent area densities on the properties of friction and wear were explored corresponding to different dent center distance and spot diameter. Some important conclusions can be drawn as follows.

The tribological properties of high speed steel W9Mo3Cr4V treated by laser shock peening are affected by dent area density. The friction coefficients of textured surfaces with dent area densities less than 6% are lower than smooth surface, and exists a best optimum value of dent area density with the least friction coefficient. Friction coefficient increases rapidly with the increasing dent area density

when the dent area density is more than the best value. Increments of friction coefficients are not obvious when the dent area density is more than the value of inflection point.
Wear mass losses of textured surfaces are less than the smooth surface's when the dent area densities are less than 4.9%. Wear mass losses increase gradually with the increasing dent area densities. When the dent area densities are less than 5.7%, the width and depth of wear scars are less than the smooth surface's. The results of wear losses are consistent with friction coefficients under dry fiction.
Abrasive wear and adhesive wear, oxidative wear appear in the wear process. The micro dents have the ability of capturing abrasive dust. Reasonable control of geometric parameters of surface texturing induced by laser shock processing is helpful to improve friction performance.

Acknowledgements
The authors are grateful for the support provided by National Natural Science Fund for Youth (NO.51405207), Nature Science Foundation of Jiangsu Province for Youth (NO.BK20140510), China Postdoctoral Science Foundation (NO.2014M561573), Postdoctoral Science Foundation of Jiangsu Province (NO.1301012B), The practice innovation project of the professional degree graduate students in Jiangsu Province (SJLX16_0454)

References
[1] G. Hoyle, High speed steels, Buttrtworth, London 1988.
[2] B. C. Li, Y. Li, The development of coating materials for cutting tool, *Journal of Machinery*, 50 (2012) 37-39.
[3] R. D. Han, J. Wu, Discussion on green cutting technology , *Journal of Tool technology*, 40(12)(2006) 8-11.
[4] X. L. Li, N. He, L. Li, MQL of green cutting technology, *Journal of Aero precision manufacturing technology*, 41(2)(2005) 24-35.
[5] Q. L. Zang, C. LI and J. H. Yao, Research of strengthening and toughening technology for HSS Cutters, *Journal of New technology and new process*, 2(2005) 56-58.
[6] Kawasegi N. Sugimori H. Morimoto H. et al, Development of cutting tools with microscale and nanoscale textures to improve frictional behavior, *Journal of Precision Engineering*, 33(2009) 248-254.
[7] Y.B. Guo, R. Caslaru, Fabrication and characterization of micro dent arrays produced by laser shock peening on titanium Ti–6Al–4V surfaces, *Journal of Materials Processing Technology*, 729-736(2011). (in Chinese)
[8] T. C. Hu, L. T. Hu and Q. Ding, Tribological Properties of Laser Textured Surfaces of 45# Steel under Dry Friction , *Journal of Tribology*, 30(2010) 46-52.
[9] W. Zhu, Research on laser shock processing bio-corrosion and frictional properties of Ti13Nb13Zr alloy, Jiangsu University Master's degree paper, 2012.

IOP Conf. Series: Materials Science and Engineering **230** (2017) 012011 doi:10.1088/1757-899X/230/1/012011

Corrosion Behavior of Brass In Tio_2 Nanofluids

Kai Wu, Hong-Hua Ge*, Feng Wang and Hong-Wang Zhou

Shanghai University of Electric Power, Shanghai Engineering Research Center of Energy-Saving in Heat Exchange Systems, Shanghai Key Laboratory of Materials Protection and Advanced Materials in Electric Power, Shanghai 200090, China
E-mail: gehonghua@shiep.edu.cn

Abstract. Corrosion behavior of brass electrode in TiO_2 nanofluids was analyzed using electrochemical impedance spectroscopy. The experimental results show that TiO_2 nanoparticles promote corrosion of brass. The corrosion resistance of brass electrode decreases with the increase of the temperature of the TiO_2 nanofluids. Sodium dodecyl benzene sulfonate (SDBS) is a dispersant in nanofluids and also appears corrosion inhibition to brass, and the corrosion inhibition enhances with the increase of SDBS concentration. The corrosion resistance of brass in TiO_2 nanofluids would decrease when the concentration of dispersant SDBS exceeds a certain value.

1. Introduction

With the development of thermal science and technology, the traditional heat transfer medium in the manufacturing, metallurgy, energy, transportation and microelectronics has been difficult to meet the increasing heat transfer load and heat transfer intensity, so the efficiency of the heat transfer system put forward higher requirements. An effective method for improving the heat transfer efficiency of the liquid is to add metal, non-metal or polymer solid particles to the base liquid. Because the thermal conductivity of solid particles is several orders of magnitude larger than that of liquids, the thermal conductivity of the two-phase fluid with solid particles is much larger than that of pure liquids. However, there are many problems in practical application, such as large particles, easy settlement, easy to produce clogging and abrasion, which limits its application in industry.

In 1995, Choi et al.[1] at Argonne National Laboratory in the United States, first proposed the concept of nanofluid, which refered to the formation of nano-sized metal or non-metallic particles in liquid medium in a certain way and proportion to form stable, high thermal conductivity of the new heat transfer medium, which was used in thermal engineering nanotechnology in this traditional field of innovative research. In recent years, nano-fluid had become one of the hot spots in heat transfer and material research because of its high thermal conductivity[2-4], small wall friction and good optical properties[5].

In recent years, people began to pay attention to the corrosion of metals in nano-fluids. However, the results from different systems were not consistent, nanoparticles inhibited the corrosion of metals in some systems and promoted the corrosion of metals in other systems, and most of the above researches were taken in distilled water system.

Deionized water or organic solvent as the base fluid were often used in the past to research the stability of nanofluids[6,7]. However, the application of nanofluids in cooling water systems in the industrial production has a good prospect in the future because of the significant energy-saving efficiency. TiO_2 nanofluid is an economical and effective nanofluid, and it is one of the main researched nanofluids. So TiO_2 nanofluids were prepared using a simulated cooling water and corrosion of brass in the TiO_2 nanofluids were analyzed in this paper.

2. Experimental

2.1. Test solution and materials

Simulated cooling water (SCW), SDBS and nano TiO_2 were used to produce the nanofluids. Simulated cooling water consisted of 20 mg/L Ca^{2+}, 6 mg/L Mg^{2+}, 360 mg/L SO_4^{2-}, 300 mg/L Cl^-, 379 mg/L Na^+, 122 mg/L HCO_3^-[8]. The TiO_2 nano particles (diameter of 5-10 nm, purity > 99.8%) and SDBS (purity > 99.8%) were purchased from Aladdin Industrial Corporation (Shanghai, China).

2.2. Nanofluids preparation

SCW based TiO_2 nanofluid were prepared by two-step. TiO_2 was added to the solution, and then it was ultrasonically agitated and homogeneously dispersed in the solution.

2.3. Measurement of EIS

The working electrodes (1cm×1cm) were prepared from brass (type HSn70-1) and embedded in epoxy resin. Before the tests, all electrodes were mechanically polished by silicon carbide (SiC) paper progressively up to 2000 grid, degreased with alcohol and rinsed with distilled water. A saturated calomel electrode (SCE) and a latinum electrode were used as the reference and the auxiliary electrodes, respectively. Electrochemical Workstation CHI660E was used to measure the electrochemical impedance (EIS). All the experiments were performed at 30 ± 1℃.

3. Results and discussion

3.1. Corrosion behavior of brass in TiO_2 nanofluids

3.1.1. The effect of temperature and TiO2 nanoparticles

The influence of temperature on the corrosion behavior of brass was studied in both SCW and TiO_2 nanofluids. Nyquist plots of brass electrode in both systems were showed in figure 1. The fitting of electrochemical impedance spectroscopy was based on equivalent circuit in figure 2, where R_s is the solution resistance, R_{ct} denotes the charge-transfer resistance and CPE is the constant phase angle element which explains the depression of the capacitance semi-circle. The fitting results display in table 1.

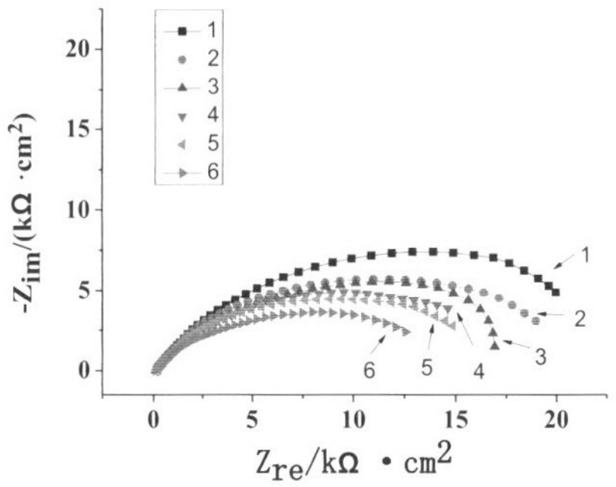

Figure 1. Nyquist plots of brass electrode in simulated cooling water and in TiO_2 nanofluids at different temperatures. SCW: (1) 30℃; (2) 40℃; (3) 50℃. TiO_2 nanofluids: (4) 30℃; (5) 40℃; (6) 50℃.

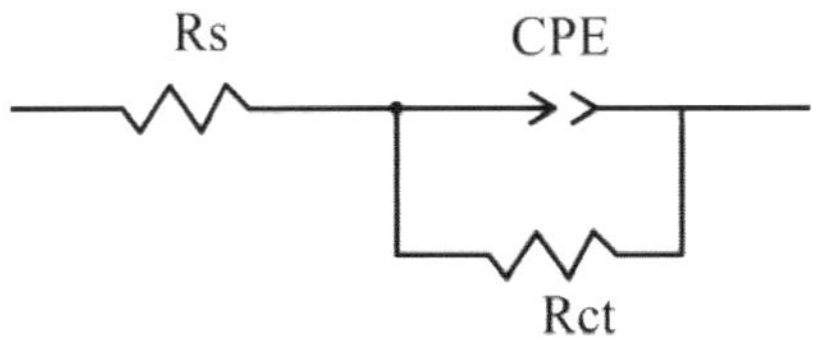

Figure 2. Equivalent circuit

Table 1. Fitting results of figure 1

Test systems	R_s ($\Omega\cdot cm^2$)	R_{ct} ($k\Omega\cdot cm^2$)	CPE	
			$Q(\mu\Omega^{-1}\cdot cm^{-2}s^n)$	n
1	158.5	23.57	68.99	0.68
2	122.4	20.01	61.59	0.67
3	100.6	17.66	58.60	0.68
4	167.0	16.82	88.32	0.65
5	129.4	15.83	74.78	0.66
6	110.0	13.18	91.31	0.63

From table 1, it can be found that charge transfer resistance was largest at 30℃ in both SCW and TiO_2 nanofluids. With the increase of temperature, the charge transfer resistance of the electrode/electrolyte interface decreased, indicating that corrosion resistance of brass reduced with the increase of temperature, which might be due to the accelerated diffusion of ion and destroyed the protection film of hydroxide ion. Comparing the data in table 1, the R_{ct} value of brass in TiO_2 nanofluids were smaller than in SCW, which means that TiO_2 nanoparticles promote the corrosion. This is due to the Brownian motion between the nanoparticles and the local convective motion of the liquid produced by Brownian motion, which enhances the mass transfer rate of O_2[9]. Therefore, TiO_2 nanoparticles exist alone in simulated cooling water, the probability of contact between O_2 and the brass surface increases, which may promote the corrosion of brass. The local convection caused by nanoparticles also promoted the diffusion of corrosion products.

3.1.2. Comparison the corrosion behavior of brass in different solution system

Due to the poor dispersibility of nanoparticles in aqueous solutions, SDBS is usually added to improve the dispersion of nanoparticles in solution. In this work, SDBS as disperser was entered into SCW. The experimental conditions were reported in table 2. Nyquist plot of brass in the three test systems at 30 ℃ were showed in figure 3. The fitting results were showed in table 3.

Table 2. The composition of the three test systems

Test systems	TiO_2 (wt.%)	SDBS (wt.%)
1-SCW	0	0
2- SCW + TiO_2	0.1	0
3- SCW + SDBS	0	0.05

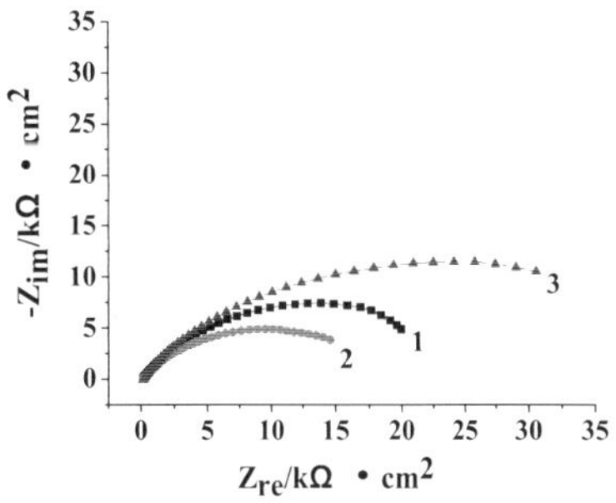

Figure 3. Nyquist plot of brass electrode in three test systems

Table 3. The fitting results of figure 3

Test systems	R_s ($\Omega\cdot cm^2$)	R_{ct} ($k\Omega\cdot cm^2$)	CPE	
			$Q(\mu\Omega^{-1}\cdot cm^{-2}s^n)$	n
1	167.0	16.82	88.32	0.65

2	158.5	23.57	68.99	0.68
3	185.7	36.23	57.60	0.68

It showed in figure 3 and table 3 that R_{ct} of brass in the TiO_2 nanofluids was minimal, indicating that corrosion of brass in TiO_2 nanofluids was most serious, probably due to the promotion of the mass transfer of O_2 and corrosion products. Brownian motion between TiO_2 nanoparticles and local convection of liquid produced by Brownian motion could enhance the mass transfer rate of O_2, which had mentioned above. When SDBS was added into the simulated water, the Rct value of the brass electrode increased from 16.82 kΩ·cm^2 to 36.23 kΩ·cm^2, indicating that SDBS had good corrosion inhibition effect on brass. As a commonly used surfactant, SDBS could adsorb on metal surface to form a protection layer against corrosion [10].

3.2. Effect of dispersant SDBS on brass

In order to further explore the corrosion inhibition effect of SDBS, different SDBS concentrations were performed in this paper. Nyquist plots of brass at 30 ℃ in TiO_2 nanofluids with different concentrations of SDBS were displayed in figure 4, and the fitting result in table 4.

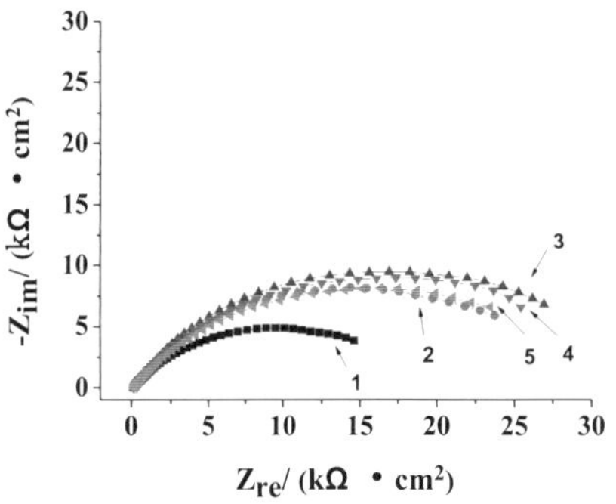

Figure 4. Nyquist plot of brass electrodes in TiO_2 nanofluids with different concentrations of SDBS
1) SCW + TiO2;
2) SCW +TiO2 + 0.005%SDBS;
3) SCW + TiO2 + 0.01%SDBS;
4) SCW + TiO2 + 0.02%SDBS;
5) SCW + TiO2 + 0.04%SDBS

Table 4. Fitting results of figure 4

Test systems	R_s (Ω·cm^2)	R_{ct} (kΩ·cm^2)	CPE	
			Q(μΩ^{-1}·cm^{-2}s^n)	n
1	167.0	16.82	88.32	0.65
2	162.2	26.81	63.79	0.68
3	156.0	30.89	53.96	0.57
4	150.7	29.43	62.30	0.68
5	151.6	27.42	72.93	0.66

It could be seen from figure 4 and table 4 that the addition of dispersant SDBS increased the R_{ct} of brass electrode in the TiO_2 nanofluids, indicating that SDBS had a good corrosion inhibition effect on brass in the water-based nano-fluid. When the concentration of SDBS reached 0.01%, the impedance value of brass electrode decreased, which indicated that when the amount of dispersant SDBS exceeded a certain value, the corrosion inhibition to brass decreased, which may be related to the threshold effect of SDBS[11].

4. Conclusions

(1)The corrosion resistance of brass electrode in nanofluids decreased with the increase of temperature. The increase of temperature could promote the corrosion of brass in SCW and TiO_2 nanofluids.
(2) The addition of TiO_2 nanoparticles in the nanofluids promoted the corrosion of the brass, and the SDBS could effectively inhibit the brass corrosion.
(3) The inhibition effect of SDBS on brass in TiO_2 nanofluids was enhanced with the increase of SDBS concentration. When the amount of dispersant SDBS was more than a certain value, the corrosion resistance of brass would be reduced.

References

[1] U.S. Chio 1995 *Asme Fed* **66** 99-105.
[2] Rashidi A, Amrollahi A, Lotfi R, Javaheryzadeh H. Rahimi H, & Rahimi A R and Jorsaraei A 2013 *Mater. Res. Bull.* **48** 4438-4443.
[3] Krishnamurthy S, Bhattacharya A P, Phelan P E and Prasher R S 2006 *Nano Lett.* **6** 419-23.
[4] Prasher R, Phelan P E and Bhattacharya P 2006 *Nano Lett.* **6** 1529-1534.
[5] Haddad Z, Abid C, Oztop H F and Mataoui A 2014 *Int. J. Therm. Sci.* **76** 168-189.
[6] Rashidi A M, Packnezhad M, Moshrefi-Torbati M and Walsh F C 2014 *Mcrofluid. Nanofluid.* **17** 225-232.
[7] Yang L, Du K, Zhang X S and Cheng B 2011 *Appl. Therm. Eng.* **31** 3643-3647.
[8] Xue-Min X U, Hong-Hua G E, Jia-Ni W U, He-li X U and Zuo-zhen L 2012 *Corrosion & Protection*
[9] Nagy E, Feczkó T and Koroknai B 2007 *Chem. Eng. Sci.* **62** 7391-7398.
[10] Xin-jing M, Dong-chen J and Hong-hua G 2004 *Chemistry World* **8** 473-474.
[11] Tavakoli H, Shahrabi T and Hosseini M G 2008 *Mater. Chem. Phys.* **109** 281-286.

Acknowledgements

The authors acknowledge the financial support of the National Natural Science Foundation of China (51471104) and Science and Technology Commission of Shanghai Municipality (16dz0503302)

Experimental Parameter Effect on Ceramic Coating Film on LY12 Alloy by Micro-arc Oxidation

H ZHAO[1], J H YANG[1] and X H WANG[2]

[1] College of Materials Science and Engineering, Shenyang Ligong University, China
[2] College of Materials Science and Engineering, Shenyang University of Technology, China
E-mail: zhaohui_1968@sina.com

Abstract. Ceramic coatings were fabricated on the surface of LY12 aluminum alloy by micro-arc oxidation (MAO) method in $NaSiO_3$-NaOH solution. Effect of experimental parameter on the MAO film was investigated. The results show that the ceramic layers formed on the surface of the alloy show different surface morphologies, and thickness of MAO film increases with the increase of voltage and oxidation time. These films contain two layers, an outer porous layer and an inner barrier layer, which consists mainly of γ-Al_2O_3 phase and a small amount of α-Al_2O_3 phase..

1. Introduction

A microarc oxidation (MAO) is a novel surface technique being wide used in depositing thick and ultra-hard ceramic coatings on valve metals substrate [1-3]. A key feature of the process is a plasma discharge that occurs at the metal/electrolyte interface when the applied voltage exceeds a certain critical breakdown value and appears as a number of discrete short-lived micro-discharges moving across the metal surface [4]. The MAO film has a lot of characteristics, such as high micro-hardness, high electric insulation resistance and good adhesion with the substrate [5].

Much Research has been focused on aluminum and aluminum alloys in recent years due to their low density, high specific strength and great machinability [6], but they are sometimes limited in applications because of their moderately poor corrosion resistance and tribological properties [7, 8]. Therefore, surface modification techniques of aluminum alloys, such as MAO, have been paid much attention. It is necessary to investigate the formation of ceramic film on aluminum alloy prepared by MAO.

MAO coatings were prepared on LY12 alloy by microarc oxidation in this study. Different experimental parameters, such as different voltage, different oxidation time, are employed. The microstructure, surface morphology and phase composition of MAO coatings on LY12 alloy under different conditions were analyzed in details.

2. Experimental

LY12 alloy plates in the size of 60×30×2 mm were used as the substrate. The MAO coatings were prepared by using pulse microarc oxidation system. The constant voltage mode was employed, and the constant pulse width and interval was selected as 500 μs and 1000 s. An aqueous solution, comprising mainly $NaSiO_3$ (4g/L), NaOH (2g/L) and some other substances to adjust its alkalinity and increase electric conductivity, was used as the electrolyte. Stainless steel plate was used as a cathode while LY12 alloy specimen as an anode. The temperature of the solution was controlled below 25 ^{o}C and the

IOP Conf. Series: Materials Science and Engineering **230** (2017) 012012 doi:10.1088/1757-899X/230/1/012012

oxidation time is 15 min, 20 min and 25 min respectively. Different voltages (400V, 440 V, 480 V) were used respectively.

The thickness of the coatings was measured using a coating thickness gauge (TT230). SEM (HORIBA S-3400) was used to investigate the morphology of surface and the cross-section of the specimen. Rigaku Vltima IV diffractometer was used to investigate the phase constituent of the coatings.

3. Results and discussion

3.1 Voltage effect on the MAO film of LY12 alloy

Figure 1 shows the surface morphologies of the MAO coatings on LY12 alloy under different voltages. When the coating is formed at lower voltage, the particles on the surface coatings are smaller, fine and uniform micro-pores distribute randomly on the surface (Fig. 1(a)). With the increase of voltage, MAO reaction become to be violent, more and larger oxide particles are produced. Therefore, the number of micro-pores decreases quickly, while the size of which increases significantly (Fig. 1(b)). Moreover, the surface of coatings is roughened with the increase of voltage. At this time, melting particles are deposited on the surface and obstruct the discharging channels as a result of much higher discharging energy induced by intense micro-arc discharging. At the voltage of 480 V, rough ceramic surface is formed and much more oxides are developed under higher breakdown voltage (Fig. 1(c)).

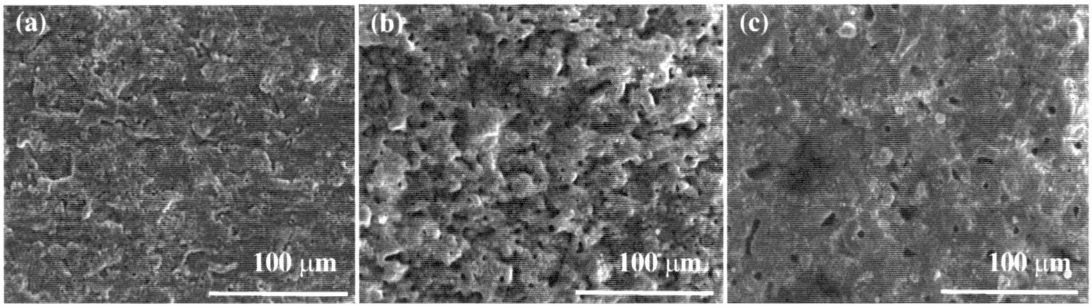

Figure 1. Surface morphologies of the ceramic coatings on LY12 alloy under different voltages. (a) 400 V, (b) 440 V, (c) 480 V.

The variation of the thickness of the MAO coatings on LY12 alloy is shown in Fig. 2 as a function of voltage. With an increasing voltage, the thickness of the coatings increases. An approximately linear growth of the coatings has been observed. The growth rate of the coatings is very fast, and the coating thickness goes up from about 2 to 10 μm.

Figure 3 shows the XRD patterns of the coatings on LY12 alloy under different MAO voltages. It can be determined that the MAO coating is composed of γ-Al_2O_3 mainly and a small amount of α-Al_2O_3. With the increase of voltage, more and more aluminum hydrate anions accumulated at the interface between the alloy and the electrolyte. Aluminum substrate was transformed into γ-Al_2O_3 and α-Al_2O_3, and bright surface of aluminum alloy became to be dark. Therefore, XRD peaks show decreased aluminum phases and increased γ-Al_2O_3 and α-Al_2O_3 phases with the increase of voltage.

3.2 Oxidation time effect on the MAO film of LY12 alloy

Surface morphologies of the MAO coatings on LY12 alloy under different oxidation time were shown in Fig. 4. At the beginning, uniformly dispersed micro-discharging channels are easier to occur on the surface of alloy, so that the size of micro-pores is significantly small, and the coatings are compact. With an increasing oxidation time, the coatings are characterized by micro-pores with different sizes.

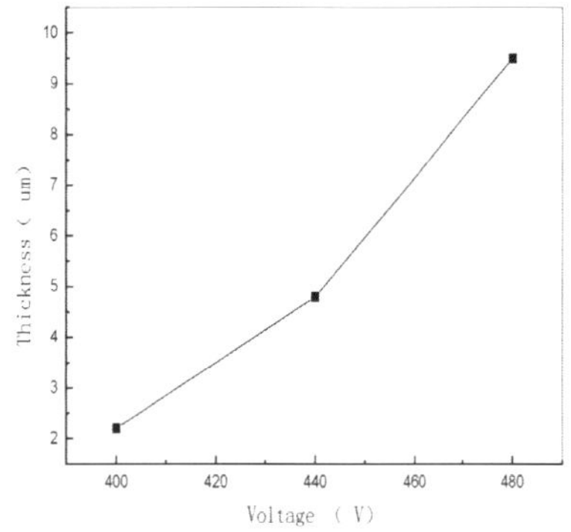

Figure 2. Variation of the thickness of MAO coatings on LY12 alloy with voltage.

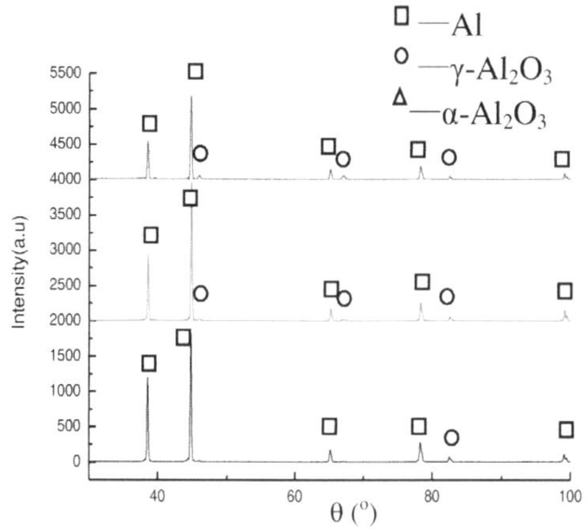

Figure 3. XRD patterns of MAO coatings on LY12 alloy under the voltage of (a) 400 V, (b) 440 V, (c) 480 V

Enlarged micro-pores were observed, and the coatings become porous and loose, as shown in Figs. 4(a) and (b). The porous feature strongly depends on discharging nature involved in MAO process [9]. However, with the further increase of time, the micro-pores on the surface seemed to be much less, and the coatings became much dense. The best treatment time is about 25 min for LY12 alloy to obtain the dense coatings.

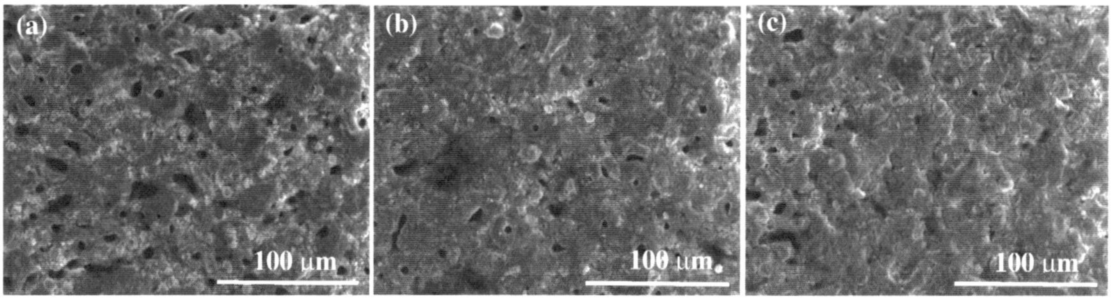

Figure 4. Surface morphologies of the ceramic coatings on LY12 alloy after different MAO treatment time. (a) 15 min, (b) 20 min, (c) 25 min.

Figure 5 shows the variation of the thickness of the MAO coatings on LY12 alloy as a function of oxidation time. With an increasing oxidation time, the thickness of the coatings increases rapidly. An approximately linear growth of the coatings was also observed. The growth rate of the coatings is very fast at the initial stage, and the coating thickness can reach 8.7μm at 15 min. However, in the period of 20 to 25 min, the thickness only goes up from 9.6 to 11 μm.

XRD patterns of the MAO coatings on LY12 alloy are shown in Fig. 6 under different oxidation time. There are γ-Al_2O_3 and α-Al_2O_3 in the MAO coatings, and the intensity of these diffraction peaks increases with an increasing time. Moreover, the peaks of aluminum substrate become to be weak when the thickness of the coatings increases. It has been indicated that the composition of the coatings may have a lot of γ-Al_2O_3 phase and a small amount of α-Al_2O_3. Obviously, γ-Al_2O_3 phase occurred initially and then transformed into α-Al_2O_3 phase.

3.3 Cross sectional structure of MAO film on LY12 alloy

Fig. 7 shows the SEM micrographs of the cross-sections for the LY12 alloys subjected to the MAO treatment. It can be noted that the MAO coatings have a good adhesion with the alloy substrate. It was found that the coating/metal interface was not very clear, and the oxygen content increases rapidly at the interface. The thickness of the oxide coating is about 15 μm. The coating contains two layers, an outer porous layer and an inner barrier layer. In general, barrier layer have great influences on the

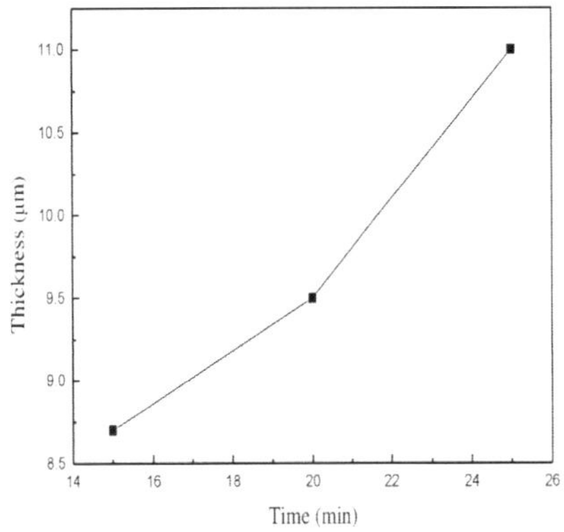

Figure 5. Variation of the thickness of MAO coatings on LY12 alloy with oxidation time.

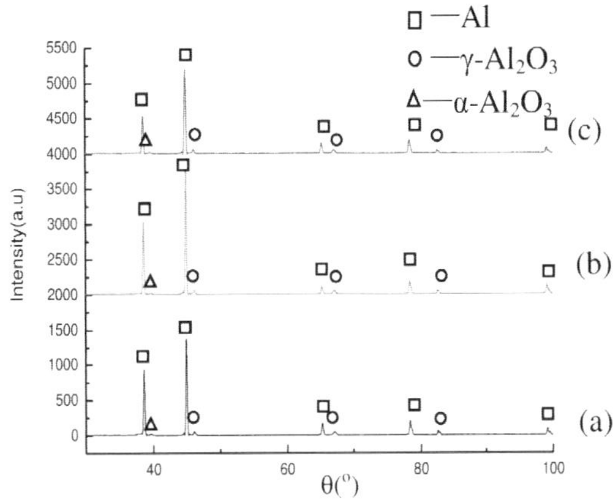

Figure 6. XRD patterns of MAO coatings on LY12 alloy. (a) 15 min, (b) 20 min, (c) 25 min.

corrosion barrier of MAO coatings [9]. With an increase in the thickness of the coating, the outer porous layer becomes much thinner, and the inner barrier layer will be thicker, which can account for a majority of coatings.

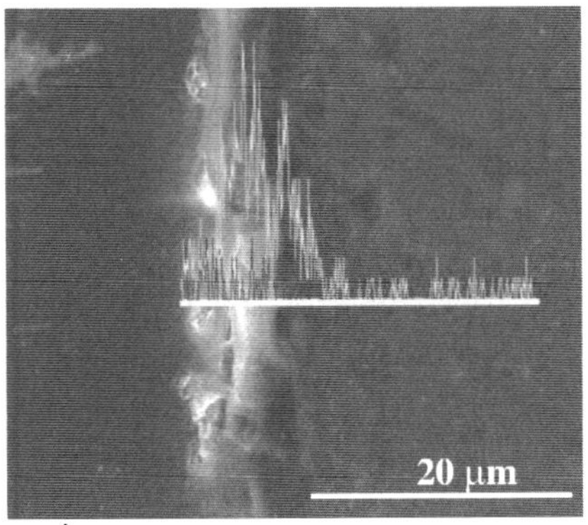

Figure 7. Back-scattered electron image of the cross-sectional MAO coating on LY12 alloy under a voltage of 480 V and an oxidation time of 25 min.

4. Conclusions

The ceramic layers formed on the surface of the LY12 alloy has been obtained by different micro-arc oxidation treatment conditions. Different surface morphologies of MAO films were obtained under different conditions. The thickness of MAO coatings on the LY12 alloy increases with an increase of voltage and oxidation time. The MAO coatings contain two layers, an outer porous layer and an inner barrier layer. These films consist mainly of γ-Al_2O_3 phase and a small amount of α-Al_2O_3 phase.

Acknowledgment

This work was supported by Shenyang Science & Technology Ministry (No.F15-200-6-03).

References

[1] Wirtz G P, Brown S D and Kriven W M 1991 Mater. Manufac. Proc. **6** L87
[2] Sundararajan G and Rama Krishna L 2003 Surf. Coat. Technol.**167** L269
[3] Hoche H, Scheerer H, Broszeit E and Berger C 2003 Surf. Coat. Technol. **174-175** L1018
[4] Yerokhin A L, Nie X and Leyland A 1999 Surf. Coat. Tech. **122** L 73
[5] Yerokhin A L, Lyubimov V V and Ashitko R V 1999 Ceram. Int. **122** L 1
[6] Kojima Y Mater. Sci. Forum 2000 **350/351** L3

[7] Liang J, Guo B G, Tian J, Liu H W, Zhou J F and Xu T 2005 Appl. Surf. Sci. **252** L345
[8] Gray J E and Luan B 2002 J. Alloys Compd. **336** L88
[9] Cai Q Z, Wang L S, Wei B K and Liu Q X 2006 Surf. Coat. Technol. **200** L3727

Effects of Friction Stir Processing on Mechanical Properties and Damping Capacities of AZ31 Magnesium Alloys

F Y Lan[1] , H M Chen[2], W P Guo[3], J Zhang[4] and Y X Jin[5]

[1] Provincial Key Lab of Advanced Welding Technology, Jiangsu University of Science and Technology, Zhenjiang, 212003, China
[2] School of Metallurgy and Materials Engineering, Jiangsu University of Science and Technology, Zhangjiagang, 215604, China
E-mail: hmchen@just.edu.cn

Abstract. AZ31 base alloy were subjected to friction stir processing (FSP) with different rotating speed. The effect of FSP on the microstructure, mechanical properties and damping capacities of AZ31 alloy was discussed in this paper. The fine and equiaxed grains were obtained in the SZ for FSPed samples with the rotating speed from 1300rpm to 1700rpm with the average grain size of 11.82μm, 12.93μm and 14.50μm, respectively. The FSPed samples had higher ductility and damping capacities compared with the AZ31 base alloy. The ductility and damping values $Q_0^{-1}(\varepsilon=10^{-4})$ and $Q_H^{-1}(\varepsilon=10^{-3})$ of 1500rpm FSPed sample were increased by 121.44%, 5.02% and 7.49%, respectively.

1. Introduction

Magnesium alloys are gaining increasing importance in aerospace, defence and automobile applications owing to their low density, high strength to weight ratio , high damping capacity and high recycling capability [1,2]. Friction stir processing (FSP) was developed by adapting the concept of FSW to obtain a fine grain size in a stirred zone [3]. After the successful advent of FSP, the application of arena of this technology has been growing rapidly [4]. Some if its most recent applications involve surface modifications, enhancement of surface and bulk properties and generation of in-situ and ex-situ nanocomposites in number of aluminum and magnesium-based alloys [1,5-9].

However, few studies have been conducted to excogitate the effect on damping properties of FSPed magnesium alloys possessing the modification of grain size and grain orientation which is an important factor that affects the damping capacity of magnesium alloys [10,11]. Therefore, there is a great possibility of resolving the contradiction between the mechanical properties and damping capacity of magnesium alloys by FSP.

2. Materials and Methods.

The material used in this study was AZ31 magnesium alloy of hot rolling state with composition of (in wt.%): Al-2.14, Zn-0.967, Mn- 0.307, Si-0.096, Mg-balance. The base material was cut into specimen of 250 mm× 170 mm×4 mm. The shoulder diameter, pin diameter and pin height were 16.0 mm, 4.0 mm and 3.8 mm, respectively. The FSP tool was rotated in 1300, 1500, 1700 rpm in the clock-wise direction, severally. The traverse speed was 50 mm/min. The tilted angle was 2.5°.

For optical examination, the samples were sectioned, cold mounted, polished by silicon dioxide paste with grain size of 1μm and finally etched in a solution of oxalic acid(2g), nitric acid (1mL) and distilled water (98mL), and then examined by using a ZEISS optical microscope. The cross-section were subjected to micro hardness testing employing Vicker's hardness indentation at 100g load. Specimens for tensile test and damping test were machined from the center region in the SZ. Tensile testing was performed on a SANS CMT5205 material testing machine at a stretching rate of 2mm/min at ambient conditions. Damping samples were machined to dimensions of 35 mm×10 mm×1 mm. Damping capacity was measured by dynamic mechanical analyzer (NETZSCH DMA-242C) in single cantilever deformation mode. For the measurements of strain amplitude dependence of damping capacity, the range of strain amplitude was from 1 μm to 200 μm, and the measurement frequency was 1 Hz.

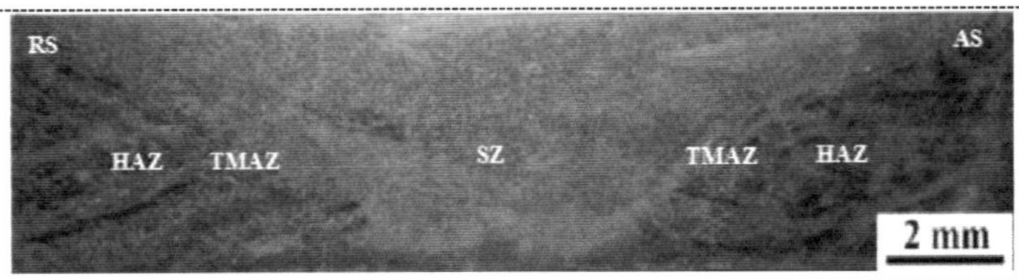

Figure 1. Electronic cross-sectional macro-image of FSPed sample with a rotating speed of 1500 rpm.

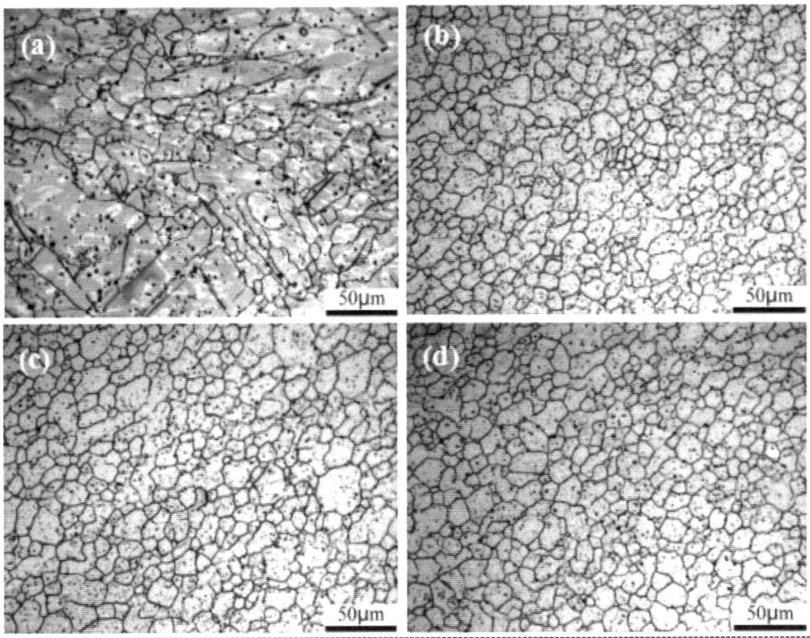

Figure 2. Microstructure of AZ31 base alloy and SZ of FSPed samples.
(a) AZ31 base alloy (b) 1300rpm (c) 1500rpm (d) 1700rpm

3. Results and Discussion

3.1. Macrostructure and microstructure

Figure 1 shows the cross-sectional macrostructure of the FSPed sample with a rotating speed of 1500 rpm.Three distinct regions are present in the final macrostructure: heat affected zone (HAZ), thermo-mechanically affected zone (TMAZ) and stir zone (SZ)[12]. Compared with the retreating side (RS), the advancing side (AS) has a much clearer and sharper boundary which can be visible to the naked eye.

The microstructure of AZ31 magnesium alloy consisted of elongated, uneven distributed grains along the rolling directions (as shown in figure 2 (a)). The equiaxed grains were found in the SZ after FSP. Compared with the AZ31 base alloy, the FSPed samples had significantly refined grains. At a constant traverse speed, the average grain size was respectively 11.82μm, 12.93μm and 14.50μm (as shown in figure 2 (b),(c),(d)), getting larger with the increasing of rotating speed.

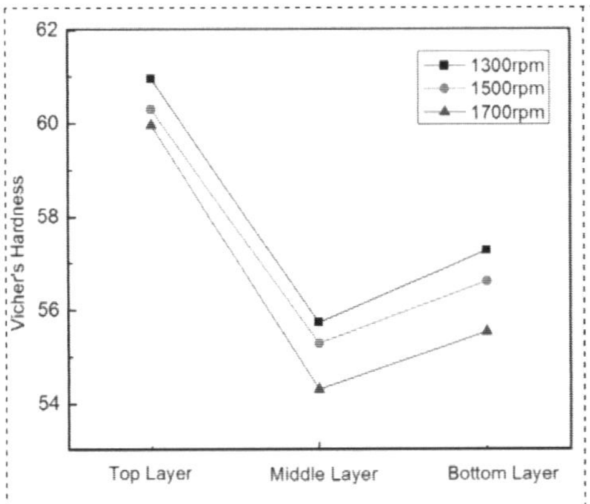

Figure 3. The Vicher's hardness of different layers of FSPed AZ31 alloys.

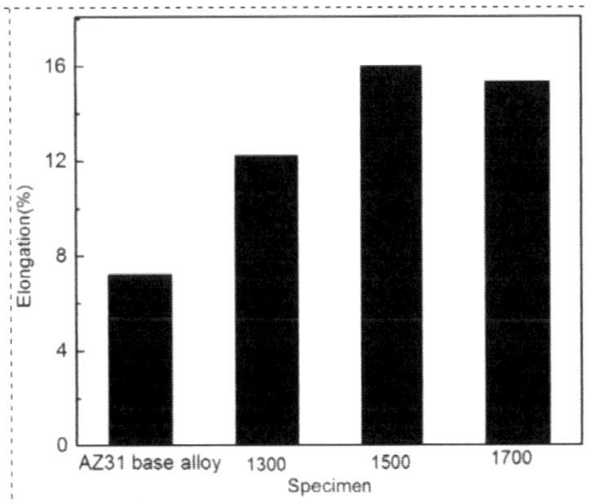

Figure 4. The ductility of AZ31 base alloys and FSPed samples.

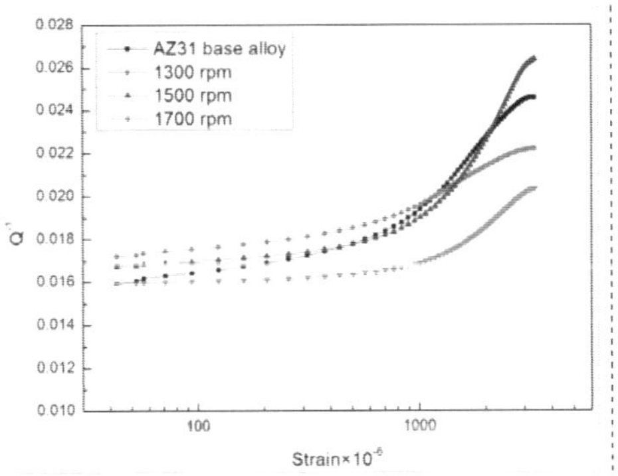

Figure 5. Strain amplitude dependence damping capacity of AZ31 base alloy and FSPed samples.

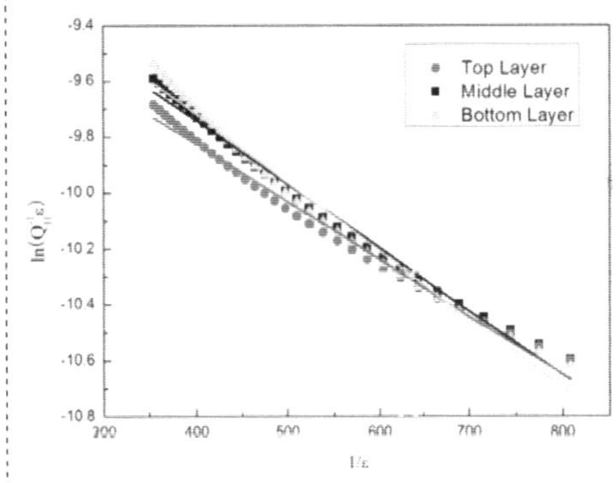

Figure 6. G-L plots of SZ in different layers after FSP of 1500rpm.

3.2. Mechanical properties

Figure 3 shows the Vicher's hardness of different layers of FSPed AZ31 alloys with the rotating speed of 1300rpm, 1500rpm and 1700rpm, respectively. Compared to the AZ31 base alloy, the FSPed had lower hardness values significantly. As the rotating speed increased, the grain size increased, while the hardness values decreased at the same time which were relatively in good accordance with the Hall-Petch Relationship. The Vicher's hardness of FSPed samples of different rotating speed had the similar trends, decreased at the middle layer and then rebounded at the bottom layer. This is because the softening caused by the dynamic recrystallization in the SZ was much stronger than the strengthen caused by the grain refining.

The tensile strength did not get promoted though the average grain size of the SZ of fine grained materials is much less than the AZ31 base alloy. The elongation has been significantly improved as shown in figure 4. The reason for this phenomenon may be related to the changes of texture of the material after FSP. Hung and Shih [13] in their investigation identified that the grain orientation has changed after FSP, causing the increase of the residual stress. AZ31 magnesium alloy has a HCP structure which has less slip systems at room temperature. Therefore, the effect of the texture was much more than the average grain size on the strength of the AZ31 alloy.

3.3. Damping capacities

Figure 5 shows the strain amplitude dependence damping capacity of AZ31 base alloy and FSPed samples. The damping values $Q_0^{-1}(\varepsilon=10^{-4})$ of AZ31 base alloy at low strain amplitude was 0.01594, and that of the samples subjected to 1300rpm FSP (0.01721), 1500rpm FSP (0.01674) and 1700rpm (0.01595) were increased by 7.97%、5.02% and 0.06%, respectively. The damping values $Q_H^{-1}(\varepsilon=10^{-3})$ at high strain amplitude of AZ31 base alloy and FSPed samples were 0.02456, 0.0222(1300rpm FSP), 0.0264(1500rpm FSP) and 0.02037(1700rpm FSP), respectively. Compared to the AZ31 base alloy, only the 1500rpm FSPed sample had a increase by 7.49 %. When the damping capacity is bigger than 0.01, it is generally considered that the alloys exhibit high damping capacity [14]. The 1500rpm FSPed samples obtained higher damping capacity.

According to the G-L theory[15, 16] ,

$$C_1/C_2^2=\rho L_N^3/6F_B. \tag{1}$$

Where C_1 and C_2 are the material constants; ρ is the dislocation density; F_B is the binding force between dislocation and weak pinning points; L_N is the average dislocation distance between strong pinning points.

Table 1. Values of C_1 and C_2 according to G-L plots.

Treatment	Top Layer	Middle Layer	Bottom Layer
$C_1\times10^{-4}$	1.23	1.46	1.60
$C_2\times10^{-3}$	2.06	2.27	2.45
C_1/C_2^2	28.98	28.33	26.66

Strain amplitude dependence damping capacity of SZ in different layers after FSP of 1500rpm abided by the G-L model from Figure. 4(b). The values of C_1 and C_2 calculated according to G-L plots[17] are shown in Table 1.

From equation 1, the C_1/C_2^2 value is proportional to ρL_N^3. The C_1/C_2^2 values of SZ in different layers after FSP of 1500rpm were decreased due to the decrease of the average grain size from the top layer to the bottom layer while the L_C was increased according to the change of C_2.

4. Conclusions

The fine and equiaxed grains were obtained in the SZ for FSPed samples with the rotating speed from 1300rpm to 1700rpm with the average grain size of 11.82μm, 12.93μm and 14.50μm, respectively.

The FSPed samples had higher ductility and damping capacities compared with the AZ31 base alloy. The ductility and damping values $Q_0^{-1}(\varepsilon=10^{-4})$ and $Q_H^{-1}(\varepsilon=10^{-3})$ of 1500rpm FSPed sample were increased by 121.44%, 5.02% and 7.49 %, respectively.

References.

[1] Reddy G M, Rao A S, Rao K S. *Friction Stir Processing for Enhancement of Wear Resistance of ZM21 Magnesium Alloy[J].* Transactions of the Indian Institute of Metals, 2013, 66(1):13-24.

[2] Sanchez C, Nussbaum G, Azavant P and Octor H, Mater Sci Eng 221A (1996) 48.

[3] R.S. Mishra and M.W. Mahoney: Mater. Sci. Forum, Vol. 357-359(2001), pp.507-514.

[4] Arora H S, Singh H, Dhindaw B K. *Parametric Study of Friction Stir Processing of Magnesium-Based AE42 Alloy[J].* Journal of Materials Engineering & Performance, 2012, 21(11):2328-2339.

[5] Takayama Y, Nagai T, Shibayanagi T, et al. *Superplasticity and Crystallographic Orientation Distribution in Friction Stir Processed Sheet of an Al-Mg-Mn Alloy[C].* Ibaraki district conference. The Japan Society of Mechanical Engineers, 2004:67-68.

[6] Xin R L, Li B, Liu Q. *Microstructure and Texture Evolution during Friction Stir Processing of AZ31 Mg Alloy[C].* Materials Science Forum. 2010:1195-1200.

[7] Khodabakhshi F, Simchi A, Kokabi A H, et al. *Microstructure and texture development during friction stir processing of Al-Mg alloy sheets with TiO_2, nanoparticles[J].* Materials Science & Engineering A, 2014, 605(6):108-118.

[8] Arora H S, Singh H, Dhindaw B K. *Wear behaviour of a Mg alloy subjected to friction stir processing[J].* Wear, 2013, 303(s 1-2):65-77.

[9] Chang C I, Wang Y N, Pei H R, et al. *Microstructure and Mechanical Properties of Nano-ZrO_2 and Nano-SiO_2 Particulate Reinforced AZ31Mg Based Composites Fabricated by Friction Stir Processing[J].* Key Engineering Materials, 2007, 351:114-119.

[10] Sug imo to K, Niiya K, Okamo to T, et al. *A study of damping capacity in magnesium alloys [J].* Transactions of Japan Institute Metals, 1977, 18:277 288.

[11] Sugimoto K, Matsui K, Okamo to T, et al. *Effect of crystal orientation on amplitude dependent damping in magnesium [J].* Transactions of Japan Institute Metals, 1975, 16:647-655.

[12] P. Asadi, R.A. Mahdavinejad, S. *Tutunchila. Simulation and experimental investigation of FSP of AZ91 magnesium alloy.* Materials Science and Engineering A 528 (2011) 6469-6477

[13] Hung Fei-Yi, Shih Chien-Chih, Chen Li-Hui, et al. *Microstructures and high temperature mechanical properties of friction stirred AZ31-Mg alloy [J].* Journal of Alloys and Compounds, 2007, 428:106-114.

[14] Wan Diqinga, Wang Jincheng, Yang Gencang. *A study of the effect of Y on the mechanical properties, damping properties of high damping Mg-0.6%Zr based alloys.* Materials Science and Engineering A, 517 (2009) 114-117.

[15] A. Granato, K. Lüker. *Theroy of Mechanical Damping Due to Dislocation[J].* Appl. Phys. 1956, 27 (6): 583-593.

[16] A. Granato, K. Lüker. *Application of Dislocation Theory to Internal Friction Phenomena at High Frequencies[J].* Appl. Phys. 1956, 27 (7): 789-805.

[17] Zhang Zhenyan, Peng Liming, Zeng Xiaoqin. *Effects of Cu and Mn on mechanical properties and damping capacity of Mg-Cu-Mn alloy.* Trans. Nonferrous Met. Soc.China. 18(2008): s55-s58.

Acknowledgments.

This work was supported by National Natural Science Foundation (ProjectNo: 51301077), Natural Science Foundation of Jiangsu Province (Project No: BK20130470) , Undergraduate Innovation Projects of Jiangsu Province and Jiangsu University of Science and Technology(Project No: SJLX16_0459 and 1062161515) and Priority Academic Program Development of Jiangsu Highuer Education Institutions.

IOP Conf. Series: Materials Science and Engineering **230** (2017) 012014 doi:10.1088/1757-899X/230/1/012014

Present Development Status of Anti-creep Magnesium Rare-Earth Alloys

Yunwei Gui[1], Quanan Li[1, 2] and Xiaoya Chen[1]

[1]School of Materials Science and Engineering, Henan University of Science and Technology, Luoyang 471023, China
[2]Collaborative Innovation Center of Nonferrous Metals, Henan Province, Luoyang 471023, China
[a]guiyunwei1@163.com,

Abstract. The research status of creep resistance of rare earth magnesium alloy at home and abroad is reviewed, and the mechanism of high temperature creep resistance and the way of improving the creep resistance of magnesium alloy were also discussed. The problems of high temperature resistance and creep resistance of cast magnesium alloy are pointed out, and its future development direction is forecasted. The purpose of this paper is to provide the idea and basis for the development of creep resistant and heat resistant magnesium alloy.

1. Introduction

Metal magnesium and its alloys are among the lightest metal structural materials currently available. It has a series of advantages such as low density, high specific strength, high specific rigidity, high specific elastic modulus, excellent damping shock absorption performance, good thermal conductivity, good electromagnetic shielding effect, excellent machinability, stable dimension and easy recovery, which has aroused more and more attention[1-4]. Nowadays magnesium alloy has been widely used in aerospace, automotive, communications electronics, computers and other industrial fields. However, the heat resistance of magnesium alloys is poor, when the temperature increases, its strength and creep resistance dramatically reduced, which limits the further application of magnesium alloys. Compared with the general magnesium alloy, rare earth heat-resistant magnesium alloy has a higher high temperature strength and high temperature creep resistance[1-5]. Therefore, how to improve the high temperature creep properties of magnesium alloys has become a hot topic in recent years.

2. Research progress on creep resistance of rare earth magnesium alloys

From the 20th century, 30 years, the addition of rare earth elements in magnesium alloys to improve the creep resistance of magnesium alloys has begun, until the 70's rare earth magnesium alloy began to get practical application. At present, the heat-resistant magnesium alloy systems are mainly Mg-Al alloy, Mg-Zn alloy and Mg-RE alloy[6-9]. The following describes several types of new anti-creep Mg-RE magnesium alloy.

2.1 Binary rare earth magnesium alloy

Suzuki M et al[10] studied the relationship between the creep rate and the stress of the Mg-Y binary alloy and found that the creep mechanism is a dislocation-climbing mechanism when the Y content is between 0.2% and 1.1%. When the Y content is 1.1% -2.4%, the creep mechanism is mainly sticky sliding.B. L. Mordike[11] studied Mg-Gd binary alloys with different Gd contents. Compared with the alloys of WE43 and QE22, the creep resistance of Mg-Gd alloy is better than that of other alloys. With the increase of Gd content, the creep resistance of the alloy gradually increased.YAN J L et al[12] studied the microstructure and creep resistance of Mg-2Nd alloy, the results show that the as-cast microstructure of Mg-2Nd alloy is composed of α-Mg matrix and Mg12Nd eutectic phase distributed at grain boundaries. The tensile strength and yield strength of as-cast Mg-2Nd alloy decrease with increasing temperature, however the magnitude of decrease is not big, it has better high temperature stability. Elongation increased with temperature increased significantly. The as-cast Mg-2Nd alloy exhibits good creep resistance. The stress exponent of the alloy in the range of 150 ~ 250 ℃, 30 ~ 110 MPa is 3.3 ~ 8.0, and the creep activation energy is in the range of 108 ~ 142 kJ / mol. The creep mechanism of the alloy is due to the dislocation climbing control and the grain boundary slip play a certain role.ZHU B B et a[l][13] studied microstructure and mechanical properties of Mg-Sm alloy, it is found that the comprehensive mechanical properties of Mg-3Sm alloy are better. The maximum solid solubility of Sm in magnesium was 5.7% (mole fraction), so it has good solid solution strengthening and aging strengthening effect. T6 state of binary Mg-6Sm alloy creep resistance and WE43 is equivalent.Mordike's[11] research shows that the creep resistance of simple binary Mg-Sc alloy is poorer than WE43 alloy, and must be combined with other elements to produce composite strengthening. After addition of Mn, the creep resistance of Mg-Sc-Mn alloy was found to be two orders of magnitude higher than that of alloy WE43 at 350 ℃.XUE S et al[14] studied microstructure and properties of Mg - 2La binary rare earth magnesium alloy. At 175℃ and 70MPa, the stress exponent n and creep activation energy Q of the alloy are 2 and 67 kJ / mol, respectively, the creep mechanism of alloy is grain boundary sliding mechanism. Because the eutectic Mg17La2 phase does not have a high melting point (about 600 ℃), the thermal stability is poor, so the creep resistance of the alloy is low.

2.2 Multiple rare earth magnesium alloy

FU S L et al[15] studied high temperature creep behavior of Mg-12Gd-2Y-0.5Zr alloy, the results show that the creep resistance of the alloy is generally better than that of the WE54 alloy. In the stress range of 50-70MPa, the stress index n is between 2.5-3.8 at 200℃, n is between 4.2-6.1 at 300℃, the creep of this series of alloys is mainly controlled by dislocation climbing control to dislocation cross slip control mechanism. The creep activation energy Qc is 66kJ/mol at 50MPa and 87.2kJ/mol at 70MPa under the temperature range of 200-300℃. With the increase of temperature and stress, the grain boundary diffusion activation energy (80kJ/mol) of pure magnesium is getting closer and closer.ZHANG Q et al[16] studied creep behavior of Mg-5Y-3Sm-0.8Ca-0.5Sb alloy, the results show that Mg-5Y-3Sm-0.8Ca-0.5Sb alloy has similar creep properties to the commercial heat-resistant magnesium alloy WE43 at 200 ℃. With the increase of creep stress, the grain size of the alloy changed little and the grain boundary became not obvious. The calculated values of the stress index show that the creep is controlled by both diffusion and grain boundary sliding mechanisms.FU S L et al[17] studied high temperature creep behavior of Mg-12Gd-2Y-2Sm-0.5Zr magnesium alloy, the results show that the aged Mg-12Gd-2Y-2Sm-0.5Zr alloy has excellent creep resistance at 200 ℃, 300 ℃ / 50MPa and 70MPa. Under the stress of 50 MPa and 70 MPa, the stress exponent n=3.8 at 200℃ and n=5.6 at 300℃. The creep activation energy Qc = 50.8kJ / mol at 50MPa and the creep activation energy Qc = 64.8kJ / mol at 70MPa under 200 ℃ and 300 ℃.TONG Y et al[18] studied microstructure and properties of Mg-13Gd-3Y-0.4Zr alloy, the steady-state creep rate of the T6 alloy is 3.61×10-9s-1, and the creep strain rate is 1.18% under the condition of 250℃ and 80MPa.At 300℃ and 50MPa, the fracture time of alloy specimen is 30h, the strain rate is 12.47%, and the steady state creep rate is 2.41×10-8s-1.The precipitation phase of high melting point alloy effectively prevents the slip and climbing of the dislocation, and improves the creep resistance of the alloy.SHA G Y et al[19] studied compressive creep behavior and micro - mechanism of Mg-4Y-2Nd alloy , at 300 ℃, 100 MPa conditions, the alloy still has extremely excellent creep properties, the steady-state creep rate is 1.505×10-6s-1. Sliding and twinning are the basic ways of creep deformation of alloys, precipitation strengthening and grain boundary strengthening are the main ways to improve the creep resistance of this alloy.

3. Creep mechanism of magnesium alloys

Creep refers to the material in the sustained force, with the time of the slow and sustained non-elastic deformation. For creep theory research, a large number of scholars[20,21], the formation of a variety of creep theory, the more commonly used theory dislocation creep, diffusion creep, Nabarro-Herring creep, Coble creep, Harper -Dom creep, the third power viscous sliding creep and low temperature creep and so on. It is pointed out that magnesium alloys have many creep deformation mechanisms under different creep conditions, in which dislocation slip and grain boundary slip are the main deformation mechanisms of magnesium alloys. Dislocation slip creep is the deformation of material under the action of stress, slip through the dislocation to achieve the deformation, and slip of the essence of intraocular dislocation movement[20-22]. In the process of creep, the dislocations slide on the sliding surface. When the dislocation slip is blocked, dislocations accumulate to form dislocation agglomerates. At low temperatures, dislocations accumulate to form strain hardening. At high temperatures, dislocations Dislocations can be dislocated and crossed[20-22]. Diffusion creep generally occurs at high temperature and low stress creep conditions, where low creep stresses do not permit a large number of dislocations to slip, and creep deformation occurs through the diffusion of matter[21, 22]. The grain boundary sliding is mainly due to the diffusion of the atoms at the grain boundaries at high temperature. The grain boundaries are moved and the creep deformation occurs under the action of stress. As the temperature increases and the creep stress decreases, The more the contribution of creep deformation, and the finer the grain size, the more grain boundaries provide the channel for diffusion, and the contribution of grain boundary slip to creep deformation increases. The Raj and Ashby studies have shown that if there is enough slip in the creep process to meet the deformed Von Mises condition, grain boundary slip is not needed to coordinate grain deformation, but in diffusion creep, Is indispensable coordination mechanism of grain deformation[20-22].

4. The way to improve the creep property of magnesium alloy

For the creep mechanism of magnesium alloy, creep strengthening of magnesium alloy should be to limit dislocation movement, to prevent the grain boundary sliding principle, so far to improve the creep resistance of magnesium alloy is mainly solid solution strengthening, precipitation strengthening, dispersion strengthening and grain boundary.

4.1 Solid solution strengthening

Solid solution strengthening refers to adding solute elements in the alloy to improve its homogenization temperature and elastic modulus, slow down the diffusion and self-diffusion process, reduce the rate of dislocation climbing, so as to improve the high temperature creep properties of the alloy. Solid solution strengthening is the result of the interaction between solute atoms and dislocations. In terms of their properties, they can be divided into elastic, chemical, electrical and geometric types. Solute atoms can be segregated around the dislocation to form a variety of air mass, can also be irregularly distributed in the matrix Therefore, adding alloy elements with high solid solubility to the magnesium alloy can improve the aging hardening ability, and can obviously improve the creep resistance of the alloy. In the process of high temperature creep, the role of solid solution strengthening mainly in the recovery process, can delay the recovery process, thereby reducing the creep rate of magnesium alloy. In the process of high temperature creep, the role of solid solution strengthening mainly in the recovery process, can delay the recovery process, thereby reducing the creep rate of magnesium alloy[20-24].

4.2 Precipitation strengthening

Precipitation aging strengthening is the precipitation phase in which the solid solubility of alloying elements decreases with the decrease of temperature during the aging process. The interaction between precipitates and dislocations leads to the increase of the yield strength of the alloy. As the magnesium atoms larger, usually with the magnesium matrix formation of non-co-precipitated complex phase, the interface energy is high, easy to coarsen at high temperatures, it is difficult to effective grain boundary pinning role Therefore, the key to improve the creep resistance of magnesium alloy is to improve the crystal structure of the precipitation phase with the appropriate alloying elements to reduce the mismatch of the lattice constant with the magnesium matrix, to improve the precipitation mode and morphology of precipitated phase, Phase to reduce its diffusivity, so as to improve the creep properties of magnesium alloy[20-24].

4.3 Dispersion strengthening

Dispersion strengthening is much less affected by temperature changes than precipitation strengthening. It has high melting point, good thermal stability, little solubility in the matrix and dispersed in the grain boundaries and grains. It can effectively control the grain deformation, grain boundary sliding and dislocation movement, so it can be produced in the alloy Diffusion strengthening phase to improve the alloy in different temperature range of the ability to adapt to the alloy and the mechanical properties, especially magnesium alloy creep resistance[20-23].

4.4 Grain boundary strengthening

The grain size has great influence on the high temperature properties of the metal materials. The grain size of the magnesium alloy can be controlled by controlling the preparation way, solidification process and adding appropriate alloying elements In general, the creep at low temperature and high stress is mainly due to the dislocation of intracrystalline dislocation climbing, grain boundary will become barriers to dislocation climbing, so the smaller grain size of magnesium alloy creep beneficial; But at the same temperature and above, the grain boundary sliding increased, appropriately reduce the number of grain boundaries, increasing the grain size, you can improve the creep resistance of magnesium alloy.Grain boundary strengthening is mainly achieved through purification and microalloying methods. The purpose of purification is to reduce the harmful impurities in the grain boundary of the segregation. Microalloying improves the thermal strength of the alloy by improving the shape and distribution of the second phase of the grain boundary and the organization of the region near the grain boundary by segregation of the microalloying element at the grain boundary[20-24].

5. Conclusions

In recent years, the main industrial countries put a huge financial and human resources for the development of anti-creep heat-resistant magnesium alloy,which achieved initial results. Mg-Al and Mg-Zn systems are inexpensive at present, but the heat resistance is not satisfactory and the space for improvement is small .The Mg-RE system has high heat resistance, but it is expensive and difficult to be commercialized and marketized..It Should be further optimized by alloying elements, effective alloy design and casting magnesium alloy to solve high temperature, creep resistance and other problems.

It is the good choice to have a depth study of magnesium alloy creep behavior and creep mechanism, and the use of rare earth elements developed to meet the various requirements of the new high-performance, low cost creep resistant magnesium alloy .And it will be an important research direction in Magnesium alloy.

6. References

[1] M D Xin, Z S Ji. *Research situation and application prospects of rare earth in foundry magnesium alloy*[J]. Journal of the Chinese Society of Rare Earths, 2010, 28(6): 643-653.

[2] M K Wang,J X Sun, X C Liu. *Development of research and application of cast magnesium alloys*[J]. Nonferrous Metals, 2012, (2): 56-59.

[3] H.R. Jafari Nodooshan, W C Liu, G H Wu. *Effect of Gd content on microstructure and mechanical properties of Mg–Gd–Y–Zr alloys under peak-aged condition*[J]. Materials Science & Engineering A, 2014, 615(2014): 79–86.

[4] L L Rokhlin, N I Nikitina. *Aging kinetics and methanical properties of Mg-Gd-Sm system alloys*[J]. Fizika metallov i metallovedenie, 1996, 82(4): 113-118.

[5] J Chen, Q A Li, Q Zhang. *Effect of yttrium on microstructure and properties of AZ61 magnesium alloy*[J]. Journal of the Chinese Society of Rare Earths, 2015, 33(4): 449-454.

[6] H Wang, Q D Wang, C J Boehlert. *The impression creep behavior and microstructure evolution of cast and cast-then-extruded Mg–10Gd–3Y–0.5Zr (wt%)*[J]. Materials Science and Engineering: A, 2016,649:313-324.

[7] J D Robson, S J Haigh, B Davis. *Grain boundary segregation of rare-earth elements in magnesium alloys*[J]. Metallurgical and Materials Transactions A-Physical Metallurgy and Material, 2016, 47A: 522-530.
[8] D S Li, X N Cheng. *The development of creep resistant magnesium alloys*[J]. Materials Review, 2006,5(20):424-427.
[9] Y Jono, M Yamasaki, Y Kawamura. *Quantitative evaluation of creep strain distribution in an extruded Mg–Zn–Gd alloy of multimodal microstructure*[J]. Acta Materialia, 2015,82:198-211.
[10] M Suzuki, H Sato, K Maruyama. *Creep deformation behavior and dislocation substructures of Mg–Y binary alloys*[J]. Materials Science and Engineering A, 2001, 319: 751-755.
[11] B L Mordike. *Creep-resistant magnesium alloys*[J]. Materials Science and Engineering A, 2002, 324: 103-112.
[12] J L Yan, Y S Sun, F Xue. *Microstructure and creep resistance of Mg-2Nd alloy*[J]. Foundry, 2007, 56(8): 805-808.
[13] B B Zhu, Y S Sun, D Jia. *Microstructure and mechanical properties of Mg-Sm alloys*[J]. Journal of Southeast University (Natural Science Edition), 2009, 39(3): 610-614.
[14] S Xue, Y S Sun, T B Zhu. *Investigation on the properties of Mg-La and Mg-Nd binary magnesium-rare earth alloys*[J]. Foundry, 2005, 54(9): 888-891.
[15] S L Fu, Q A Li, X T Jing. *High temperature creep behavior of Mg-12Gd-2Y-0.5Zr magnesium alloy*[J]. Special Casting & Nonferrous Alloys, 2013, 33(12): 1090-1092.
[16] Q Zhang, Q A Li, J Chen. *Creeping behavior of Mg-5Y-3Sm-0.8Ca-0.5Sb alloy*[J]. Rare Metals and Cemented Carbides, 2014, 42(4): 52-54.
[17] S L Fu, Q A Li, X T Jing. *High temperature creep behavior of Mg-12Gd-2Y-2Sm-0.5Zr magnesium alloy*[J]. Ordnance Material Science and Engineering, 2014, 37(3): 5-7.
[18] Y Tong, Q D Wang, Y Gao. *Heat treatment optimizing and property of Mg-13Gd-3Y-0.4Zr alloy*[J]. Light Metals, 2007, 3: 45-49.
[19] G Y Sha, E H Han, T Yu. *Greep behavior and microscopic mechanism of Mg-Y-Nd alloy*[J]. Acta Metallurgica Sinica, 2003, 39(10): 1025-1030.
[20] K Sawada., M Tabuchi, K Kimura. *Analysis of long-term creep curves by constitutive equations*[J]. Materials Science and Engineering: A, 2009, 510-511: 190-194.
[21] F Von Buch, B Mordike. *High-temperature properties of magnesium alloys*[J]. Magnesium-Alloys and Technology, 2004, 106-129.
[22] W F Xu, Y Zhang, L M Peng. *Linear precipitate chains in Mg-2.4Gd-0.1Zr alloy after creep*[J]. Materials Letters, 2014,137:417-420.
[23] N Liu, Z Zhang, L Peng. *Microstructure evolution and mechanical properties of Mg-Gd-Sm-Zr alloys*[J]. Materials Science and Engineering: A, 2015,627:223-229.
[24] A F Abd El-Rehim. *Effect of cyclic stress reduction on the creep characteristics of AZ91 magnesium alloy*[J]. Acta Metallurgica Sinica (English Letters), 2015,28(8):1065-1073.

Acknowledgments
This work was supported by the National Natural Science Foundation of China (Nos 51571084 and 51171059), and Henan Province Key Scientific and Technological Projects (Nos 152102210072). The authors are grateful to Q A Li and X Y Chen for their intellectual discussions and assistance.

IOP Conf. Series: Materials Science and Engineering **230** (2017) 012015 doi:10.1088/1757-899X/230/1/012015

Effects of Combining Na and Sr additions on Eutectic Modification in Al-Si alloy

G L ZHU[1], N J GU[1] and B J ZHOU[1]
[1]Chalco Institute Research of science and technology Co.LTD, Beijing, China
E-mail: my2150@163.com

Abstract. Experiments were designed to investigate the effects of strontium and sodium modified on the eutectic silicon for Al-Si alloy. It was found that combining addition of Na and Sr did not appear to cause deleterious interactions of modification, at at the same time, Sr-Na was fairly constant with holding time and without obvious modification fading. Addition of Na-Sr modifier could take effect quickly and decrease incubation period.

1. Introduction

Al-Si alloy is one of the commercial Eutectic aluminum silicon alloys widely applied in automotive engines, because of their good mechanical properties at elevated temperature, excellent wear resistance and low coefficient of expansion[1-3]. However, eutectic Si in untreated exhibits poor ductility even at room temperature[3]. In order to improve mechanical properties, especially fracture toughness and elongation, sodium or strontium is added to Al-Si alloy to transform the eutectic silicon from faceted acicular flakes to a fine fibrous structure[4-7]. It has been reported that the addition of sodium to Al–Si alloys has some drawbacks, such as obvious effect of porosity and sodium fading. On the other hand, the addition of Sr results in increase of hydrogen content and a long incubation[6, 8].Therefore, in present work, attempts were made to investigate any possible interaction by determining the characteristic modification for the eutectic silicon containing Na and Sr. Furthermore, the effect of combining addition of Na and Sr on incubation and modification fading were studied by metallographic analysis.

2. Experimental

Chemical composition of the Al-Si alloy used in this study is shown in table 1 The alloy was melted in a resistance furnace at 750°C. After the alloy had melted, the melt was degassed for 15 min using Ar. The modification of the melt were add separately by adding Al-10Sr master alloys, a mixture of Na-salts (two thirds of Na-fluoride and one third of Na-chloride). To investigate the influence of sequential additions, the experiments were divided in two series. In the first series, Sr or Na was first added alone. In the second series, Sr and Na were added simultaneously.The melt was poured at 700℃ into the graphite mould with a cooling rate in the centre of the samples of about 10K/s. Metallographic samples were sectioned horizontally, and were mounted and polished. The ground specimens were then subjected to a final polishing with colloidal silica suspension. An optical microscope (Olympus BX60F5) was used to analyze the microstructures.

For the purpose of illustrating the quantitative modification, the AFS chart known as modification rating was employed in the study. The eutectic silicon morphologies of all samples were categorized into five qualitative levels ranging from coarse needle-like eutectic silicon to fine fibrous eutectic silicon[9]: (1) large plate eutectic Si; (2) plate eutectic Si; (3) longer fibrous eutectic Si; (4) fibrous and needle eutectic Si; (5) fine fibrous eutectic Si.

Table 1. Chemical composition (wt %) of Al-Si alloy

Si	Cu	Mg	Ni	Al
11.5	0.85	0.95	0.85	Bal.

3. Results

3.1Modification with sodium

Figure 1(a) shows the typical microstructure of Al-Si alloy produced by unmodified process. Unmodified alloy reveals coarse plate-like silicon morphology. In contrast, figure (b) clearly shows the fine fibrous morphology of the eutectic Si in a fully modified structure.

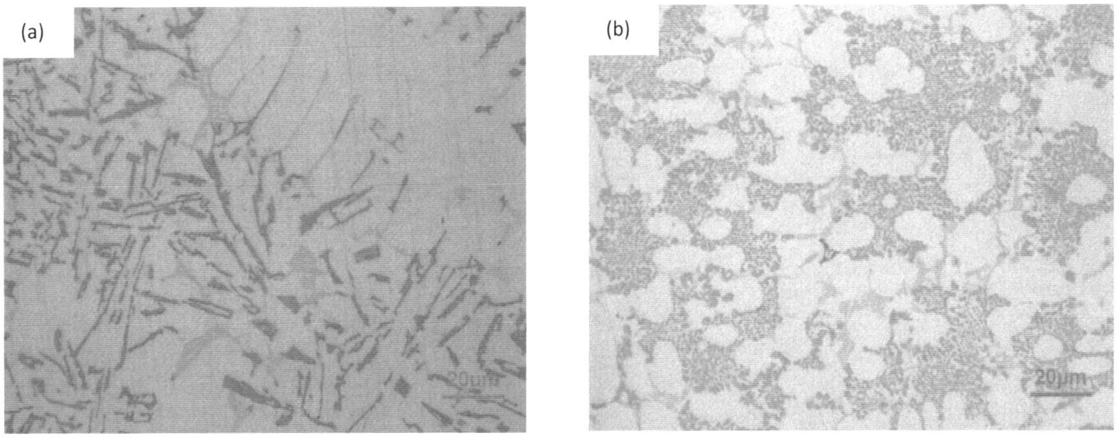

Figure 1. The typical microstructure of Al-Si alloy (a) unmodified alloy;(b) Na-modified (100ppm)

Figure 2 shows modification rating as a function of Na concentration. The results show that there is a strong relationship between Na concentration and modification rating. At the first stage, modification rating increases with the increase of Na concentration, however, it decreases with further increase in Na concentration at the late stage. There is a maxima in the rating–Na curve. The increase in modification rating implies that there is under-modified during initial stage, while the decrease in modification rating suggested that there is over-modification during late stage. The over-modified microstructures of Al-Si alloy, i.e. the over-modification band is obvious infigure 3.

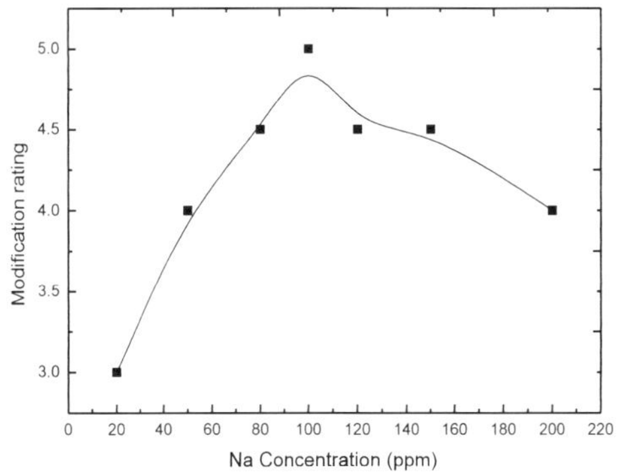

Figure 2. Modification rating a function of Na concentration

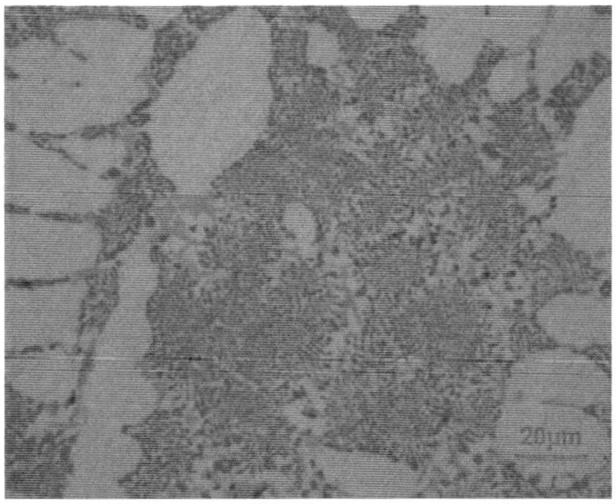

Figure 3. Over-modified structures of Al-Si alloy

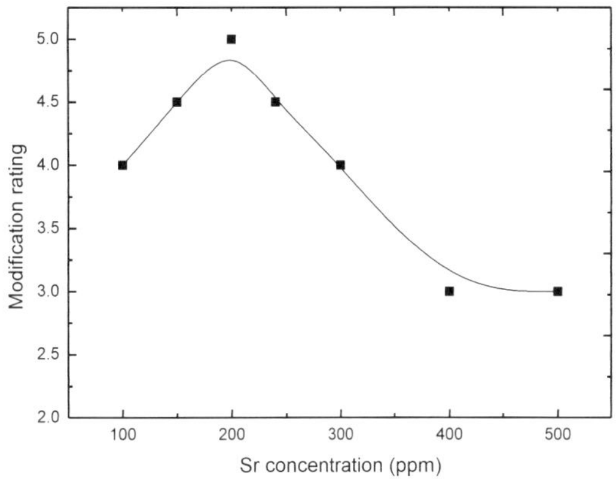

Figure 4. Na concentration a function of holding time

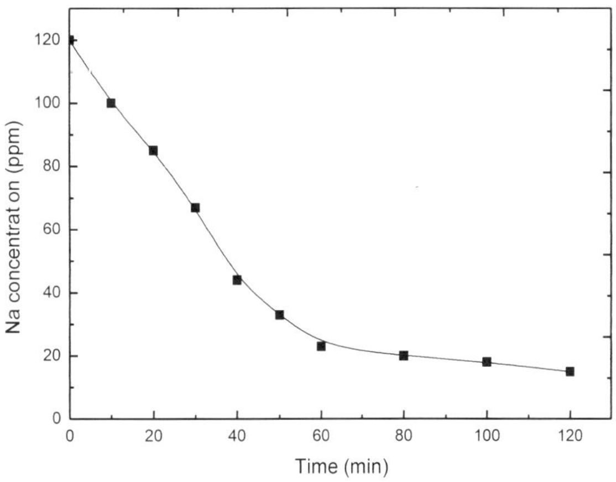

Figure 5. Modification rating a function of Sr concentration

Figure 4 shows Na concentration as a function of holding time with 120ppm Na. Chemical composition analysis indicates that Na concentration decreases with the increase in holding time. A comparison between figure 2 and figure 4 indicates that there is a distinct Na fading. Loss of Na is believed to be responsible for the disappearance of the modification effect during holding time. The lifetime of Na-modified Al-Si alloy is about 20-30 min.

3.2Modification with strontium

Figure 5 shows the correlation between Sr concentration and modification rating. Modification rating first increases and then decreases with increase of Sr concentration. The under-modified structure is observed as at low Sr concentration, which is similar to figure 2. A fibrous or fully modified structure is only obtained when Sr addition is near to 200 ppm. With increase of Sr concentration, the Al_2Si_2Sr intermetallic compound is also observed. Consequently, modification rating is high when Sr concentration takes place at appropriate concentration for Al-Si alloy.
Figure 6 shows Sr concentration as a function of holding time. Sr concentration in the melt decreases slowly with holding time. It is clear that Sr has a slower loss rate than that of Na and a well-modified structure in a long time. A comparison between figure 5 and figure 6 indicats that Sr could has a better modification with increase of holding time, especially for 60 min, or longer.

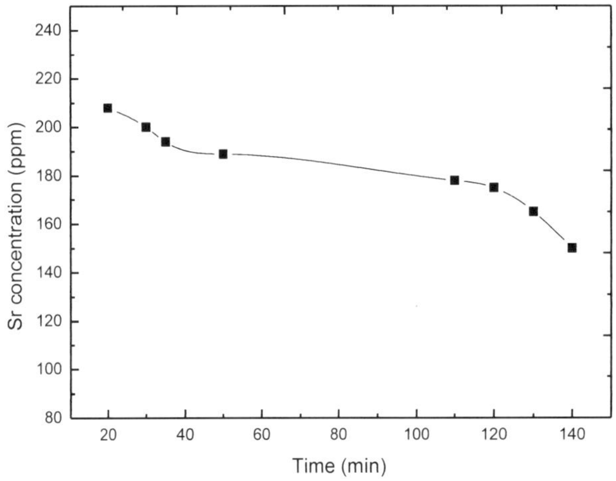

Figure 6. Sr concentration a function

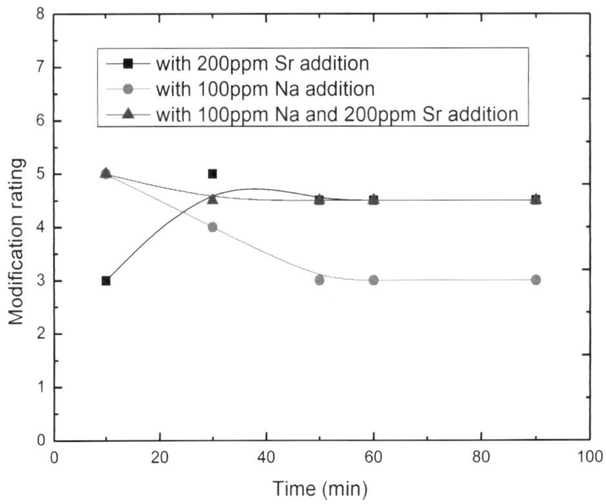

Figure 7. Modification rating a function of holding time of holding time

3.3.Modification with sodium and strontium combined

Figure 7 shows the modification rating as a function of holding time at 100ppm Na and 200ppm Sr. The results show that it has a good modification over 60min. At 10 min, modification rating of Al-Si is a full modification. A full modification is at 30min when Sr is added alone, It is obvious that there is a shorter incubation for adding Sr and Na compared with Sr. Moreover, it does not show obvious effect of modification fading for combining addition of Na and Sr.

4. Discussion

Na and Sr are commonly used in Al–Si alloys to modify the morphology of eutectic Si and enhance its mechanical properties. Microstructure examinations show that reasonable concentration Sr or Na is sufficient for well-modified of the eutectic silicon. However, Na fades considerably more quickly than expected in general. Na fades due to oxidation reactions with the surrounding environment and modification effect on the eutectic Si is lost. Sr modifier has some weakness, which is a long incubation. with combining addition of Na-Sr, full modification is fairly constant with increase of holding time and without obvious modification fading. It can take effect quickly and decrease the time of incubation. At the same time, no evidence of Sr-Na intermetallics is found in the samples from the bottom of the melt in this study. It is therefore believed that no deleterious reaction is founded between Na and Sr.

L. Lu found that Na addition was likely to promote Sr vaporization and/or oxidation kinetically, leading to a progressive loss of both modifiers[10]. Figure 7 shows that modification rating has been kept constant over 120min. It is concluded that Sr does not cause a large loss during this period. The possible reason for this difference is the atmosphere of modification. This relative quick Sr fading may be related to large surface to volume ratio of the small laboratory furnces. As L. Lu had already discussed, due to its high solubility in liquid Na and limited solubility in molten Al, Sr is expected to migrate upwards from the molten Al into the liquid Na layer. In the open space, Sr is continuously depleted with Vaporization and/or oxidation until its depletion[10]. However, in a relatively closed environment, there is no further loss of Sr from the liquid Na layer. The lifetime of Na-Sr modified Al-Si alloy is different, depending on the initial content of Na-Sr in the melt, melt circumstances, surface to volume ratio of the furnces. For example, the turbulence of melt and bubbling during degassing will make Na-Sr easily to fade.

Based on the above discussion, it can be concluded that the effect of Na is more transient whereas Sr is more persistent. Combining addition of Sr and Na appeares to reduce incubation time compared for Sr alone. It is not deleterious reaction between Na and Sr, but rather that a combination is beneficial in giving prolonged modification (less fading). Consequently, modification with sodium and strontium combined can produce a well-modified microstructure over a long period.

5. Conclusion

1. The appropriate concentration of Sr and Na is sufficient for well-modified of the eutectic silicon.
2. It is not deleterious reaction between Na and Sr, but rather that a combination is beneficial in giving prolonged modification (less fading).
3. Modification with sodium and strontium combined provides a constant long term modificationand in a shorter incubation.

6. References

[1] Haque M M 1995 *J. Mater. Process. Tech.* 55 193
[2] Nogita K, McDonald S D and Dahle A K 2003 *Mater. Trans.* 44 692
[3] Wang W X and Gruzleski J E 1989 *Mater. Sci. Technol.* 5 472
[4] Dahle A K, Nogita K, McDonald S D, Dinnis C M and Lu L 2005 *Mater. Sci. Eng.A.* 413-414 243
[5] Ho C R and Cantor B 1995 *J. Mater. Sci.* 30 1921
[6] Dinnis C M, Dahle A K, Taylor J A and Otte M O 2004 *Metall. Mater. Trans. A.* 35 3531
[7] Hegde S and Prabhu K N 2008 *J. Mater. Sci.* 43 3009
[8] Fat-Halla N 1989 *J. Mater. Sci.* 24 2488

[9] Apelian D, Sigworth G K and Whaler K R 1984 *AFS Trans.* 92 297
[10] Lu L, Nogita K and Dahle A K 2005 *Mater. Sci. Eng.A.* 399 244

Acknowledgments
The authors thank for SINR their support and facilities.

IOP Conf. Series: Materials Science and Engineering **230** (2017) 012016 doi:10.1088/1757-899X/230/1/012016

Research of flammability of fireproof materials in ship safety

Yizhou Jiang[1,2] , Duanfeng Han[2] and Ziwei Zhang[3]

[1]China Academy of Transportation Sciences, Beijing 100029, China
[2]College of Shipbuilding Engineering, Harbin Engineering University, Harbin 150001, China
[3]South China University of Technology, Guangzhou 510641, China
E-mail: jiang.yizhou@outlook.com

Abstract. This paper analyzes the classification, performance and application of ship fireproof and heat insulating materials, and describes the test standard and performance evaluation criteria of the non-combustibility, low flame-spread characteristics and smoke and toxicity of marine fireproof materials in detail. So the paper has certain reference value and guidance significance for the selection of heat insulating materials with fire divisions and the use of flammable materials on board in accordance with requirements.

In recent years, ship fire accidents have happened continuously. The issue of ship fire prevention has aroused the concern of all countries. There has been a trend in ship cabin interior materials that non-combustible material displacing combustible material. In this aspect, all countries in the world have done a lot of research work. Combustible material are clearly defined in The International Convention for the Safety of Life at Sea (SOLAS) and the relevant amendments [1]. The International High-Speed Craft Safety Code (HSC Code), enacted by the International Maritime Organization (IMO), entered into force on 1 January 1996, which assert claims from prevention, suppression and escape to ensure the safety of life at sea and prevent shipwreck. Fire-proof insulation materials should have a variety of conditions such as light mass, fire proof, heat insulation, sound proof, beauty and so on. One of the most important is the extensive use of a variety of fire-resistant material to constitute fire divisions, which strengthens the fire protection capacity of ship structure.

This paper makes in-depth analysis of the performance, advantages and disadvantages and application on the vessel of some ship fireproof and heat insulating materials, and conduct a comparative study in the main performance of several materials.

2. Overview of ship fireproof and heat insulating material

As Euro-American corporations involve in China's marine development business, there is a higher requirement for fire divisions which are not defined in the Convention so as to safeguard the safety of life at sea better. Fire happening when the ship in operation will cause great loss probably because the fight conditions at sea is worse than that on land. If fire happens when the ship is sailing in heavy sea, fighting difficulties will increase. Therefore, in accordance with the requirements of the SOLAS Convention, the study of fireproof and heat insulating materials with good flame resistance and low fame-spread characteristics is extremely important [2].

2.1. Development of ship fireproof materials in China

IOP Conf. Series: Materials Science and Engineering **230** (2017) 012016 doi:10.1088/1757-899X/230/1/012016

China began to study ship fire proof materials in the late 70s of last century, mainly referring to the international popular class of fire divisions, which is generally divided into two types of organic materials and inorganic materials. Organic materials include black cork board, polyurethane foam board, yellow cork board and so on; inorganic materials include asbestos foam board, glass wool insulation board, ceramic fiber and so on. There are many enterprises in China producing inorganic fireproof and heat insulating materials. Fire-resistant materials have been used for on board are mainly PS-2 mineral cotton board, moisture-proof ultra-fine glass wool felt, double-sided aluminum-plastic composite rock wool board, stainless steel composite rock wool board, polyethylene foam board and so on.

2.2. Green marine fireproof material

Green materials, also known as eco-materials, refers to the materials which are in harmony or coexistence with chemical ecology environment in raw material collection, product manufacturing, use or recycling, waste processing and other aspects. Such green fireproof and heat insulating materials are not much produced in China. However, Western countries have a variety of similar green refractory fiber products on the market, such as Firemaster Marine Lite from Thermal Ceramics; Insulfrax, Fiberfrax from Unifrax.

3. Classification of fireproof materials

The SOLAS Convention classifies materials into non-combustible materials and combustible materials. As for deck covering and associated low flame-spread surface materials, the classification is based on a specific performance difference. Non-combustible materials in accordance with the main components, application location and fire-resistant rate are classified as follows:

3.1. Classification according to the main components

Classification of non-combustible material according to the main components see the Table 1a and 1b.

Table 1a. Classification of non-combustible material according to the main components.

Name	Characteristics	Main Uses
Sodium silicate binder	Sodium silicate as the base material, and a certain filler, curing agent and other semi-liquid material.	As adhesive of heat and sound insulating material and so on.
Cement materials	Cement mixed with sand and lightweight aggregate.	As flooring and deck covering with non-combustibility high hardness.
Asbestos products	Natural inorganic fiber asbestos mainly.	Heat insulation, but not recommended for harmful effect on human health.
Glass cotton products	Manmade inorganic fibre products, weak high-temperature resistance.	General heat insulation and sound absorption material.
Rock wool products	Manmade inorganic fibre products, higher high-temperature resistance.	Heat insulation, composite plate as components of bulkhead and deck.
Ceramic wool products	Manmade inorganic fibre products, very high high-temperature resistance.	Heat insulation, fire-resistant separation composition, sound absorption material.

Table 1b. Classification of non-combustible material according to the main components.

Name	Characteristics	Main Uses
Calcium silicate board	Artificial synthetic board, high intensity.	Heat insulating material of bulkhead plate and ceiling.
Pearlite plate	Artificial synthetic board, moderate intensity.	Heat insulating material of bulkhead plate and ceiling.
Vermiculite plate	Artificial synthetic board, low intensity.	Heat insulating material of bulkhead plate and ceiling.

3.2. Classification according to application location and fire-resistant rate

Flammability classification is based on the ability to react to fire of materials to determine its fire rating. Flammability ratings for materials or products are traditionally expressed as flammability ratings or flammability index, which is the overall performance of the product is integrated into one or more characteristic parameters. These parameters describe the combustion characteristics of the product during the test. The description of the material reaction to the fire depends on the combustion characteristics of the material and the test method.

There are three kinds of divisions' types defined by SOLAS, A-Class, B-Class and C-Class [3]. Class-A is a division made of steel plates and other equivalent materials, which is a structure that can stop smoke and flame from passing through within 1 h. While according to insulation requirements, Class-A division is divided into four grades, and during the time corresponding to each grades, the elevated temperatures without exposure to the surface of the flame must be within the limits specified in the Convention: grade 'A-60' 60 min, grade 'A-30' 30 min, grade 'A-15' 15 min and grade 'A-0' 0 min; insulation value of B-Class, in the following two standard fire test time, the average temperature of the side which is back to the flame rises no more than 139 degrees, and the temperature at any point including any of the joints rises no more than 139 degrees; Constructions of grade 'B-15' 15 min and grade 'B-0' 0 min should prevent the passage of flame at the end of the 0.5 h standard fire test; Class-C divisions are divisions constructed of non-combustible materials. They do not have to meet the requirement of preventing the passage of smoke and flame and limiting heating as long as they meet the relevant requirements. The following is the classification in accordance with the application location and fire-resistant rate of different materials, see Table 2a and Table 2b.

Table 2a. Classification in accordance with the application location and fire-resistant rate.

Application location	Fire-Resistant rate	Common Materials
Bulkhead	A-60, A-30, A-15, A-0, B-15, B-0, C	Steel, heat insulating material, non-combustible materials.
Deck	A-60, A-30, A-15, A-0, B-0	Steel, heat insulating material, non-combustible materials.
Ceiling	B-15, B-0, C	Non-combustible materials.
Lining board	B-15, B-0, C	Non-combustible materials.
Door, doorframe	A-60, A-30, A-15, A-0, B-15, B-0, C	A-Class for the steel material, B-Class and C-Class for the non-combustible materials.

Table 2b. Classification in accordance with the application location and fire-resistant rate.

Application location	Fire-Resistant rate	Common Materials
Floor	--	Cement grout and Lightweight aggregate concrete.
Heat insulating material	--	Ceramic wool, rock wool, glass wool board.
Through position of cable	Without destroying the fire integrity of the through position.	Combine composite rockwool panel or ceramic plate with materials preventing flame spread.
Through position of fire air valve	Without destroying the fire integrity of the through position and slightly higher than fire divisions class.	Marine fire air valve should be conducted a fire division test.

Heat insulating materials with 'A', 'B' and 'C' class fire-resistant fire divisions used in ship must be non-combustible materials; damp-proof layers and adhesives used with insulation and insulation for pipe fittings of cooling systems need not be non-combustible materials, while they should maintain the practicable minimum, and their exposed surfaces should have low flame-spread characteristics. Paints used in exposed surfaces, varnish and other decorative materials and deck coverings should not produce excessive substances with smoke and toxicity. Judging whether the marine fireproof materials have non-combustibility, low flame spread characteristics, smoke and toxicity are determined by the specified test.

4. Flammability of materials

4.1. Non-combustibility

Non-combustible material refers to a material heated to about 750 ℃, neither burning, nor giving off sufficient flammable boil-off gas [4]. International marine non-combustible materials are determined by the test procedure specified by FTP Code. The marine material is specified that any material other than non-combustible material is combustible. According to the SOLAS Convention, the material is classified into non-combustible materials and combustible materials. Differently, the building materials are classified into non-combustible materials, fire- retardant materials, combustible materials and flammable materials by China national standard GB8624-1997 [5]. According to China National Standard 1501182-1990 Fire Test-Test Method of Non-Combustibility of Building Materials and the application rules of the procedures of international fire resistance test, the conditions for determining the non-combustibility of the construction material and marine material are shown in Table 3.

In addition to considering the changes of the material's own performance during the whole test, the current standard should also consider the change of its environmental conditions, which is qualitative technical index. Material non-combustibility is determined by the prescribed test process. The material required to achieve the standard test conditions can be judged as non-combustible materials, and for some non-single-component material, even if the inorganic and organic composite materials, it is said that it has non-combustibility as long as it can meet the criteria for evaluation. Compared with non-combustible materials for buildings, there are higher requirements for marine non-combustible materials, and the average surface temperature rise of the sample is also required indicators.

Table 3. The non-combustibility of the construction material and marine material.

Determinant conditions of non-combustibility	Average furnace temperature rise	Average burning time of sample	Average surface temperature rise of sample	Average mass loss rate of sample
Non-combustibility of the construction material	≤50°C	≤20s	--	≤50%
Non-combustibility of the marine material	≤30°C	≤10s	≤30°C	≤50%

Non-combustible materials are generally inorganic materials. Whether using surface coatings or chemical soaking methods, organic materials can not be changed to non-combustible materials by general processing methods [6]. Organic materials are mainly used for thermal insulation materials used by fire divisions in ships or insulating materials of ventilation pipes, etc. The furnace temperature was controlled and stabilized at 750±5°C by heating resistance wire. Three thermocouples were placed in the furnace and the sample to measure the temperature of the furnace, the surface temperature of the sample and the center temperature of the sample, and the curves of temperature change and time are output through data acquisition.

Select the soft fiber, rigid plate type, semi-rigid categories of three non-combustible materials for use, the non-combustibility curves of the test are shown in Fig. 1. From the comparison of the non-combustibility curves of various materials, it can be seen that the furnace temperature of the fiber material and the surface temperature of the sample reach a constant value in a short time. There is an obvious peak in the center temperature of the rigid plate material, and the final temperature takes a long time to reach a constant. For high density materials, the central thermocouple temperature will slowly reach the highest value, which has a greater impact on the furnace temperature and the temperature difference of the sample surface temperature. The amount of internal combustibles exist in the material that meets the requirements for non-combustibility will directly affect the performance of the material. Combustion and duration of the combustible will directly lead to the furnace temperature and sample surface temperature rising, making the temperature difference of the final equilibrium temperature too high [7].

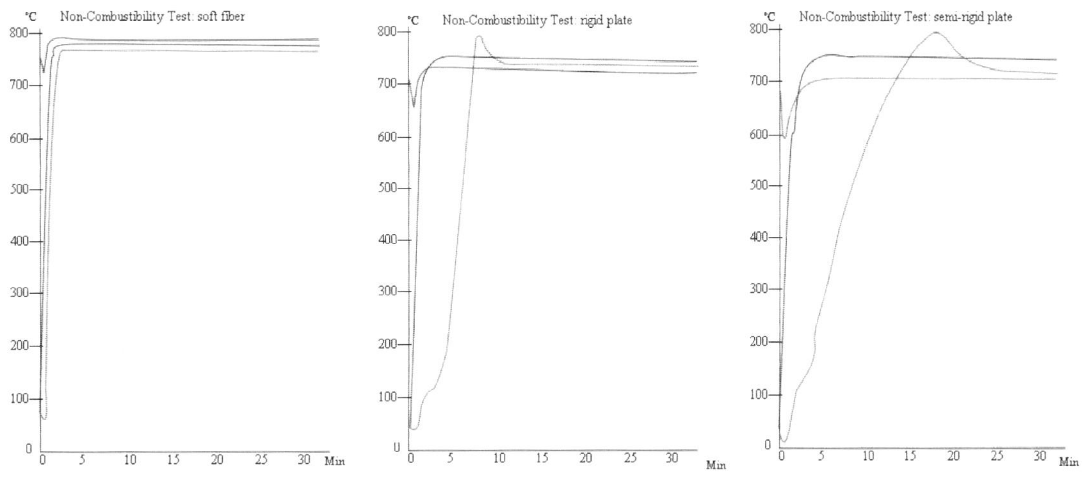

Figure 1. The non-combustibility curves of the test.

In general, products made only of glass, concrete, ceramic products, natural stone and commonly used metals and alloys can be considered non-combustible materials which can be used onboard without test and approval.

4.2. Low flame-spread characteristics

Low flame-spread characteristics refers to the surface material can effectively limit the spread of the flame. This characteristics is assessed by the FTP Code. surface material flammability test, tested by the fire test procedures for the surface flammability of bulkheads, ceilings and deck decorative materials in accordance with Resolution A653 (16) passed by IMO. Specific marine low-flame-spread materials are bulkheads, linings, ceilings and the like, such as film, decoration layer and so on; surface paint, varnish, coating and so on; moisture barrier and adhesives used with insulation and the exposed surface of the insulation for pipe fittings of cooling system; deck covering, floor covering(carpet), ventilation pipe and so on.

The general low-flame-spread test apparatus provides a method of evaluating the combustion characteristics of a sample in the vertical direction. The sample is exposed to a graded radiation flow field provided by a gas-fired radiant panel, and the radiation surface is close to the ignition source. Meanwhile, observe the ignition time, flame spread and extinguishment of the sample, and measure the millivolt voltage signal of thermocouple in the combustion process. Heat for ignition, sustain burning heat, critical heat flux during extinguishment and the value of thermal release can be obtained by the temperature and information collection module. Marine flammable materials to meet the performance requirements of low flame-spread characteristics are shown in Table 4.

Table 4. The performance requirements of low flame-spread characteristics.

Low flame spread characteristics performance criteria	Critical heat flux during extinguishment	Sustain burning heat	Total value of thermal release	Peak value of thermal release
Bulkhead, lining board, damp-proof layers and adhesives, the exposed surface of the insulation for pipe fittings of cooling system and plastic pipe	≥20KW/m2	≥1.5MJ/m2	≤0.7MJ	≤4KW
Deck base covering and floor covering	≥7KW/m2	≥0.25MJ/m2	≤1.5MJ	≤10KW

4.3. Smoke and toxicity

It is necessary to reduce the life risk caused by smoke and toxic substances on board in the event of fire. Therefore, flammable materials, surface coatings included, flammable gases and toxic substances released from the fire should be restricted to effectively control potential smoke and toxicity. Smoke generation tests should be carried out in accordance with smoke and toxicity test in FTP Code Part Z and International Standard 1505659: 1994.

Marine flammable materials which are needed doing the smoke and toxicity test involve paints, varnishes and other decorative coating for exposed surfaces as specified in n-2, section 6 of the SOLAS Convention. According to Appendix 2 of FTP Code, when the decorative material and the deck covering cannot simultaneously satisfy the total heat release value Q of not more than 0.2MJ and the heat release rate peak qp of not more than 1.0 kw, further testing of smoke density and toxicity is required for validation. These include: decorative materials for bulkheads, linings or ceilings, such as films and decorative veneers; deck coverings materials, such as deck base coverings; materials used on the surface of floors, such as carpets; plastic pipes and cables; paints, varnishes and other decorative coating for exposed surfaces.

The general test equipment can collect and measure the amount of smoke generated on the surface of a horizontally placed sample in a closed combustion chamber with or without the use of ignition conditions under specified heat radiation conditions. The combustion chamber is equipped with a photometric test device, which can measure attenuation of beam through the smoke and represent smoke volume in the form of a specific optical density.

Tests are carried out by accredited laboratories in accordance with FTP rules Part Z. The test should be carried out in three states: (1) With test flames, the irradiation is 25kw/m2; (2) Without test flames, the irradiation is 25kw/m2; (3) Without test flames, the irradiation is 50kw/m2.

5. Conclusion

Flammability of ship fireproofing material is one of the important attributes to evaluate the safety performance of ship. Based on the flame spread and the spread speed, the heat generated when burning and smoke volume and concentration, toxicity, etc. people can judge whether it meets the requirements of marine fireproof materials. As a mobile steel structural buildings on the water, ships will often be subjected to vibration and collision. Sailing in waters with diverse climates is different from the service conditions on land.

Therefore, the approach used on land cannot be simply copied and the particularity of the ship must be considered, that is, to meet the requirements of the ship. In the meantime, when the ship is in operation, the firefighting condition is worse than that on land, so it is more necessary to judge the flammability of fireproof materials. When ship fire breaks out, by passive fire proof of materials, people can maintain the integrity of the ship structure in a certain time, control the smoke and flames in local range, avoid the spreading of the fire, and win time for evacuation and extinguishment in order to minimize casualties and property damage.

References

[1] Information on http://www.imo.org
[2] Z.G. Wang *Wuhan Shipbuilding*. Vol.1 **26-27.**
[3] J.M. Qian *Fire Science and Technology*. Vol.22 **91-93.**
[4] X.N. Xu *Journal of the Chinese People 's Armed Police Force Academy*. Vol.18 **29-31.**
[5] Lei Chen *Fire Science and Theory*. Vol.21 **12-15.**
[6] W.G. He *Shipbuilding Technology*. Vol.3 **14-19.**
[7] Q. Sun, Y.C. Zhu *Journal of Anhui Institute of Architecture (Natural Science)*. Vol.12 **9-12.**

IOP Conf. Series: Materials Science and Engineering **230** (2017) 012017 doi:10.1088/1757-899X/230/1/012017

Hot Compression Deformation Behavior of Mg-Gd-Nd-Zr Heat-Resistant Magnesium Alloy

Yanhui Xin[1], Quan-an Li[1,2]

[1]School of Materials Science and Engineering, Henan University of Science and Technology, Luoyang 471023, China
[2]Collaborative Innovation Center of Nonferrous Metals of Henan Province
E-mail: xinyanhui@163.com

Abstract. The hot compression deformation of Mg-12Gd-2.5Nd-0.5Zr heat-resistant magnesium alloy was carried out on Gleeble-1500D thermo-mechanical simulator with the strain rate range of 0.002-1 s^{-1}, the deformation temperature range of 623-773 K and maximum strain of 50 %. The results show that true stress decreases with the increase of deformation temperature at a constant strain rate, and the true stress increases with the increase of strain rate at a same temperature. The deformation constitutive equation was established, and the hot deformation activation energy of 292.679 kJ/mol is determined.

1. Introduction

Magnesium alloys are the lightest material used in modern industry, which has the advantages of low density, high specific strength, and other advantages, and magnesium alloys are widely used in automotive industry[1]. It has been shown that the high temperature creep resistance of the alloy can be improved obviously by adding rare earth element gadolinium and other heavy rare earth elements in magnesium alloy[2].

At present, most magnesium alloy products are produced by casting. However, the defects, such as voids and inclusions in the casting process have caused great harm to the mechanical properties of the alloy. Compared with cast magnesium alloys, wrought magnesium alloy products produced by extrusion, rolling and forging have excellent properties[3].

In this work, the hot compression deformation behavior of Mg-12Gd-2.5Nd-0.5Zr magnesium alloy was performed on Gleeble-1500D thermo-mechanical simulator. the relationships between flow stress and deformation temperature as well as strain rate of the tested alloy were analyzed. Furthermore, the deformation activation energy of plastic deformation is obtained by numerical calculation, and the constitutive equation is also established, which provides theoretical and experimental basis for optimizing the hot deformation processing conditions.

2. Experimental

The nominal composition of investigated alloy is Mg-12Gd-2.5Nd-0.5Zr (wt.%). The tested alloy was prepared from high purity Mg (99.9%) and Mg-30%Gd, Mg-25%Nd, Mg-30%Zr master alloys (wt.%) by melting in a medium-frequency induction furnace with a Al_2O_3 crucible under the protection of mixture gas of SF_6 and CO_2.

The ingot was homogenized at 798 K for 8 h, and then the ingot were machined to cylindrical specimens with dimension of Φ10mm×15mm. The specimens were compression deformed at the temperature range from 623 K to 773 K and the strain rate range from 0.002 s^{-1} to 1 s^{-1} on the Gleeble-1500D thermo-mechanical simulator.

3. Results and analysis

3.1. Microstructure of the as-cast alloy

Figure 1(a) shows the optical microstructure of the as-cast Mg-12Gd-2.5Nd-0.5Zr alloy. The as-cast structure of the alloy is composed of the α-Mg matrix and the eutectic phase distributed at the grain boundaries. In this picture, the black non-continuous phase in the grain boundary is a eutectic structure formed by non-equilibrium solidification, which is composed of α-Mg and magnesium rare earth intermetallic compound. According to the X-ray diffraction (XRD) spectrum analysis of the as-cast alloy in figure 1(b), it is found that there are mainly α-Mg, Mg_5Gd and $Mg_{41}Nd_5$ phases in the alloy.

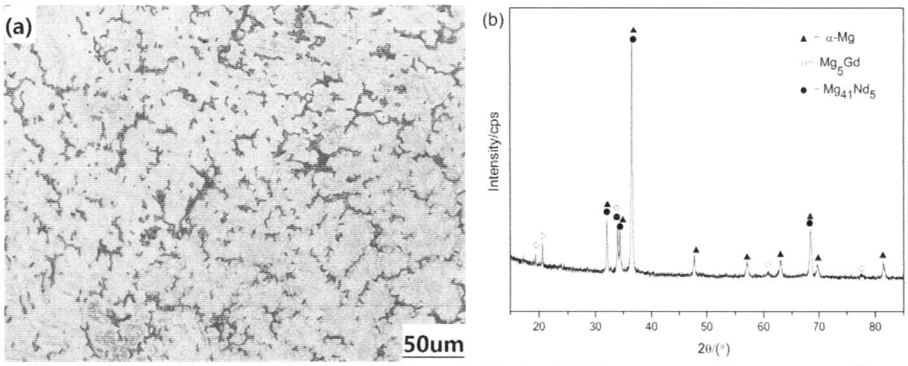

Figure 1. (a) The optical micrograph and (b) the XRD pattern of the as-cast alloy.

3.2. Microstructure of the homogenized alloy

The microstructure of the alloy after homogenizing treatment at 798 K for 8 h is shown in figure 2. dendritic segregation has been basically eliminated, and most of the eutectic compounds distributed at grain boundaries have disappeared and dissolved into the matrix.

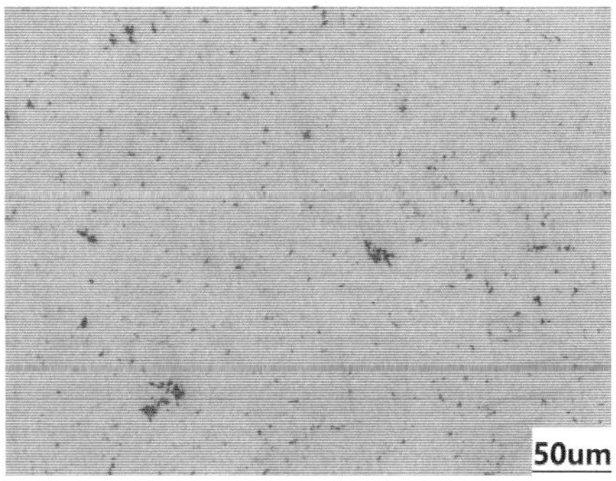

Figure 2. The optical micrograph of the homogenized alloy

3.3. Flow stress behavior

The high temperature plastic deformation behavior of Mg-12Gd-2.5Nd-0.5Zr alloy at different deformation conditions was studied by hot compression simulation test. The true stress-true strain curves measured at different deformation temperatures and different strain rates are shown in figure 3.

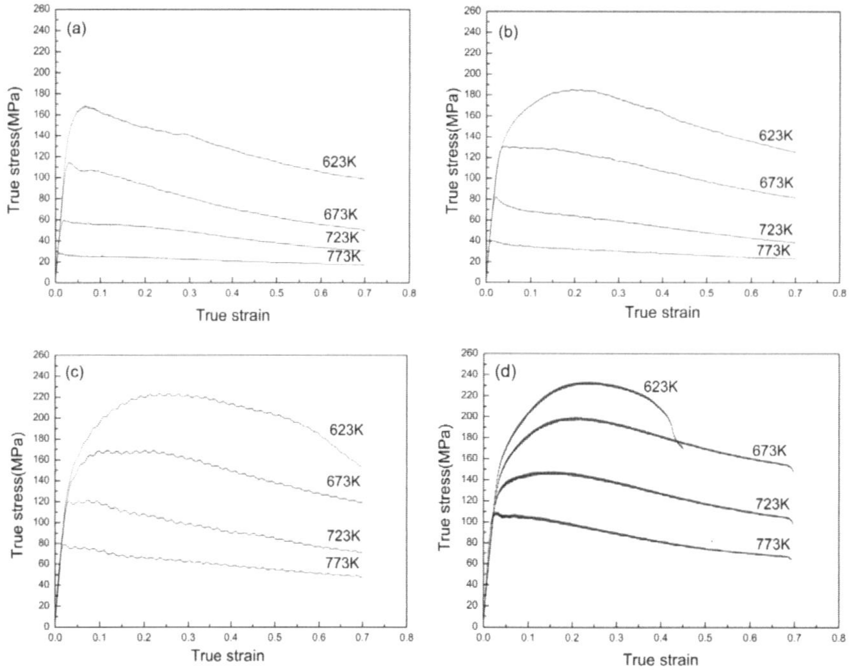

Figure 3. True stress-true strain curves of the alloy during hot compression: (a) 0.002 s^{-1}, (b) 0.01 s^{-1}, (c) 0.1 s^{-1}, (d) 1 s^{-1}.

As can be seen from the figure 3. The variation trend of flow stress is that the flow stress firstly increases sharply to a maximum value and then decreases gradually to a steady state as the strain increases. This flow behavior is a work hardening accompanied by dynamic recrystallization softening behavior[4]. At the initial stage of strain, hardening rate is higher than the softening rate, thus the flow stress rises sharply; After it attains the peak stress, all stress-strain curves are almost in flat shape, this is because a balance is established between work hardening and softening. Alternatively, as shown in figure 3(d), when the specimen was deformed at 623 K and strain rate of 1 s^{-1}, the true stress-strain curves of the alloy appears obvious instability phenomenon, the specimen is early cracked at a strain of 0.45, which implies that strain hardening dominated the plastic deformation at 623 K and strain rate of 1 s^{-1}.

Meanwhile, at a constant strain rate, the flow stress decreases with the increase of deformation temperature; at a same temperature, the flow stress increases with the increase of strain rate. This is because the higher the deformation temperature, the stronger the effect of the thermal activation energy, and the kinetic energy of the atoms in the alloy increases, and the critical resolved shear stress to decrease, and the dislocation is easier to move. Furthermore, with the increase of temperature, the dynamic recrystallization is easier to carry out, and the softening effect is enhanced, which result in the decrease of flow stress[5]. In the case of constant temperature, the greater strain rate, the dynamic recrystallization cannot be fully carried out, and rapid increment and accumulation of dislocations, which result in the increase of flow stress.

3.4. Constitutive equation

In order to further investigate the high-temperature plastic deformation behavior of the alloy, it is necessary to establish the constitutive equation of the alloy.

The relationship among the flow stress, the strain rate and the deformation temperature is usually expressed by the following three forms[6, 7]:

$$\dot{\varepsilon} = A_1 \sigma^{n_1} \exp(-\frac{Q}{RT}) \qquad (1)$$

IOP Conf. Series: Materials Science and Engineering **230** (2017) 012017 doi:10.1088/1757-899X/230/1/012017

$$\dot{\varepsilon} = A_2 \exp(\beta\sigma)\exp(-\frac{Q}{RT}) \tag{2}$$

$$\dot{\varepsilon} = A[\sinh(\alpha\sigma)]^n \exp(-\frac{Q}{RT}) \tag{3}$$

Where σ is the flow stress (MPa), $\dot{\varepsilon}$ is the strain rate (s^{-1}), Q is the activation energy for deformation (kJ/mol), R is the ideal gas constant (8.314 J mol^{-1} K^{-1}), T is the absolute temperature (K), A_1, A_2, A, n_1, n, α and β are the material constants, with $\alpha=\beta/n_1$.

For the low stress level ($\alpha\sigma$<0.8) and high stress level ($\alpha\sigma$>1.2), the relationship between flow stress and strain rate can be described by Equation (1) and Equation (2), respectively. Equation (3) is a hyperbolic sine law proposed by Sellars and Tegart[8], which can be applied to all the stress levels.

Get the natural logarithms of both sides for Equations (1) and (2). Then we have

$$\ln\dot{\varepsilon} = \ln A_1 + n_1 \ln\sigma - \frac{Q}{RT} \tag{4}$$

$$\ln\dot{\varepsilon} = \ln A_2 + \beta\sigma - \frac{Q}{RT} \tag{5}$$

The values of n_1 and β can be determined according to Equations (4) and (5) from the slopes of the $\ln\dot{\varepsilon}-\ln\sigma$ and $\ln\dot{\varepsilon}-\sigma$ plots, respectively. By linear regression of the relations of $\ln\dot{\varepsilon}-\ln\sigma$ and $\ln\dot{\varepsilon}-\sigma$, α was determined as 0.0076. The linear relationships of $\ln\dot{\varepsilon}-\ln\sigma$ and $\ln\dot{\varepsilon}-\sigma$ at different temperatures were fitted as shown in figure 4.

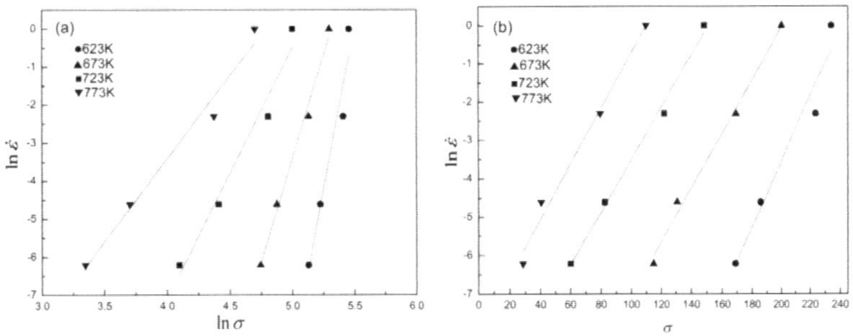

Figure 4. Linear relationship fitting: (a) $\ln\dot{\varepsilon}-\ln\sigma$ and (b) $\ln\dot{\varepsilon}-\sigma$.

Taking the natural logarithm of both sides of the above Equation (3), then we have

$$\ln\dot{\varepsilon} = \ln A + n\ln[\sinh(\alpha\sigma)] - \frac{Q}{RT} \tag{6}$$

The deformation activation energy Q can be determined by differentiating Equation (6):

$$Q = R\left[\frac{\partial\ln\dot{\varepsilon}}{\partial\ln[\sinh(\alpha\sigma)]}\right]_T\left[\frac{\partial\ln[\sinh(\alpha\sigma)]}{\partial(1/T)}\right]_{\dot{\varepsilon}} \tag{7}$$

The linear relationships of $\ln\dot{\varepsilon}-\ln[\sinh(\alpha\sigma)]$ and $\ln[\sinh(\alpha\sigma)]-1/T$ were fitted as Equation 5. The mean value of $\left[\frac{\partial\ln\dot{\varepsilon}}{\partial\ln[\sinh(\alpha\sigma)]}\right]_T$ can be obtained as 6.87434 from the slopes of $\ln\dot{\varepsilon}-\ln[\sinh(\alpha\sigma)]$ plots(figure 5(a)), and the average of $\left[\frac{\partial\ln[\sinh(\alpha\sigma)]}{\partial(1/T)}\right]_{\dot{\varepsilon}}$ can be determined as 5.12094 from the slopes in plots of $\ln[\sinh(\alpha\sigma)]-1/T$ (figure 5(b)). Finally, The Q value of the alloy is calculated as 292.679 kJ/mol.

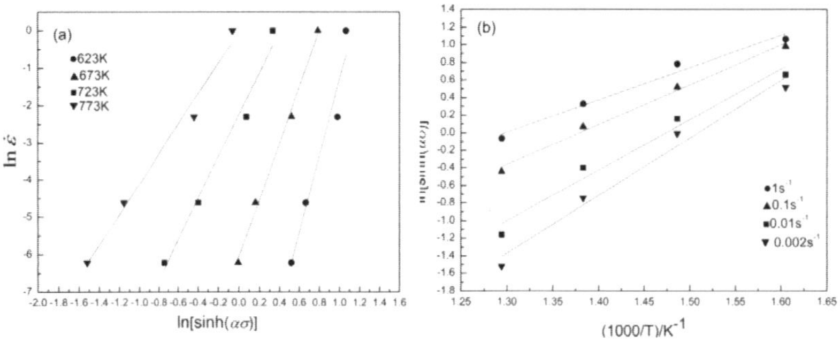

Figure 5. Linear relationship fitting: (a) $\ln\dot{\varepsilon}$–$\ln[\sinh(\alpha\sigma)]$, (b) $\ln[\sinh(\alpha\sigma)]-1/T$.

The effects of temperature and strain rate on the flow stress can be expressed by Zener-Hollomon parameter[9], and the hyperbolic sine function incorporated with the Z parameter at all the stress levels as follows:

$$Z=\dot{\varepsilon}\exp(\frac{Q}{RT})=A[\sinh(\alpha\sigma)]^{n} \quad (8)$$

Where Z is the temperature compensated strain rate factor; A is the structural factor; n is the stress exponent; α is the stress level parameter.
Taking the natural logarithm of both sides of Equation (8) gives

$$\ln Z=\ln A+n\ln[\sinh(\alpha\sigma)] \quad (9)$$

Based on Equation (9), it can be concluded that ln$Z$ and ln[sinh($\alpha\beta$)] should exhibit linear relationship. The relationship between lnZ and ln[sinh($\alpha\beta$)] is plotted in figure 6 through linear regression. Therefore, n and ln$A$ in Equation (9) are determined as 6.87434 and 47.17858, respectively. And then, the value of A can be obtained as 3.08601×10^{20}.

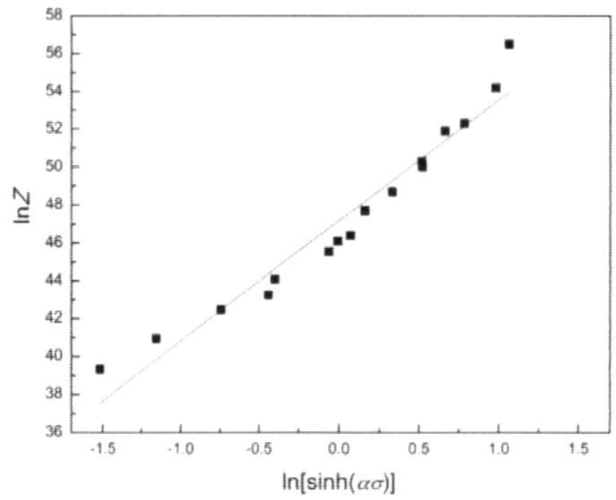

Figure 6. Relationship between ln[sinh($\alpha\sigma$)] and ln$Z$

Finally, the calculated values of Q, α, n and A were substituted into the Equation (3), and the constitutive equation for the hot compression of Mg-12Gd-2.5Nd-0.5Zr alloy was obtained as follows:

$$\dot{\varepsilon}=3.08601\times10^{20}[\sinh(0.0076\sigma)]^{6.87434}\exp(-292679/8.314T) \quad (10)$$

4. Summary
There is a dendritic segregation in the as-cast Mg-12Gd-2.5Nd-0.5Zr alloy, and most of the precipitates are distributed at the grain boundaries. The alloy after homogenizing treatment at 798 K for 8 h, the majority of segregation structure can be eliminated.

The flow stress increases to a maximum and then decreases slowly to a steady state with the increasing of strain. Furthermore, the flow stress decreases with the increasing of deformation temperature and decreasing of strain rate.

The hot deformation activation energy of the alloy is 292.679 kJ/mol, and the constitutive equation of the alloy is $\dot{\varepsilon} = 3.08601 \times 10^{20}[\sinh(0.0076\sigma)]^{6.87434}\exp(-292679/8.314T)$.

5. References

[1] Mordike B L and Ebert T 2001 *Mater. Sci. Eng. A* 302(1) 37–45.
[2] Anyanwu I A, Kamado S and Kojima Y 2001 *Mater. Trans* 42(7) 1212–1218.
[3] Chen Z H 2005 *Wrought magnesium alloy* (Beijing: Chemical Industry Press) pp 1–15.
[4] Xia X S, Chen Q, Li J P and Zhao Z D 2014 *J. Alloys Compd* 610(30) 203–211.
[5] Zhang Z G, Xia C Q, Qiao X, Peng X M, and Liu Y Y 2009 *Min Metall Eng* 29(5) 105–108.
[6] Mcqueen H J and Ryan N D 2002 *Mater. Sci. Eng. A* 322(1–2) 43–63.
[7] Shi H, Mclaren A J, Sellars C M, Shahani R and Bolingbroke R 1997 *Mater. Sci. Technol* 13(3) 210–216.
[8] Qin Y J, Pan Q L, He Y B, Li W B, Liu X Y and Fan Xi 2010 *Mater. Sci. Eng. A* 527(10) 2790–2797.
[9] Yu H, Kim Y M, Yu H S, You B S and Min G H 2012 *Acta Metall Sin* 48(9) 1123–1131.

Acknowledgments

The project is supported by the National Natural Science Foundation of China (No.51571084, No.51171059), Science and technology research project of henan province (No.152102210072).

Research on Protective Coating on Inner Surface of Alloy Tube

Y C Zhang[1, 2], Y H Liu[1,3], Z J Zhou[2], M M Zheng[1], S Y Kong[1], H H Xia[1] and H L Li[1]

[1]State Power Investment Corporation Science and Technology Research Institute, Beijing 102209, China
[2]School of Materials Science and Engineering, University of Science and Technology Beijing, Beijing 100083, China
E-mail: liuyanhong@spic.com.cn

Abstract. Materials are one of the most important factors which limit reactor development. Molten salt not only used as the coolant but used as application in which fissile materials and fission products are dissolved in Molten Salt Reactors (MSRs). Therefore the corrosion resistance of structure materials is the one of most important aspects for application in MSRs. Compatibility and chemical stability with the molten salt should be considered for some common structural alloys such as Incoloy-800H. In this research, the pure nickel coating was obtained by electroplating on the inner surface of nickel alloy to improve the corrosion resistance. However, there are some problems for plating on the inner surface of tube. For example the current is shielded and the anode is easy to passivate. The inner anode was used for solving these problems in this study. Pure nickel coating was obtain and the microstructure and properties of coating were analysed using this method. The thickness, hardness and microstructure of coating were observed by metallographic microscope, micro hardness tester and field emission scanning electron microscope, and the influence of deposition duration and annealing treatment duration on properties were analysed. Thermal shock performance was investigated as well. The results showed that the coating thickness increased linearly with the increasing of plating durations and the size of grain increased with the durations as well, the surface of coating became inhomogeneous correspondingly. The hardness of coating changed as the change of durations of annealing treatment. The thermal shock test showed that bonding strength of coating with substrate was good.

1. Introduction

Nuclear energy is recognized as a safe, economical and low-carbon energy source all over the world. It is also the best choice to solve the problems of low-carbon economy, emission reduction and environmental stress [1]. After the Fukushima Daiichi nuclear accident, the safety of nuclear power and nuclear power plant how to deal with sudden natural disasters have caused widespread concern [2], meanwhile, researchers accelerated the pace of research on higher security nuclear power technology. Molten salt reactor (MSR) has become one of research hot spots because of its characteristics such as high inherent safety and recyclability of nuclear fuel. However, materials corrosion has been recognized as an issue in molten salts [3]. According to Fabre et al. [4], the corrosion rates of Ni, Mo and W were smaller than that of Fe in LiF-NaF. The corrosion performance of a number of nickel-based alloys in molten FLiNaK salt at 850 °C was study by Olson [5], the results showed that the Ni-201 alloy, which did not contain Cr, was resistant to corrosion, and the corrosion weight loss of Cr-content alloy correlated with the Cr-content of the corresponding alloy. At present, nickel-base alloy with good high-temperature resistance is studied for structural material of MSR, but the corrosion resistance of molten salt is not ideal, because most of them contain Cr, Fe and other elements. According to common alloying constituents to corrode in molten salts increased in the following order [6]: Ni, Co, Fe, Cr, Al, we can see that it can be increased the nickel-based alloy corrosion resistance in molten salts to obtain a layer of pure nickel on the alloy matrix [7].

Pulse electroplating is an effective technology for obtaining high properties coating, this process involves the swift alternating of the voltage or current between two different values resulting in a series of pulses of equal amplitude, duration and polarity, separated by zero current. By changing the pulse amplitude and width, it is probable to change the deposited film's composition and thickness [8]. A large number of experiments showed that pulse electroplating is better than direct electroplating. The deposition layer obtained by pulse electroplating is uniform and compact, and the corrosion resistance of the layer is better than the layer obtained by dc method. In addition, the layer has a lower porosities, higher brightness and conductivity using pc method [9,10]. Y H Liu et al.[11] used different plating solutions to prepare nickel coatings on plate substrates, it was found that the nickel coating prepared by the nickel sulfamate system had better performance. The effects of duty ratio on the corrosion resistance of nickel coating prepared on cooper substrate in NaCl solution was studied by L Zhou et al. [12], who found that the smaller duty ratio, the better brightness and corrosion resistance coating was obtained. The similar research reported by Sivasakthi P et al. [13], they found that the micro-hardness of nickel coating increased with the duty ratio and pulse frequency decreased.

Most of corrosion resistant structural materials are used in pipeline system of MSR, the studies of nickel coating on the surface of tubular substrate prepared by electroplating are rarely reported. At present, there are some technical difficulties in electroplating coating on the inner surface of tubular substrate. If the anode is arranged in the outside of tube, the current in tube will be blocked, and this will result in obstacles to plating coating on the inner surface of tube; if the anode is placed inside the tube, the area ratio of cathode to anode is large, the current density of anode will be far more than the cathode, it will result in anodic passivation. To solve these problems, the experimental apparatus was designed and the electroplating time was controlled respectively, the nickel rod with diameter of 4 mm and length of 95 mm was used as the internal anode, the 800H nickel-base alloy tube with diameter of 25 mm and length of 50 mm was used as substrate, and the nickel coating on inner surface of tube was prepared by pulse electroplating. In this study, the thickness, hardness and surface morphology of coating were observed respectively, and the thermal shock test was carried out simultaneously.

2. Experimental

2.1. Electroplating solution preparation

Sulfamate solution was used in this research, and table 1 shows the composition of electroplating solution. The solution of nickel sulfamate, boric acid and nickel chloride were configured respectively, then mixed the three solutions and adjusted the pH to acidic, finally added sodium dodecyl sulfonate in the solution, and mixed well.

Table 1. The composition of electroplating solution

Composition	Concentration ($g \cdot L^{-1}$)
$Ni(NH_2SO_3)_2 \cdot 4H_2O$	400~500
H_3BO_3	30~40
$NiCl_2 \cdot 6H_2O$	10~20
$C_{12}H_{25}SO_3Na$	0.05~0.15

2.2. Sample pretreatment

The anode was a nickel rod with purity of 99.9%, and the cathode was a nickel-based alloy tube, the anode and cathode were pretreatment to remove surface defects, impurity and oxide film. The process of pretreatment was as follows: sandpaper grinding → washing → alcohol degreasing → washing → chemical degreasing → washing → pickling → washing → activation → washing →drying.

2.3. Experimental procedure

Figure 1 below shows the experimental apparatus. The anode and cathode were fixed on the pendant and placed in the solution, the cathode. The position of anode and cathode was adjusted respectively to ensure that the anode was in the center of nickel-based alloy tube and the cathode was put completely into the solution. Finally the wires were connected to the pulse power. According to previous work [11, 12, 14, 15], the parameters were set as shown in table 2 below. After electroplating, the tube was taken out and washed by ultrasonic cleaning, then annealed at 500 °C for 2 h, 4 h and 6 h respectively. Afterwards, the sample was cut into rings shape with thickness of 1 cm, and rectangular shape along the axial direction of the tube. The thickness, hardness and microstructure of the coating were investigated. In addition, the thermal shock test was investigated using the ring samples. The parameters of test as follows: 500 °C for 5 min followed by quenching. The pulse power supply used in this experiment was HP-MCC, and the SX-G08133 muffle furnace was used for annealing treatment and thermal shock test. The hardness of the samples was tested by HMV-G21 microhardness tester, the loading was 980.7 mN holding for 10 s. The thickness of coating was observed using a DMILM inverted metallographic microscope. The surface morphology of coating was observed by JSM-7001F field emission scanning electron microscopy (FESEM), and the elemental characterization was carried out using energy dispersive spectrometer (EDS).

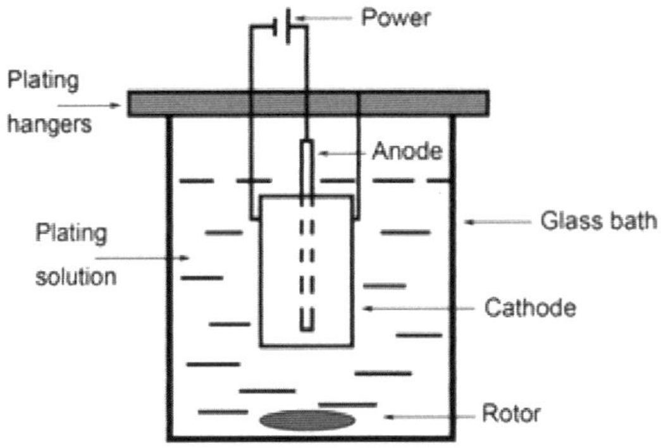

Figure 1. Diagram of experimental apparatus

Table 2. Experiment parameters

Experiment parameter	Value
Current density ($A\cdot dm^{-2}$)	3
Duty factor	20%
Cycle (ms)	10
Temperature (°C)	50
Duration of deposition t [h]	1-3

3. Results and discussion

3.1. Effect of electroplating duration on the thickness of coating

Thickness is an important index of the quality for coating. For molten salt reactor protective coating, in order to ensure the effect of protection, thickness should be more than 40 μm [16]. Electroplating duration is the most important parameter when others are fixed. Figure 2 shows the thickness of the coating in different electroplating durations. When the electroplating durations were 1 h, 1.5 h, 2 h, and 3 h respectively, the thickness of coatings was 32.8 μm, 42.0 μm, 75.0 μm and 138.4 μm accordingly. Figure 3 (a) displays the relationship between the thickness of coating and electroplating duration. It can be seen from graph, the thickness of coating increases with the extension of electroplating duration, and the change is an approximately linear relation. The following formula for the relationship between the thickness and duration was used in Ref. [17]

$$d = \frac{J_k \eta_k K t}{1000\rho} \tag{1}$$

where d is thickness of coating (mm); J_k is cathode current density ($A{\cdot}dm^{-2}$), η_k is current efficiency, K is electrochemical equivalent [$g{\cdot}(A{\cdot}h)^{-1}$)] t is electroplating duration (h), and ρ is density of coating ($g{\cdot}cm^{-3}$)

In this research, J_k was fixed, and K and ρ were constant value for electroplating nickel, η_k was generally between 85% and 100%. Therefore, according to formula (1), d was a linear function of t, and the result of this study conformed to this law. Due to the occurrence of side reactions such as hydrogen evolution during the process of electroplating and stirring in the tube was weakened, the current efficiency may be unstable and the growth rate of the coating may be affected. In addition, when the electroplating duration was controlled within 3 h, the current efficiency was not reduced, indicating that the anode was not passivated at this time.

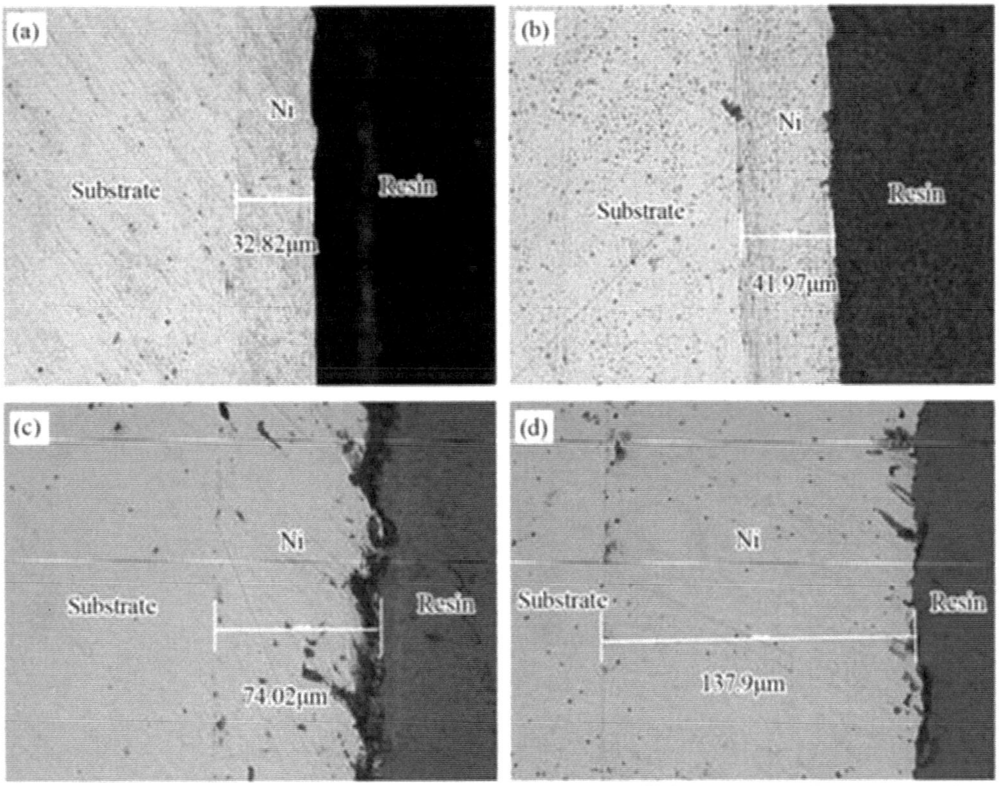

Figure 2. Thickness of coating in different electroplating duration

(a) 1 h; (b) 1.5 h; (c) 2 h; (d) 3 h

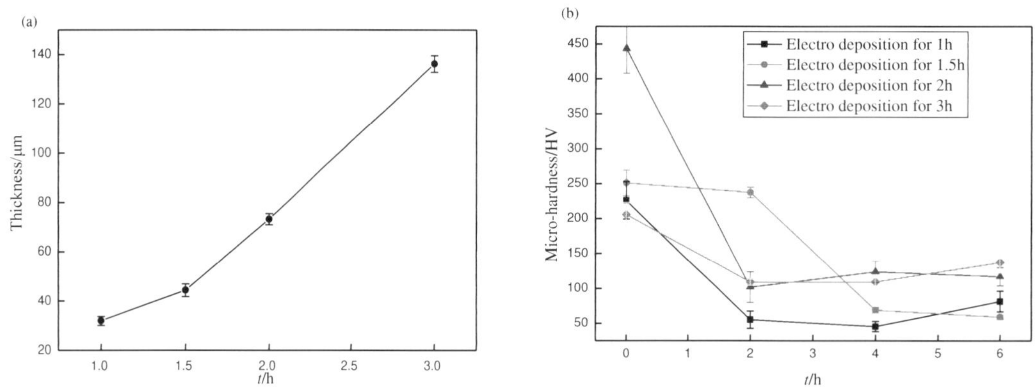

Figure 3. (a) Curve of thickness with electroplating duration; (b) Curves of coating hardness with annealing treatment duration

3.2. Effect of annealing time on hardness of coating

Figure 3 (b) shows the curve of Vickers hardness of coating with the annealing time, and it also shows the effect of different electroplating duration on the hardness of coating. Without annealing, the hardness of coating increased first and then decreased with the extension of the electroplating duration, when the duration was 3 h, the hardness of coating was minimum and was 206 HV; when the duration was 2 h, the hardness was maximum and was about 443 HV. The most probable explanation of this phenomenon is that the structure of coating prepared by 2 h was more compact and uniform, another reason might be that the high hardness of nickel alloy substrate (the hardness of nickel alloy greater than that of pure nickel hardness), and the decreasing influence of the alloy substrate on the coating as the coating thickness increase. Moreover there might be more carbon, silicon and other elements or some impurity ions in coating for 2 h, which caused the increase of hardness. When the annealing treatment duration was 2 h, the hardness of coating prepared at the different time was obviously decreased, except the coating prepared by 1.5 h, and the hardness increased with the extension of annealing treatment duration, but the change was not obvious. After 6 h annealing treatment, the final hardness were 81 HV, 59 HV, 117 HV and 137 HV respectively. The study by Luo Y et al. [18] showed that the recovery and recrystallization of coating would occur above 450 °C, it caused residual stress of coating release. Therefore the hardness would decrease after annealing treatment. With the prolongation of annealing treatment duration, the change of hardness was not obvious. The hardness of Ni-based alloy substrate before annealing was 335 HV and 312 HV after that, which indicated that the coating had no effect on the mechanical properties of substrate.

3.3. Surface morphology

The surface morphology of coating after 2 h annealing treatment was observed by SEM and the coating samples of different deposition durations were compared and analyzed. Figure 4 (a) and (b) show that the coatings are both uniform and compact, but the surface of coating electrodeposited for 1 h has slight scratches, it was probable that the growth of the coating tended to the surface structure of substrate after grinding. While, there are still some holes on the surface plating for 1.5 h, which may be due to hydrogen evolution reaction on the surface of cathode during electroplating, as hydrogen vesicles were electrical insulators, nickel ions cannot penetrate the bubble to discharge, resulting in the partial area of coating without nickel deposition, forming holes. From figure 4 (c) and (d), it can be seen that the grains are obvious, and the surface is not smooth. The grains of coating electrodeposited for 2 h are fine and uniform, the size of grain between 0.5 μm and 2 μm, and the grain boundaries are obvious, while, the grains of coating deposited for 3 h become coarse and the size becomes non-uniform, which means that the grain size of coating increased gradually with prolongation of electroplating duration. This can probably be explained by considering that the atoms were preferentially deposited at the tip of the large particles with the extension of electroplating duration, then the large particles grew up and the surface of coating gradually became rough.

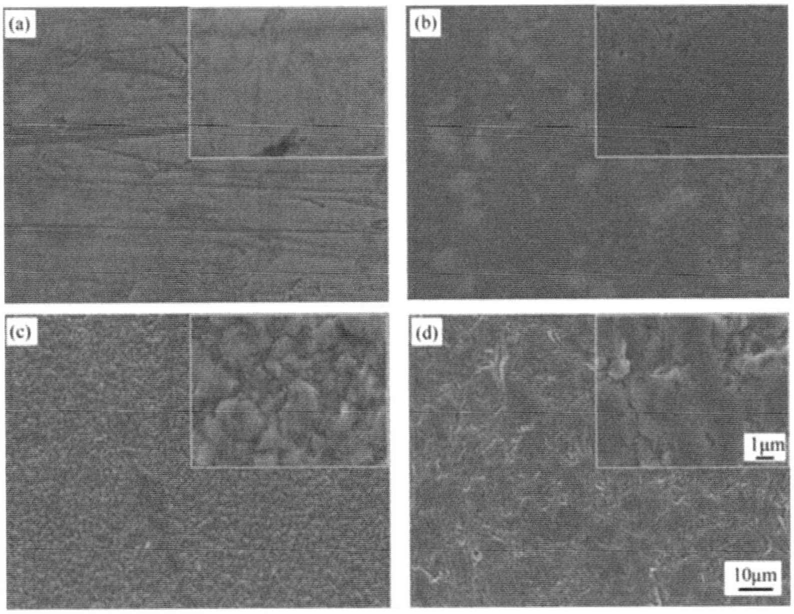

Figure 4. Microscopic morphology of coatings at different deposition duration

(a) 1 h;(b) 1.5 h; (c) 2 h; (d) 3 h

3.4. Thermal shock test

The thermal shock was investigated to reveal the ability of coating to withstand thermal stress in our research. After annealing, the sample was cut into rings with thickness of 1 cm, and then subjected to a cycle test at 500 °C for 5 min followed by quenching. After 20 cycles, surface of coating did not appear crack, blister, peeling and so on, but there was a slight partial oxidation, it is indicated that the ability of coating to withstand thermal stress was good, and coating had high bonding strength with inner surface of tube. Figure 5 shows the appearance of sample without thermal shock test (left side) and after 20 cycles of thermal shock (right side). In addition, the cross section of sample after thermal shock was observed by SEM. As shown in figure 6, there is no cracking at the boundary between coating and substrate; EDS analysis shows that there is no obvious oxygen peak appeared in the EDS pattern. The results show that coating is not oxidized, and there is nickel enrichment on the substrate near the boundary, the probably reason might be that nickel ions migrate to substrate during annealing. In addition, the peak of carbon, silicon and other elements in coating may also be atomic diffusion during annealing treatment.

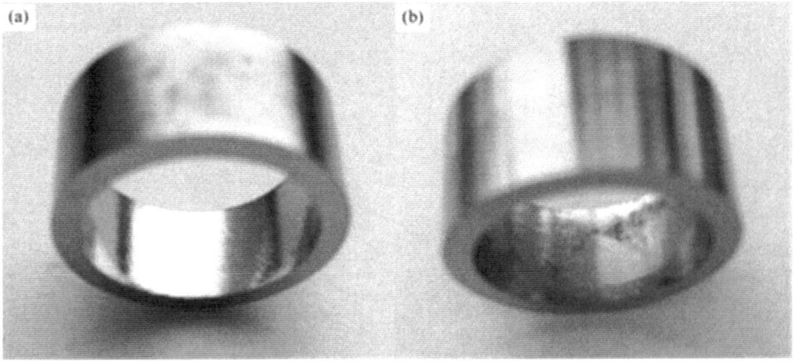

Figure 5. Photo of sample without thermal shock test and after 20 cycles of thermal shock
(a) without thermal shock test; (b) after 20 cycles of thermal shock

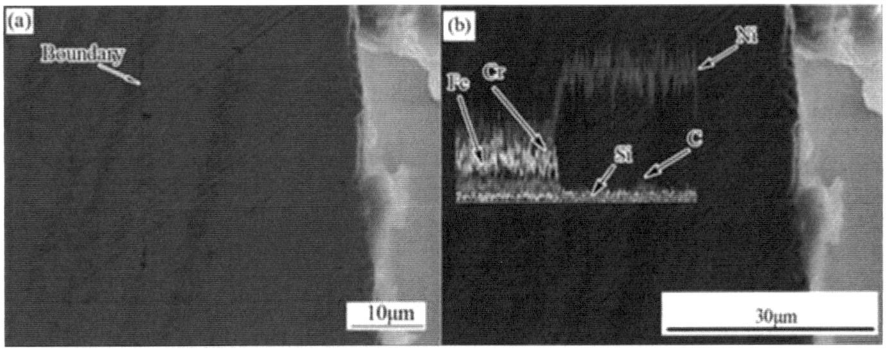

Figure 6. The SEM image and EDS analysis of cross section of coating after 20 cycles of thermal shock test
(a) SEM image; (b) EDS analysis

4. Conclusions

Pure nickel coating was successfully electrodeposited using internal anode on the inner surface of tubular substrate, and the quality of coating can be controlled by electroplating parameters and post-treatment. The thickness, surface morphology and hardness of coating can be controlled by electroplating duration and annealing treatment respectively. The coating has good bonding strength with tube shape substrate, and good oxidation resistance. The results of this study provide some necessary data for the research of the protective coating on pipeline in MSR.

5. References

[1] Ye Q Z 2016 *Adv. Climate Change Res.* 7 10-16
[2] Zeng M, Liu Y, Ouyang S, Shi H and Li C 2016 *Renew. Sust. Energ. Rev.* 58 147-156
[3] Olson L C., Ambrosek J W, Sridharan K, Anderson M H and Allen T R 2009 *J. Fluorine Chem.* 130 67–73
[4] Fabre S, Cabet C, Cassayre L, Chamelot P, Finne J, Noël D and Taxil P 2008 *Mater. Sci. Forum* 483-490
[5] Olson L C 2009 Materials corrosion in molten LiF-NaF-KF eutectic salt, Madison: University of Wisconsin-Madison
[6] Misra A K and Daniel W J 1987 Fluoride salts and container materials for thermal energy storage applications in the temperature range 973-1400 K. Intersociety Energy Conversion Engineering *Conf.* pp 188-201

[7] Muralidharan G, Wilson D F, Walker L R, Santella M L and Holcomb D E 2011 Cladding alloys for fluoride salt compatibility. Office of Scientific & Technical Information Technical Reports., America Tennessee: Oak Ridge National Laboratory
[8] Chandrasekar M S and Pushpavanam M 2008 *Electrochim. Acta* 53 3313-22
[9] Panagopoulos C N, Papachristos V D and Christoffersen L W 2000 *Thin Solid Films* 366 155-163
[10] Ul-Hamid A, Dafalla H, Quddus A, Saricimen H and Al-Hadhrami L M 2011 *Appl. Surf. Sci.* 257 9251-59
[11] Liu Y H, Yang C, Lu Y L, Zhou X and Li H L 2015 *Rare Metals* 39 865-869
[12] Zhou L, Yu J, Ma A Y 2009 *Electroplat. Finish.* 28 5-8
[13] Sivasakthi P, Sekar R and Bapu G N K R 2015 *Mater. Res. Bull.* 70 832-839
[14] Savall C, Godon A, Creus J and Feaugas X 2012 *Surf. Coat. Tech.* 206 4394–02
[15] Allahyarzadeh M H, Ashrafi A, Golgoon A and Roozbehani B 2016 *Mater. Chem. Phys.* 175 215-222
[16] Xu Y X and Zeng C L 2014 *Journal of Chinese Society for Corrosion and Protection* 34 211-217
[17] Chu R B, Wang Z X and Wu S C 2013 *Concise electroplating manual* 2nd ed. Beijing: Machinery Industry Press
[18] Luo Y, Tian W H, Shi G F and Cheng S W 2014 *J. Mater. Eng.* 4 40-45

Acknowledgments
This work has been financially supported by Employee Independent Innovation Project of the State Nuclear Power Technology Corporation (No. SNP-KJ-CX-2014-10) and the International S&T Cooperation Program of China (No. 2015DFA50510).

IOP Conf. Series: Materials Science and Engineering **230** (2017) 012019 doi:10.1088/1757-899X/230/1/012019

Photonic Bandgap Properties of One-Dimensional Graphene-Based Photonic Crystals with a Single Dielectric

Huimin Fang, Jinghui Cao, Jingzhuang Liu, Chaoyang Wang, Xifu Yue, and Jijiang Wu*

School of Science, Shandong University of Technology, Zibo 255049, China
*Corresponding author e-mail: w9513110@163.com

Abstract. The transfer matrix method was used to investigate the photonic bandgap properties of one-dimensional graphene-based photonic crystal (GPC) with a single dielectric medium. The first layer of the unit cell of the GPC consists of a traditional dielectric medium and the second is the same dielectric medium wherein graphene sheets are embedded. Numerical calculation results show that the width of the low frequency bandgap decreases as the period of the GPC, the permittivity of the dielectric, the thickness of the dielectric layer, the filling factor of the dielectric layer, and the period of the graphene-dielectric multilayer medium structure increase and increases as the thickness of the graphene-dielectric multilayer medium structure increases. The influence of the above parameters on the Bragg bandgap is different from their effects on the low frequency bandgap. Due to the graphene conductivity could be tuned by varying the chemical potential, so GPC is a tunable PC and may have many potential applications in photonics.

1. Introduction

These Photonic crystals (PCs), composed of periodic dielectric materials, have attracted great attention due to their unique ability to manipulate light. When an electromagnetic wave propagates in photonic crystal, due to Bragg scattering, wave of a certain frequency range can be inhibited resulting in the formation of photonic bandgap (PBG). This property facilitates the active control and manipulation of the flow of electromagnetic waves or photons and is used in various applications for a wide range of optical devices. Most of the PC-based optical devices are constructed by usual dielectric materials. In recent years, various non-conventional materials have also been used to construct PCs, such as superconducting materials, liquid crystal and a negative refractive index materials, etc. [1-3]. Due to these materials' special electromagnetic properties, a PC which is composed by these materials usually presents some special PBG properties.

In recent years, a new type of graphene-based photonic crystals (GPCs) has also attracted much attention [4-7]. Graphene, a flat monolayer of carbon atoms packed into a dense two-dimensional honeycomb crystal lattice, has unique thermal, mechanical, electrical and optical properties. Due to the high optical transmittance and tunable electrical conductivity, graphene has been found broad application in photonics and optoelectronics fields [8, 9].

Because of the tunability of the complex conductivity of graphene, the GPC is a very important PC with tunable PBG. El-Naggar recently proposed a new design of one-dimensional (1D) GPC [10]. In this GPC, the first layer of the unit cell of the GPC consists of a dielectric wherein graphene sheets are periodically embedded. The GPC studied by El-Naggar is made up of two dielectric materials. In this paper, we propose utilizing only one dielectric material to constitute the GPC. This will be a great significance to reduce the difficulty of preparation of the GPC. By detailed numerical calculation, it can be seen that our proposed GPC have the same optical properties as the GPC proposed by El-Naggar.

2. Theoretical model

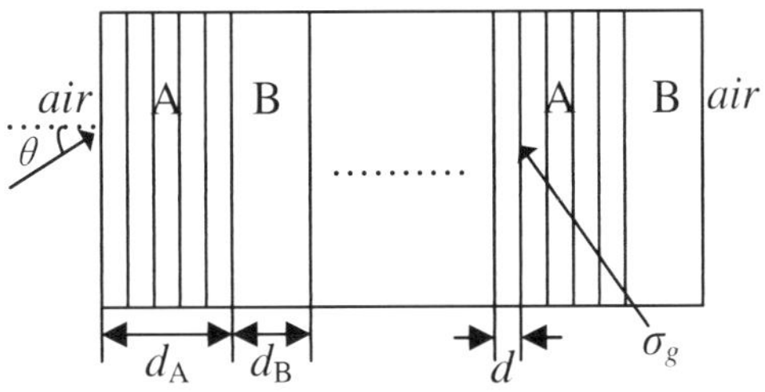

Figure 1. Schematic of a 1D graphene-based PC (AB)N

Structure of GPCs, $(AB)^N$, studied in this paper shown in figure 1, which is similar to the model discussed by El-Naggar. The first layer of the unit cell of the GPC consists of a dielectric wherein graphene sheets are periodically embedded and the second layer is a dielectric medium. The GPC contains N periods of alternating layers along the z axis of the graphene-dielectric multilayer medium and a dielectric medium, denoted by A and B, respectively. The thicknesses of the layers A and B are assumed to be d_A and d_B, respectively. Then the thickness of the unit cell of the structure is $D = d_A + d_B$. The dielectric constant of the layers A and B are assumed to be $\varepsilon_A(\omega)$ and ε_B, respectively. The frequency-dependent effective dielectric constant, $\varepsilon_A(\omega)$, of the graphene sheets separated by dielectric with a dielectric constant ε_d and a thickness d can be expressed as[11, 12]:

$$\varepsilon_A(\omega) = \varepsilon_d + i\frac{\sigma_g(\omega)}{\omega\varepsilon_0 d} \tag{1}$$

where $\sigma_g(\omega)$ is the dynamical conductivity of doped graphene for high frequencies at temperature T expressed as the sum of the intraband and interband electron transition contributions as

$$\sigma_g(\omega) = \sigma_g^{\text{intra}}(\omega) + \sigma_g^{\text{inter}}(\omega) \tag{2}$$

where

$$\sigma_g^{\text{intra}}(\omega) = \frac{e^2}{4\hbar}\frac{i}{2\pi}\left[\frac{16k_BT}{\hbar\omega}\log\left(2\cosh\left(\frac{\mu}{2k_BT}\right)\right)\right] \tag{3}$$

$$\sigma_g^{\text{inter}}(\omega) = \frac{e^2}{4\hbar}\mathrm{H}(\hbar\omega - 2\mu) - \frac{ie^2}{8\pi\hbar}\log\left(\frac{(\hbar\omega + 2\mu)^2}{(\hbar\omega - 2\mu)^2 + (2k_BT)^2}\right) \tag{4}$$

Here e is the charge of an electron, H is the Heaviside step function, k_B is the Boltzmann constant, T is the temperature in K, and μ is the chemical potential which can be tuned via electrostatic biasing.
In El-Naggar' model [10], ε_B is not equal ε_d. In this paper, we will mainly discuss the bandgap characteristic of GPC presented in figure. 1 when ε_B is equal to ε_d. The study of the bandgap properties of 1D GPCs will be carried out using the well-known transfer matrix method (TMM).

3. Numerical results and discussion

In the figure 2 we are presenting the results of the numerical calculation of the transmission spectrum in the 1D GPCs at normal incidence of light. Figure 2 (a) shows the transmission spectrum of the PC structure discussed by El-Naggar. The parameters used in the calculation are: $d_A = 2$ μm, $d_B = 6$ μm, $\varepsilon_B = 10.9$, $\varepsilon_d = 5.07$, $d = 200$ nm, the chemical potential $\mu = 0.2$ eV, the temperature $T = 300$ K. Here, the relative dielectric constant ε_B in layer B is different to the relative dielectric constant ε_d in layer A. It can be seen from figure 2 (a) that the GPC exhibits a low-frequency bandgap (LFBG), which is can't be found in the PCs constructed by the conventional dielectric materials. El-Naggar called this LFBG by graphene PBG, which is associated with a very strong dependence of the imaginary part of the graphene complex conductivity in the low frequency range. In addition to the LFBG, GPCs also have Bragg bandgap (BBG) which exists in traditional PCs.

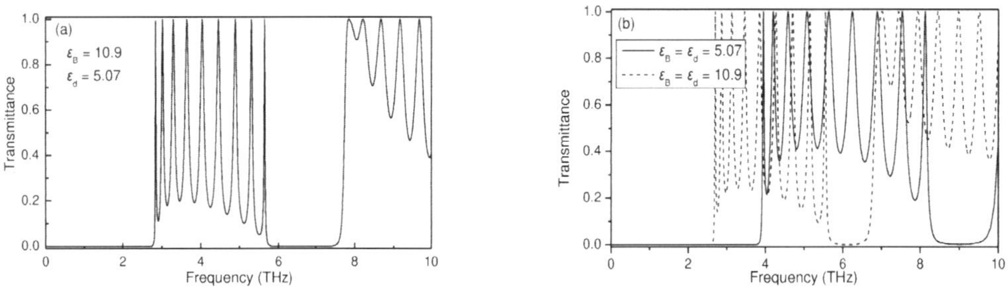

Figure 2. Transmission spectra of the GPC

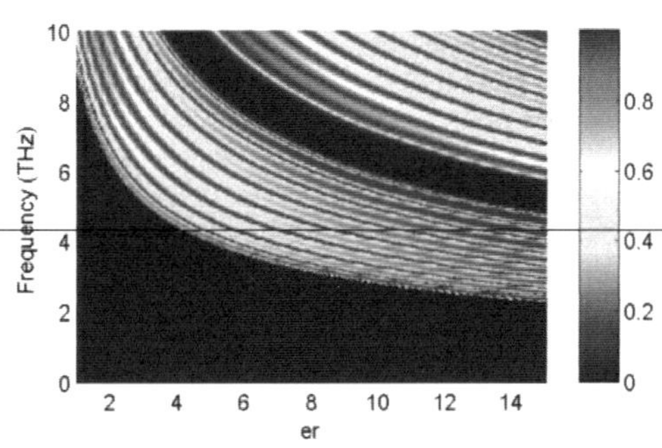

Figure 3. Transmission spectra of the GPC as a function of the permittivity of the dielectric

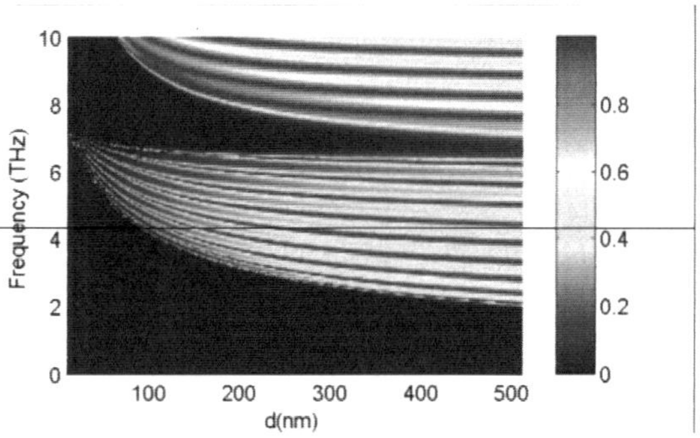

Figure 4. Transmission spectra of the GPC as a function of the period of the layer A

Figure 2 (b) show the transmission spectrum of the GPC structure with $\varepsilon_B = \varepsilon_d$. This structure is the focus of this paper. It can be seen that when $\varepsilon_B = \varepsilon_d$, the GPCs also have LFBG and BBG, and both bandgap shifts toward the lower frequency as the relative permittivity increases. The width of the two bandgap decreases with the increase of the relative permittivity increases. It can be seen that GPC constructed with the same dielectric material have the same bandgap as the GPC composed of different dielectric materials, and the GPC made of the same kind of materials is simpler in preparation.

The bandgap characteristics of the GPCs constructed with the same dielectric material will be discussed in detail below. First, we study the dependency of the PBG on the permittivity of the dielectric at normal incidence. We plot color maps of the transmittance through the GPCs versus the frequency and the permittivity of the dielectric in figure 3. It can be seen from figure 3 that with the increase of the relative permittivity, the bandgap width of the LFBG and the first BBG decreases and the center of the bandgap moves to the low frequency. Therefore, in order to get a larger PBG, relatively small relative permittivity of the dielectric material can be used to form GPCs.

Figure 4 illustrates the calculated transmittance through the GPC as a function of the period d of the graphene-dielectric multilayer medium structure A. In the calculation, $\varepsilon_B = \varepsilon_d = 8$, and the other parameters used are similar to that used in figure 2. It can be seen from figure 4 that, the bandgap width of the LFBG and the first BBG decreases with the increase of *d*. Therefore, in order to obtain a larger PBG, a graphene-based multilayer structure with a smaller period d can be used to construct a GPC.

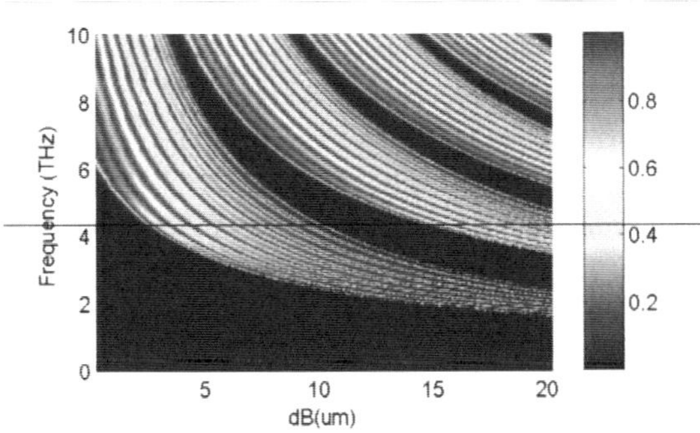

Figure 5. Transmission spectra of the GPC as a function of the thickness of layer B

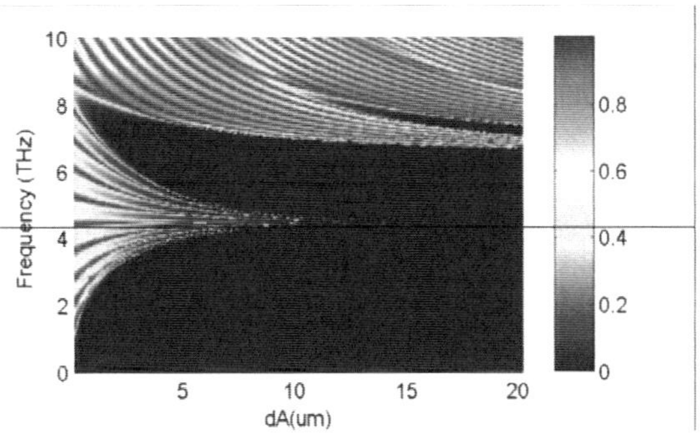

Figure 6. Transmission spectra of the GPC as a function of the thickness of layer A

Figure 5 shows the transmission spectrum of GPCs with the change of the thickness d_B of the layer B. In our calculation, d = 200 nm, and the other parameters used are similar to that used in figure 4. It can be seen from figure 5 that in our calculated frequency range, when the d_B is small, there is only LFBG, with the increase of d_B, the width of the LFBG decreases, and the other BBG appear in turn. The center frequency for all bandgaps moves to the low frequency direction with the increase of d_B.

Figure 6 shows the transmission spectrum of GPCs with the change of the thickness d_A of the layer A. In our calculation, d = 200 nm, d_B = 6 mm, and the other parameters used are similar to that used in Figure 4. It can be seen from figure 6 that, in the calculated frequency range, the band width of the LFBG increases as d_A increases. The bandgap of the other BBGs appears in turn with the increase of d_A, and the bandgap width gradually increases and the bandgap center moves to the low frequency. When d_A increases to a certain degree, the LFBG and the first BBG will overlap to form a wide bandgap.

Figure 7 shows the transmission spectrum of the GPC with the change of the filling ratio r of the layer B. r is defined as $r = d_B / (d_A + d_B)$. In our calculation, d = 200 nm, $d_A + d_B$ = 12 mm, and the other parameters used are similar to that used in figure 4. It can be seen from figure 7 that, the width of the LFBG decreases with the increase of r. All the other higher BBGs appear in turn as the r increases, and the bandgap width first increases and then decreases, while the bandgap center moves to low frequency.

Figure 8 shows the transmission spectrum of the GPC with the change of the thickness of the unit cell of the GPC D. In our calculation, r = 0.3, d = 200 nm, and the other parameters used are similar to that used in figure 4. It can be seen from figure 8 that the width of the LFBG decreases with the increase of D. All the other higher BBGs appear appears in turn with the increase of D, and the bandgap width first increases and then decreases, while the bandgap center moves to low frequency.

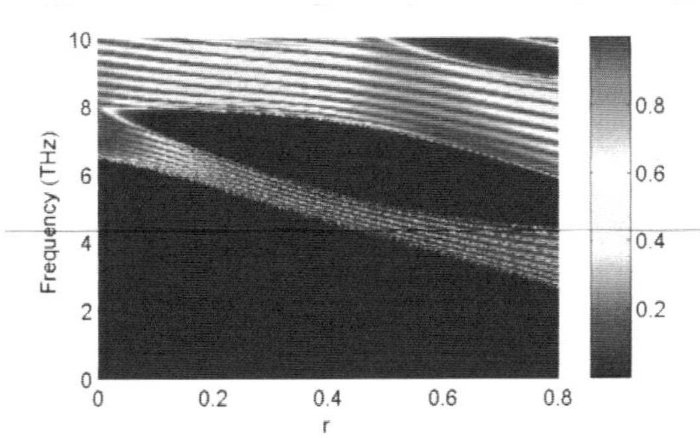

Figure 7. Transmission spectra of the GPC as a function of the filling factor

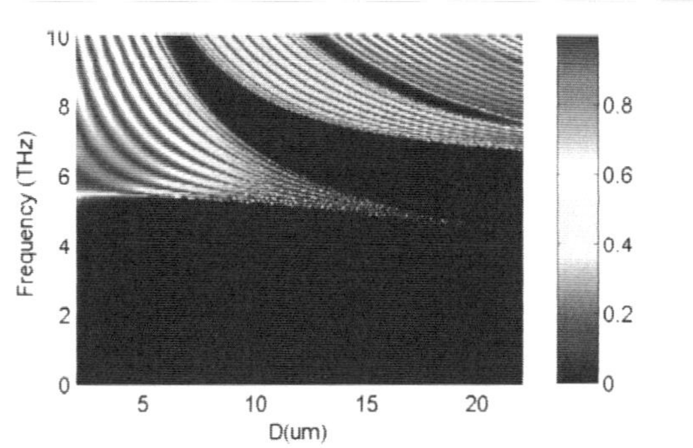

Figure 8. Transmission spectra of the GPC as a function of the period of the GPC

IOP Conf. Series: Materials Science and Engineering **230** (2017) 012019 doi:10.1088/1757-899X/230/1/012019

4. Conclusion

A 1D GPC constituted with only one dielectric material was studied by employing the TMM in this paper. The use of one dielectric material to construct PCs helps to reduce the difficulty of preparing a PC. It can be seen that GPC constructed with one dielectric material have the same bandgap as the GPC composed of two different dielectric materials. The numerical results show that the LFBG of this GPC decreases with the increase of the thickness of the unit cell D, the relative permittivity of the dielectric material, the thickness of the dielectric material, the filling ratio r and the period d of the graphene-dielectric multilayer medium structure A, but increases with increasing the thickness d_A of the graphene multilayer structure. While the BBG has a more complex response with the changes of those structural parameters above mentioned. Since the complex conductivity of graphene is a function of chemical potential, and the chemical potential can be changed by applying a magnetic field or bias, the complex conductivity of graphene is tunable. This means that the GPC is a bandgap tunable GPC, which makes it have a variety of potential applications in the field of photonics. Therefore, it is extremely important to study the GPC.

5. References

[1] Wu J J and Gao J X 2016 *Mater. Chem. Phys.* 171 91
[2] Mohamed M S, Hameed M F O, El-Okr M M and Obayya S S A 2016 *Optik* 127 8774
[3] Brandão E R, Vasconcelos M S and Anselmo D H A L 2016 *Opt. Mater.* 62 584
[4] Berman O L and Kezerashvili R Y 2012 *J. Phys. Condens. Matter* 24 015305
[5] Hajian H, Soltani-Vala A and Kalafi M 2013 *Opt. Commun.* 292 149
[6] Arefinia Z and Asgaria A 2013 *Physica* E 54 34
[7] Lin H, Xu D, Pantoja M F, Garcia S G and Yang H L 2014 *Chin. Phys.* B 23 094203
[8] Bonaccorso F, Sun Z, Hasan T and Ferrari A C 2010 *Nature Photon.* 4 611
[9] Avouris P and Freitag M 2014 *IEEE Sel. Top. Quantum Electron.* 20 6000112
[10] El-Naggar S A 2015 *Opt. Quantum Electron.* 47 1627
[11] Othman M A K, Guclu C and Capolino F 2013 *Opt. Exp.* 21 7614
[12] Zhu B F, Ren G B, Zheng S W, Lin Z and Jian S S 2013 *Opt. Exp.* 21 17089

Acknowledgment

This work has been partially supported by the Chinese Natural Science Foundation (11104165) and the National Training Programs of Innovation and Entrepreneurship for Undergraduates (201610433083)

Feasible Recycling of Industrial Waste Coal Gangue for Preparation of Mullite Based Ceramic Proppant

Guomin Li[1, a *], Haiqiang Ma[1, b], Yuming Tian[2, c], Kaiyue Wang[1, d], Yi Zhou[1, e], Yaqiao Wu[1, f], Xinwei Zou[1, g], Jianying Hao[1, h], Pinbo Bai[3, i]

[1] School of Materials Science and Engineering, Taiyuan University of Science and Technology, Taiyuan 030024, Shanxi, P. R. China
[2] Shanxi engineering vocational college, Taiyuan 030009, Shanxi, P. R. China
[3] Changqing Fracturing Proppant Co. Ltd, Yangquan 045000, Shanxi, P. R. China
E-mail: ligm@tyust.edu.cn

Abstract. Industrial waste coal gangue was successfully utilized to prepare the mullite-based ceramic proppants. The experiments involved the pelletizing technology of proppant through intensive mixer and following the sintering process under different temperatures. The crystalline phase, microstructure, density and breakage ratio of the proppants were investigated. The results showed that with the increasing of sintering temperature, the crystalline phases were transformed to rod-like mullite, which formed the cross-linked structure, improving the densification of proppants. Consequently, the breakage ratio under the closure pressure of 35 MPa exhibited declining trend and reached the minimum value of 6.8% at 1450 ^{o}C. Owing to the easy preparation, feasible design, low cost and moderate breakage ratio, the mullite-based ceramic proppant prepared by coal gangue and bauxite is promising candidate for fracturing proppants in future applications.

1. Introduction

Proppant is a crucial material used in the hydraulic fracturing operations in the petroleum and gas industry, and its main role is to "prop open" rock cracks to effectively improve the production of oil and gas wells. [1, 2] Traditionally, the high grade bauxite is the main raw material of preparing mullite-corundum based proppants, possessing excellent performance including high strength and acid resistant. However, the high grade bauxite is facing over-exploitation and exhaustion with the increasing exploitation demand of oil and gas, which results in the skyrocketing of high grade bauxite and the higher cost of proppants in return. Therefore, it is essential to seek the alternatives for the resource of high grade bauxite.

Coal gangue is the by-product discharged in the process of coal mining and coal washing, which formed and coexisted with coal seam during the course of coal formation, containing some combustibles and abundant minerals. Long-term and continuous accumulation of coal gangue not only occupies a large amount of cropland but also causes serious damage to the ecological environment. [3, 4] With the development of circular economy, industrial wastes are known as the misplaced resources. Consequently, the disposal and utilization of coal gangue has become an urgent problem. Recently, the use of coal gangue as raw materials for the preparation of cement, refractory materials and ceramics has already made certain progress. [5–8]

Despite much attention being focused on the recycling and reusing of coal gangue, there are only a few reports on its use preparing proppant. The main chemical compositions of coal gangue used in our work are SiO_2 and Al_2O_3 (see table 1), which are essential elements for the mullite formation, providing silicon source and aluminum source. Our previous study showed that coal gangue could be used as the additive to synthesize the ceramic proppant applied in coal bed methane (CBM) well. [9] In this work, the coal gangue was originally exploited as the main raw material to replace part of the II bauxite to prepare the proppants of CBM well by palletization and sintering process under different temperatures. Crystalline phase evolution, microstructure, bulk density, apparent density and breakage ratio of the proppants sintered at different temperatures were investigated to evaluate the feasibility of preparation technology. In addition, the mullite-based ceramic proppants synthesized by coal gangue and bauxite exhibit excellent anti-crushing ability, showing great potential for CBM well exploitation as low coat and high-efficiency fracturing proppants.

2. Experimental section

2.1 Preparation of proppant

The industrial waste coal gangue and natural low-grade bauxite (from Yangquan area in Shanxi Province, China) were used as the raw materials, and their components are displayed in table 1. First, the coal gangue and bauxite were ground to the desired size using a planetary ball mill. Second, the coal gangue (50 wt%) and bauxite (50 wt%) were transferred into an intensive mixer (R02, Eirich Co. Ltd, Germany) to mix homogeneously and following the pelletization of spherical green bodies by adding appropriate water discontinuously. After drying, the dried green bodies were passed through a set of sieves of meshes (aperture size of 0.90 mm–0.45 mm). Finally, the as-made samples were sintered in a multi-function electric furnace at different temperatures for 2 h, at a heating rate of 5 $^{o}C\cdot min^{-1}$ under Air atmosphere and then were passed through a set of sieves of 20/40 meshes (aperture size of 0.85 mm–0.425 mm). The final products were labeled as CP1200, CP1250, CP1300, CP1350, CP1400 and CP1450, according to the sintering temperatures.

Table 1. Chemical composition of the raw materials (wt%)

Raw materials	Al_2O_3	SiO_2	Fe_2O_3	TiO_2	CaO	L.O.I
Bauxite	62.3	12.4	5.0	2.8	0.5	17.0
Coal gangue	27.4	30.7	8.1	2.7	0.3	30.8

2.2 Characterization of proppant

The bulk density, apparent density and breakage ratio under 35 MPa pressure were determined by the Chinese Petroleum and Gas Industry Standard (SY/T 5108–2014). Apparent density was measured based on the Archimedes' principle. Breakage ratio was determined by the following formula: $\eta=m_c/m_0\times100\%$, where m_c is the weight of the crushed specimens after testing and m_0 is the weight of the specimens before testing. The composition and phase purity of the as-synthesized samples were analyzed by an X'Pert PRO diffractometer with Cu-Kα radiation (λ=1.5406 Å), in scan steps of 0.02° in a 2θ range from 10° to 90°. The microscopic morphology and structures of the samples were observed on a field emission scanning electron microscope (FESEM, S-4800).

3. Results and discussion

3.1 Crystalline phase evolution in proppants

Figure 1 shows the XRD patterns of a series of proppants treated at different temperatures. The detailed Crystalline-phase evolution of the proppant is summarized in table 2. When the sintering temperature is below 1350 ^{o}C, the main diffraction peaks of CP1200, CP1250 and CP1300 can be indexed as mullite and SiO_2, there is some Al_2O_3 coexisted as well. By calculating the ratio of Al_2O_3 to SiO_2 in our work, it is evident that the SiO_2 is sufficient. Whereas the sintering temperature is not high enough that there are still SiO_2 and Al_2O_3 coexisted. As the temperature increases to 1350 ^{o}C, the diffraction peaks of Al_2O_3 gradually disappear, the main phase composition of proppants is mullite, and there are several weak diffraction peaks that can be indexed as the crystalline planes of quartz.

With further increasing temperature, the phase composition for CP1400 and CP1450 is stable and not changed any more, implying that the reaction was completed.

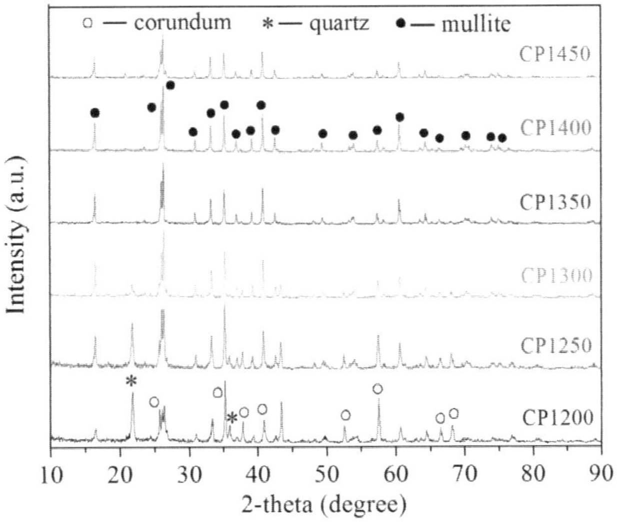

Figure 1. X-ray diffraction patterns of the proppants sintered at different temperatures, where CP1200 to CP1450 mean that the samples were sintered at temperatures from 1200 °C to 1450 °C.

Table 2. Crystalline phase, density and breakage ratio (35 MPa) of proppants

	Crystalline phase	Bulk density ($g\cdot cm^{-3}$)	Apparent density (g/cm^{-3})	Breakage ratio (%)
CP1200	mullite/quartz/corundum	1.23	2.64	12.9
CP1250	mullite/quartz/corundum	1.26	2.67	10.5
CP1300	mullite/quartz/corundum	1.28	2.71	9.9
CP1350	mullite/quartz	1.30	2.73	9.7
CP1400	mullite/quartz	1.36	2.81	9.3
CP1450	mullite/quartz	1.54	2.85	6.8

3.2 Morphology characterization of the proppants

To observe the detailed microscopic structures, SEM images provide some insights into the morphology and structure of the samples, and the representative images of cross-sections of the proppants sintered at different temperatures are shown in figure 2. As can be seen from figure 2a and b, the samples of CP1200 and CP1250 exhibit loose and flake-like structure. With the increase of sintering temperature, the needle-shaped mullite crystals (see figure 2c) come into being and begin to grow larger. For the CP1350 sample, the boundaries between the acicular mullite crystals become unclear, which may result from the formation of glass phase. Furthermore, the morphology of the products change remarkably when the sintering temperature increases to 1350–1450 °C, as seen in figure 2d–f, the acicular mullite crystals grow further and are transformed into rod-like mullite, which forms the cross-linked and dense structure. During the fracture process, this unique structure can restrain the growth of internal micro-cracks due to the fiber-reinforced effect, just like the role played by steel rebar in the reinforced concrete structure. Based on the above analysis, the crystal phase evolution in proppant during the sintering process may undergo the following reactions. [10–13]

$3Al_2O_3 + 3SiO_2 \rightarrow 3Al_2O_3\cdot 2SiO_2$ (Mullite embryo) + SiO_2 (Amorphous) (1100~1150 °C) (1)

$3Al_2O_3\cdot 2SiO_2$ (Mullite embryo) $\rightarrow 3Al_2O_3\cdot 2SiO_2$ (Acicular mullite) (1150~1200 °C) (2)

$3Al_2O_3\cdot 2SiO_2$ (Acicular mullite) $\rightarrow 3Al_2O_3\cdot 2SiO_2$ (Rodlike mullite) (1200~1300 °C) (3)

SiO_2 (Amorphous) $\rightarrow SiO_2$ (Quartz embryo) $\rightarrow SiO_2$ (Quartz cube) (1150~1300 °C) (4)

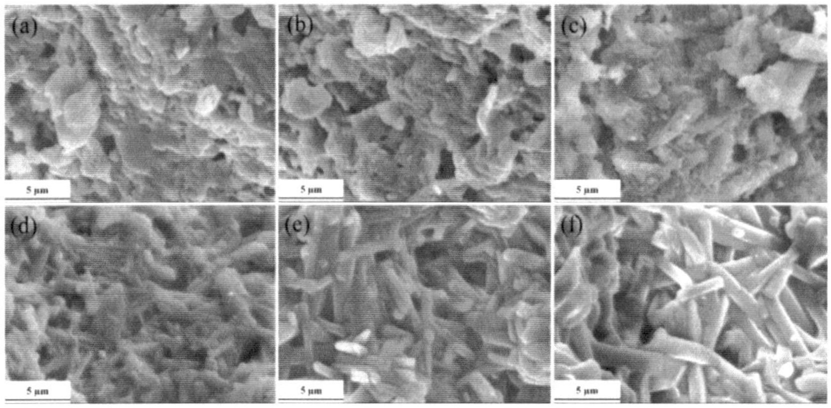

Figure 2. SEM images of cross sections of (a) CP1200, (b) CP1250, (c) CP1300, (d) CP1350, (e) CP1400 and (f) CP1450.

3.3 Density and breakage ratio

Density and breakage ratio values for the proppants are collected in table 2. Density is an important performance metric of any proppants. Variation of apparent density and bulk density with sintering temperature is shown in figure 3, from which it can be observed that the values of apparent density and bulk density incline to increase with increasing temperature. This is in agreement with the variation of breakage ratio, as shown in the inset in figure 3, which exhibits a decreasing trend with the increase of sintering temperature. Combining these results with figure 2, we know that the sintering process of proppants is actual a densification process accompanied by the elimination of pores and the formation of glass phase, because the glass phase can improve the arrangements of the rod-like mullite grains and decrease the grain boundary between the pores and gaps. This results in the increase of density and decrease of breakage ratio with the increasing of temperature. In addition, the appropriate amount of pores in proppants is beneficial to prevent the main crack from further growth and propagation in the process of fracture. The breakage ratio values of CP1300, CP1350, CP1400 and CP1450 under 35 MPa are less than 10%, meeting the standard as required by SY/T 5108–2014.

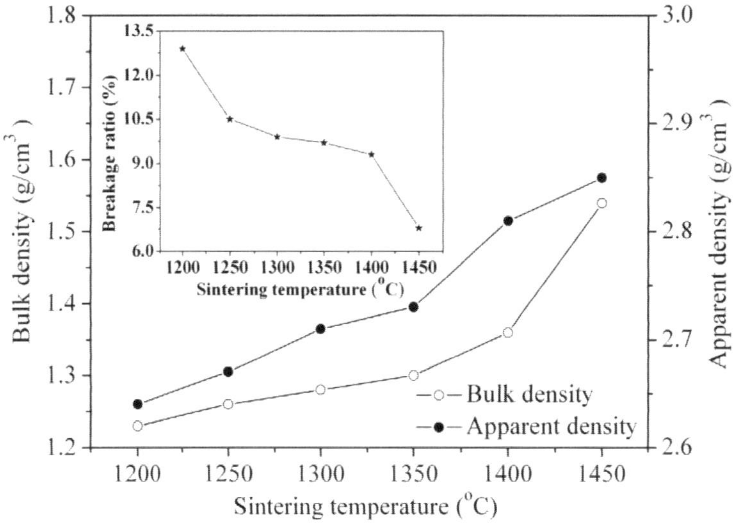

Figure 3. Bulk density, apparent density and breakage ratio (inset) under 35 MPa of proppants sintered at different temperatures.

Compared with the current products, our proppant exhibits low cost and high performance. Moreover, the roundness and sphericity of the proppants in our work all exceed 0.9, meeting the industrial standards. This further confirms the high efficiency of pelletizing technology using the intensive mixer.

4. Conclusion

The mullite-based ceramic proppants were successfully synthesized through pelletizing technology and subsequent sintering process, replacing 50 wt% bauxite by industrial waste coal gangue. Moreover, the density and breakage ratio of as-prepared products could meet the requirements of SY/T5108-2014 when the sintering temperature reached 1300 °C. This study showed that the design of replacing part of bauxite by coal gangue is feasible. The mullite-based ceramic proppants prepared by coal gangue and bauxite are promising fracturing proppants that can be mass produced and are commercially feasible for use in future applications. Furthermore, this facile synthesis strategy may open up a new avenue for the utilization of coal gangue in proppants preparation.

5. References

[1] Ma X X, Tian Y M, Zhou Y, Wang K Y, Chai Y S and Li Z G 2016 Mater. Lett. 180 127
[2] Zhao J Z, Liu Z L and Li Y M 2015 Mater. Lett. 152 72
[3] Li C, Wan J H, Sun H H and Li L T 2010 J. Hazard. Mater. 179 515
[4] Wang S B, Luo K L, Wang X and Sun Y 2016 Environ. Pollut. 209 107
[5] Cong X Y, Lu S, Yao Y and Wang Z 2016 Mater. Design. 97 155
[6] Gardner L J, Bernal S A, Walling S A, Corkhill C L, Provis J L and Hyatt N C 2015 Cem. Concr. Res. 74 78
[7] Ji H P, Fang M H, Huang Z H, Chen K, Xu Y G, Liu Y G and Huang J T 2013 Ceram. Int. 39 6841
[8] Xu H L, Song W J, Cao W B, Shao G, Lu H X, Yang D Y, Chen D L and Zhang R 2016 J. Mater. Cycles. Waste. Manag.19 1270
[9] Wang T H, Tian Y M, Zhou Y, Chai Y S, Li Z G and Bai P B 2016 J. Chin. Ceram. Soc. 35 848
[10] Chen C Y, Lan G S and Tuan W H 2000 J. Eur. Ceram. Soc. 20 2519

[11] Li N, Zhang X Y, Qu Y N, Xu J, Ma N, Gan K, Huo W L and Yang J L 2016 J. Eur. Ceram. Soc. 6 2807
[12] Schneider H, Schreuer J and Hildmann B 2008 J. Eur. Ceram. Soc. 28 329
[13] Zhu Z W, Wei Z L, Shen J X, Zhu L, Xu L, Zhang Y F, Wang S and Liu T 2017 Ceram. Int. 43 2871

Acknowledgements
This work was supported by The Doctoral Scientific Research Foundation of Taiyuan University of Science and Technology (No. 20152030), the Science Foundation of ShanXi (No. 201601D102019), the Science and Technology Innovation Projects of University in Shanxi (No. 2016-59), the Talent & Training ShanXi Programme of the joint training base for graduates (No. 2016JD37 and 2017JD33), the Fund for Collaborative Innovation Center of Shanxi Key Basic Materials and the Fund for Shanxi Key Subjects Construction (FSKSC).

IOP Conf. Series: Materials Science and Engineering **230** (2017) 012021 doi:10.1088/1757-899X/230/1/012021

Influence of Sintering Temperature on the Microstructure and Property of Low-cost Ceramic Proppants Prepared by Adding Purple Sands

H Q Ma[1], Y M Tian[2*], G M Li[1], K Y Wang[1], Y Zhou[1] and P B Bai[3]

[1] School of Materials Science and Engineering, Taiyuan University of Science and Technology, Taiyuan 030024, China
[2] Shanxi Engineering Vocational College, Taiyuan030009, China
[3] Changqing Oil Fracturing Proppant Company Limited, Yangquan 045240, China
Email:tym1654@126.com

Abstract: High-strength corundum-mullite based ceramic proppants have been fabricated from 2nd bauxite (62.3wt%Al_2O_3) and purple sands, which were used as main raw materials by solid sintering method, combined with the additive of feldspar and pyrolusite powder. The phase analysis was studied by X-ray diffraction (XRD) and the morphology of specimen was observed by the scanning electron microscopy (SEM). The result showed that the additive of feldspar was beneficial to the formation of internal liquid phase of sample with the sintering temperature increasing, which reduced the pores and promoted the densification of sample. Meanwhile the proper liquid phase resulted in the growth of mullite and promoted the close integration of mullite and corundum, forming the network structure of needle-like mullite. Ultimately the proppants exhibited the optimal performance at 1450 ^{o}C with bulk density of 1.59 g/cm^3, apparent density of 3.11 g/cm^3 and the lowest breakage ratio of 3.76% under 52 MPa pressure.

1. Introduction

Proppant, as a significant material in the hydraulic fracturing process, can increase the amount of oil and gas extraction [1]. Currently, the proppant primarily includes the quartz sand, the resin-coated product and the ceramic proppant. The quartz sand, as a traditional proppant, possesses advantage of low cost, however it is unsuitable to be applied in the deep reservoirs because of its relatively low crush resistance [2]. The resin-coated quartz sand plays a more significant influence compared to that of the quartz sand in terms of resistance crush, but it increases cost in the production process. By contrast, ceramic proppant, as an ideal proppant material, not only meets the demand of lower production cost, but also has higher resistant crushing capability [3].

In previous several literatures have focused on pursuing the higher strength. However, the high-strength ceramic proppant has been successfully synthesized using pure alumina or high-grade bauxite. Zhao et al [4] produced ceramic proppant from pure alumina and $BaCO_3$, and investigated the effect of $BaCO_3$ on the breakage ratio. It was found that the lowest breakage ratio and the apparent density were 2.5% (86MPa) and 3.48 g/cm^3 respectively. Li et al [5] prepared ceramic proppant by high bauxite. The result displayed that the breakage ratio was 4% under 69 MPa closed pressure and the apparent density was 3.27 g/cm^3. However, these above-mention materials were expensive in cost and unfavorable to the large-scale production. Therefore, to choose the low-grade bauxite or low-cost minerals as a substitute for the high-grade bauxite has become urgent.

IOP Conf. Series: Materials Science and Engineering **230** (2017) 012021 doi:10.1088/1757-899X/230/1/012021

In this work, we used the low grade bauxite (62.3 wt% Al_2O_3) and low-cost purple sands as main materials to prepare the corundum-mullite based ceramic proppants. Purple sands, as a cheap raw material, belonged to Al_2O_3—SiO_2 system compound. Therefore, adequately using the purple sands as a substitute for bauxite was feasible. Furthermore, the effects of sintering temperatures on the phase evolution, microstructure, apparent density, bulk density and breakage ratio of proppants were investigated.

2. Experimental

2.1. Raw materials

The natural low-grade bauxite and purple sands were (obtained from Yangquan city, Shanxi province, China) used as the starting material, feldspar and pyrolusite powder were selected as additives. Their main chemical components are displayed in table1.

Table 1. Chemical composition of the starting materials（wt/%)

Raw materials	Al_2O_3	SiO_2	Fe_2O_3	TiO_2	CaO	MnO_2	L.O.I
Bauxite	62.3	12.4	5	2.8	–	–	17.5
Purple sands	29.26	51	4.34	–	–	–	15.4
Feldspar	18.4	64.7	–	–	16.9	–	–
Pyrolusite powder	4.52	18.41	13.43	0.05	1.82	56.82	4.28

2.2. Preparation of proppants

Firstly, the weight ratios of purple sands, feldspar and pyrolusite powder with regard to the bauxite are of 10 wt%, 2 wt% and 4 wt% respectively. These materials were mixed homogeneously to form small granules by a strong mixing machine (R02, Eirich Co. Ltd, Germany), and then the specimens were dried in an oven. After drying, the homogeneous granules were screened by a set of sieves of 20/26 meshes to gain the desired semi-finished products of 0.85 mm~0.71 mm in diameter. Secondly, these semi-finished products were placed in a high temperature furnace (KF1700, Nanjing Bo Yun Tong Instrument Technology Co. Ltd, China) and sintered through the following process: the semi-finished products were calcinated from the room temperature to the different sintering temperatures (1300 °C, 1350 °C, 1400 °C, 1450 °C and 1500 °C) with a heating ratio of 5°C/min, and were held 2 h from1300 °C—1500 °C, then cooled to the room temperature naturally. Finally, the sintered granules were sieved by a set of sieves of 20/40 meshes to obtain finish products.

2.3. Characterization

The bulk density, breakage ratio and apparent density were tested according to the Chinese Petroleum and Gas Industry Standard (SY/T 5108-2014). The apparent density was calculated using the well-know Archimedes formula: $\rho=M/V$, where M represented the mass of test proppants and V denoted the volume of the specimens. The breakage ratio formula was expressed by $\eta=W_c/W_0$, here W_c was the mass of specimens after testing and W_0 was the mass of specimens before testing. The breakage ratio was evaluated under 52MPa closed pressure. Different phases were analyzed by XRD (X'Pert PRO, Philips Co. Ltd, Holland). The microstructures of the proppants were viewed through SEM (FESEM, S-4800, Hitachi, Japan).

3. Results and discussion

Figure1 displays the microstructures of sample at different sintering temperatures. From figure 1(a), it can be seen that the sample contains a large number of needle-like grains and the structure of sample is loose. When the sintering temperature reaches at 1350 °C, the needle-like grains are transformed to rod-like grains. For the sample with sintering temperature of 1400 °C, as can be observed that the size of rod-like grains increases further, the rod-like grains are coalesced by a small amount of liquid phase.

Furthermore, with the sintering temperature increasing to 1450 °C, the rod-like grains grow through the plate-like grain matrix to form reticular structure during to grains anisotropically grow. For the sample with sintering temperature of 1500 °C, it is obvious that a quality number of liquid phase is generated, which leads to disappearance of the grain boundary. The results illustrate that the sintering temperature can boost the growth of mullite and the moderate liquid phase is beneficial for the formation of network structure of sample.

To judge the above-mentioned microstructure of grains, the patterns of XRD of samples are represented in figure 2. A comparison of the XRD patterns observes that the phases of sample are composed of corundum and mullite respectively. It is confirmed that the needle-like grains and the rod-like grains are mullite, the plate-like grains was corundum. These morphologies have been reported in many literatures [6-9].

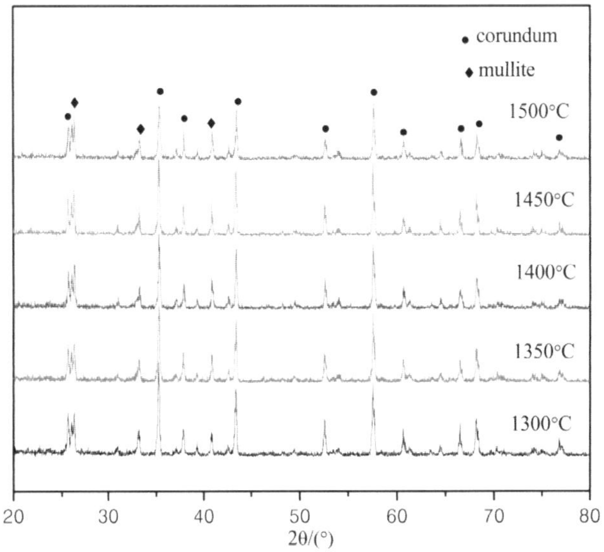

Figure 1. SEM images of samples at different sintering temperatures: (a) 1300 °C, (b) 1350 °C, (c) 1400 °C, (d) 1450 °C, (e) 1500 °C.

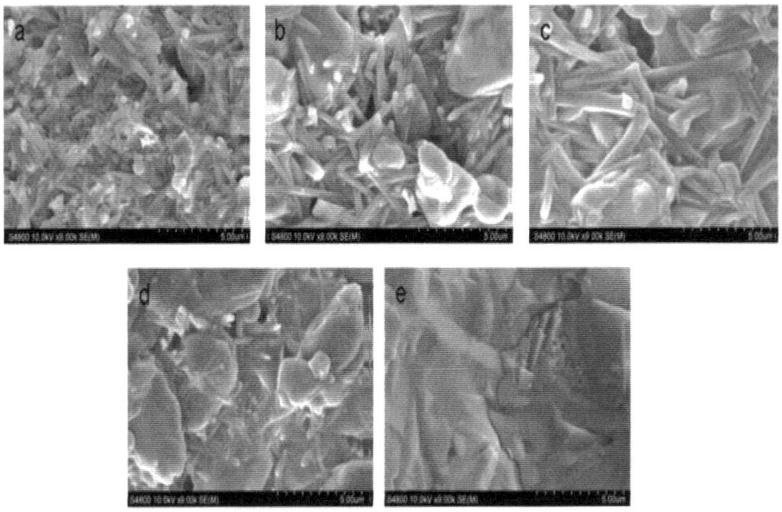

Figure 2. XRD patterns of samples at different sintering temperatures.

Figure 3 shows the relationship between bulk density, apparent density and sintering temperature. Both the bulk density and apparent density shows an upward tendency as the sintering temperature rises, which indicates the sample gradually becomes densification. Simultaneously, it is also worthwhile to note that the ascent rate of bulk density and apparent density are relatively fast at 1400—1450 °C.

As known, the bulk density is associated with the densification of sample. The densification of sample is mainly depended on liquid phase formation which is filled the pores. However, the ascent rate of bulk density is relatively slow in the sintering temperature range of 1300—1400 °C. This can be explained that the sample is sintered by solid state sintering. The diffusion mass transfer, as the main model of mass transfer of sample, forms surface tension in the sintering stage. The particles are indented distance by the surface tension, resulting in open pores reducing. However, further increasing the sintering temperature to 1400—1450 °C, it is remarkably seen that the structure trends to densification. This is due to the components of additives of feldspar and pyrolusite powder, such as Na_2O, K_2O, CaO, MgO and MnO_2, which is beneficial to liquid phase formation along with the sintering temperature increases. Meanwhile, during to alkaline-oxides can reduce the viscosity of liquid phase [10]. The mass transfer pattern of sample is transformed to flow and mass transfer. However, the rate of flow and mass transfer is faster than that of diffusion mass transfer, which accelerates densification of sample [11]. Therefore the ascent rate of bulk density is relatively fast at 1400—1450 °C.

With regard to the apparent of sample, it is correlated with the internal porosity. At a low sintering temperature, the surface temperature of sample is relatively high, which accounts for the rate of surface diffusion faster that that of volume diffusion. With the sintering temperature increasing, the rate of volume diffusion is faster than that of surface diffusion, which leads to the internal porosity constantly eliminating. Therefore the apparent density shows upward tendency. However, it can be seen that the sintering temperature is up to 1450 °C or higher temperature, the apparent density remains unchanged, which can be explained that the internal pores are nearly inexistence.

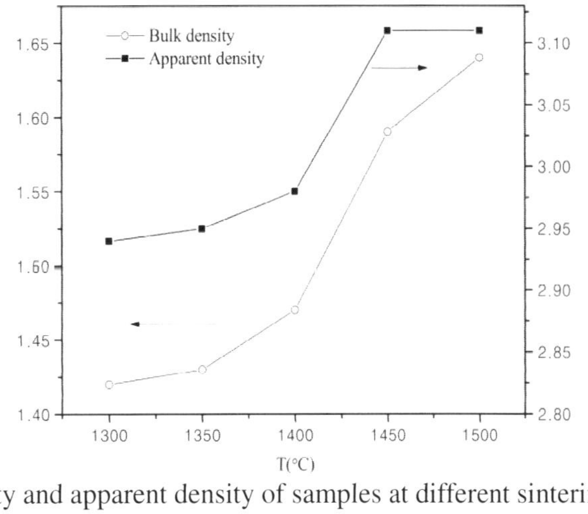

Figure 3. Bulk density and apparent density of samples at different sintering temperatures.

The relationship between the breakage ratio and temperature is shown in figure.4. The breakage ratio decreases firstly and then increases with the increasing of temperature, reaching the lowest value of 3.76 % at 1450 °C. It can be accounted for two aspects: (a) The liquid phase is generated by adding the additive of feldspar and pyrolusite powder, accelerating densification of sample. (b) The needle-like mullite anisotropically grows to form the network structure. Chen et al [10] has demonstrated the mullite whisker was beneficial to improve the strength of sample. However, it is noticeable that the sintering temperature exceeds to 1450 °C, the breakage ratio of sample decreases. It can be explained that the liquid phase is so excessive that the strength of sample reduces.

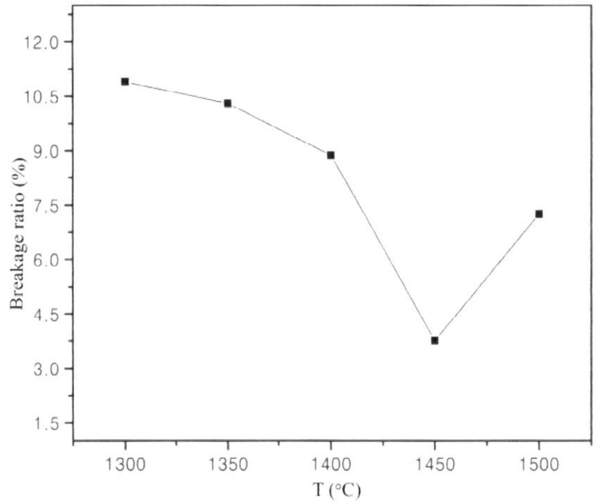

Figure 4. Breakage ratio of samples at different sintering temperatures.

4. Conclusion

High-strength corundum-mullite based ceramic proppants were synthesized by solid state sintering method. Adding feldspar is helpful to form the liquid-phase, which promotes the growth of corundum and mullite, decreasing the pore and boosting to the dense structure. The proppant sintered at 1450 ^{o}C shows an apparent density of 3.11 g/cm^3, bulk density of 1.59 g/cm^3 and the lowest breakage ratio (3.76 %) under 52 Mpa pressure, meeting the requirement of Gas industry standard SY/T 5108-2014. Compared with the proppants were fabricated by the raw material of high-alumina bauxite (>70 wt %), the proppant in this work displays the better resistance crush.

5. References

[1] Zhao J Z, Liu Z L and Li Y M 2015 *Mater. Lett.* 152 74.
[2] Ma X X, Tian Y M, Zhou Y, Wang K Y, Chai Y S and Li Z G 2016 *Mater. Lett.* 180 127.
[3] Pangilinan K D, Al Christopher C and Advincula R C 2016 *J. Petrol. Sci. Eng.* 145 155.
[4] Zhao Y R, Wu B L and Wu T T 2010 *Chin. Ceram.* 46 47.
[5] Li F Z and Zhang X X 2013 *J. Wu. Han. Univ. Technol.* 35 34.
[6] Guo H, Li W and Ye F 2016 *Ceram. Int.* 42 4822.
[7] Hou Z P, Cu B X, Liu L L and Liu X 2016 *Ceram. Int.* 42 17256.
[8] Xu X H, Li J W, Wu J F, Tang Z H, Chen L L, Li Y and Lu C L 2017 *Ceram. Int.* 43 1765.
[9] Lu Z L, Tian G Q, Wan W J, Miao K and Li D C 2016 *Ceram. Int.* 42 18855.
[10] Chen X D, Li T H, Ren Q, Wu X L, Li H, Dang A L, Zhao T K, Shang Y D and Zhang Y 2017 *J. Alloys. Compd.* 700 40.
[11] Liu Z L, Zhao J Z and Li Y M 2016 *Mater. Lett.* 171 302.

Acknowledgments

We would like to acknowledge the Scientific and Technological Innovation Programs of Higher Education Institutions in Shanxi (2016-59), the Talents Training Project of Shanxi Graduate Joint Training Base (2016JD37 and 2017JD33), the Shanxi Province Science Foundation (201601D102019), the Doctoral Scientific Research Foundation of Taiyuan University of Science and Technology (20152030) and the Fund for Collaborative Innovation Center of Shanxi Key Basic Materials.

Effect of Sintering Temperature on the Microstructure and Mechanical Properties of Low-Cost Light-Weight Proppant Ceramics

K Y Wang[1], H J Wang[1], Y Zhou[1], Y Q Wu[1], G M Li[1] and Y M Tian[2]*

[1] School of Materials Science and Engineering, Taiyuan University of Science and Technology, Taiyuan 030024, China
[2] Shanxi Engineering Vocational College, Taiyuan030009, China
E-mail: wangkaiyue8@163.com

Abstract: In this paper, the low-cost light-weight proppant ceramics were prepared with the solid wastes of coal gangue as the raw materials, and the effect of sintering temperature on the apparent porosity, bulk density, bending strength, microstructure and phase composition were investigated. The results showed that the ceramics, sintered at 1350℃, has the best performance with the bending strength of 85MPa, bulk density of 2.7 g/cm3 and apparent porosity of 18%. These properties of ceramics were very close to that of the bauxite-sintered, and thus the gangue were very probably selected for the preparation of proppants that involved in the exploitation of coalbed methane.

1. Introduction

Coalbed methane, known as "Mashgas", is a kind of high-quality chemical and energy resources. Shanxi Province is a large province of coal resources in China, and the reserves of coalbed methane are up to 10.39 trillion cubic meters, especially for two large mines of Qinshui and Hedong [1-3]. When the air concentration of coalbed methane exceeds 5%, it will explode in the case of fire. Therefore, coalbed methane has to be exploited before the coal mining. In order to achieve the industrial gas production, more than 96% of coalbed methane takes advantage of the hydraulic fracturing methods [4]. In the process of hydraulic fracturing, the proppants consist of spherical ceramic particles, which can resist the fracture closure to create the gas channel, and thus their performance directly determines the success or failure of hydraulic fracturing operations [5]. As the main raw material of proppant preparation, bauxite has become scarce and expensive [6]; thus, the need for a replacement is increasingly urgent.

The mechanical properties of the proppant are directly proportional to the content of alumina in the raw material. With the alumina content is increased, the mechanical properties of proppant are increased while the cost of proppant is greatly increased together [7]. Moreover, the density of proppant increased with the increase of alumina content, and then the proppants are not easy to be carried by the fracturing fluids which usually results in the accumulation of proppants at the gas pipeline clogging [8-9]. Compared with oil exploration, coalbed methane locates at the shallower and softer stratum, resulting in lower requirements for the volume density and the crush ratio of the proppants [10-11]. Hereby, a mineral with less Al_2O_3 is a better selection to reduce production costs. A major solid waste in the Shanxi province of China, coal gangue (~20.7 wt% Al_2O_3) has been used mainly for building cement but rarely for ceramics [7]. In this paper, the low-cost light-weight proppant ceramics were prepared with the solid wastes of coal gangue as the raw materials, and the effect of sintering temperature on the apparent porosity, bulk density, bending strength, phase composition and microstructure were investigated.

2. Experimental

2.1 Raw materials

The starting materials of coal gangue were obtained from the Changqing Fracturing Proppant Co. Ltd., Yangquan, Shanxi, China. The coal gangue consists of 20.7 wt% Al_2O_3, 65.7 wt% SiO_2, and 13.6 wt% other oxides. In this paper, the bauxite powder was also involved which were usually selected for proppants. The chemical composition analysis of the bauxite powder is 60 wt% Al_2O_3, 13.8 wt% SiO_2, and 26.2% other oxides.

2.2 Experimental procedure

The raw material powder were milled into micrometer-size by a roll crushing mill and a QLM series jet airflow mill. Following that, homogeneous mixtures could be obtained by being ball-milled in an absolute ethanol medium for 6 h and then dried in an oven at 120°C for 36 h. The dried lumps were crushed, sieved through screening (100#) to break up agglomerates, and then compressed into 30mm × 6mm × 4mm bodies by a press machine under 15MPa pressure holding for 30s. Afterwards, the black bodies were dried at 120°C for 5 h and then sintered in a KBF1700 electron furnace in air at 1200, 1250, 1300, 1350, and 1400°C at holding times of 2 h, respectively. Lastly, the samples were cooled down to 400°C at a heating rate of 5°C/min and furnace-cooled to room temperature.

2.3 Characterization

The bulk density D and apparent porosity P were determined in accordance with the American Society for Materials Standards ASTMC373, and calculated by using the formulas (1) and (2), respectively, where m0 is the mass of the samples; m1 is the mass of sample floating in the water after 24h water boiling and cooling-down to room temperature; m_2 is the mass of sample with the saturated water.

$$D = \frac{m_0}{m_2 - m_1} \times 100\% \tag{1}$$

$$P = \frac{m_2 - m_0}{m_2 - m_1} \times 100\% \tag{2}$$

The bending strength of ceramics was measured by a YDW-10 flexural testing machine. Values of the results were obtained from the average of at least 5 bars. Thermal properties of the raw materials were studied simultaneously by thermogravimetry (TG) and differential-scanning calorimetry (DSC) (METTLER TOLEDO 3+). The test was performed from 50 to 1300°C at a flowing nitrogen atmosphere with a heating rate of 10°C/min. X-ray powder diffraction (XRD, Philips Co. Ltd, Holland) was performed to identify the phase composition of the specimens by using Ni-filtered Cu Kα radiation with a scanning speed of 0.02°/step. The microscopic morphology of the specimens was observed by field-emission scanning-electron microscopy (SEM, S-4800).

3. Experimental results

The TG-DSC results of coal gangue are shown in figure 1. Weight loss of the gangue mainly occurred from 350 to 900°C. A wide endothermic peak of the DSC curve was observed in the range of 200 to 400°C, which might have been caused by the loss of free and adsorbed water. With the increase in temperature, weight loss was 21.8%, and a valley appeared at approximately 510°C, which was caused by the combustion of residual coal in the gangue [12]. Weight loss between 890 to 1067°C was slight; however, abundant reactions and phase changes occurred. Metakaolinite decomposed into amorphous Al_2O_3 and SiO_2 at 976°C [13-14], after which, rod-like mullite ($3Al_2O_3{\cdot}2SiO_2$) appeared as the temperature increased [15]. Therefore, in our research, the samples were first heated to 300°C, at 5°C/min, and were held for 30 min to remove the water vapour, and were then heated to 650°C at

2°C/min and held for 30 min to eliminate the residual coal. Moreover, the sintering temperature was defined as higher than 1067°C to obtain a high-strength high-toughness mullite phase.

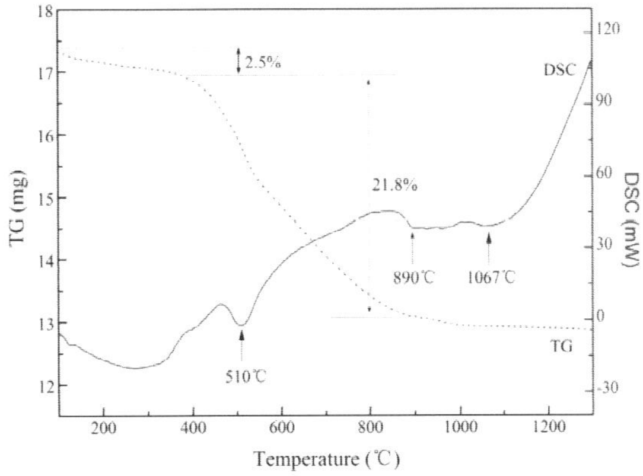

Figure 1. TG-DSC curves of coal gangue.

The sintering temperature has a great impact on the mechanical properties of ceramics [6,16]. The bending strength of ceramics sintered at different temperatures was shown in figure 2, which indicates that the highest bending strength of 85KPa was obtained at the sintering temperature of 1350°C. Figure 3 presented the bulk density and apparent porosity of ceramics sintered at different temperature. It indicated that that with the increase of sintering temperature, the apparent porosity decreased while bulk density increased, and then both of them remained steady when the temperature reached 1350°C (about 18% and 2.70 g/cm^3, respectively).

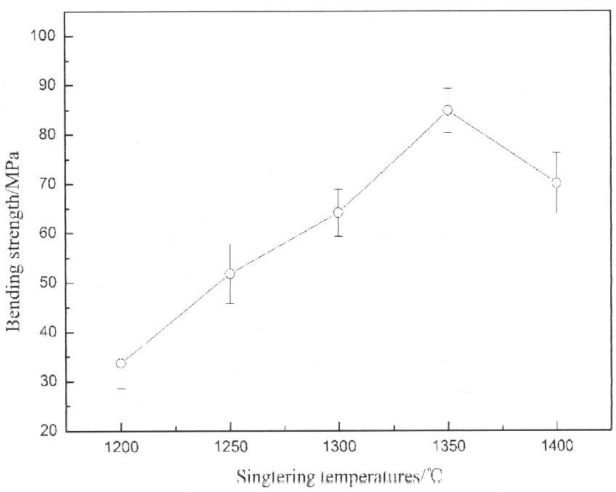

Figure 2. Bending strength of ceramics sintered at different temperatures

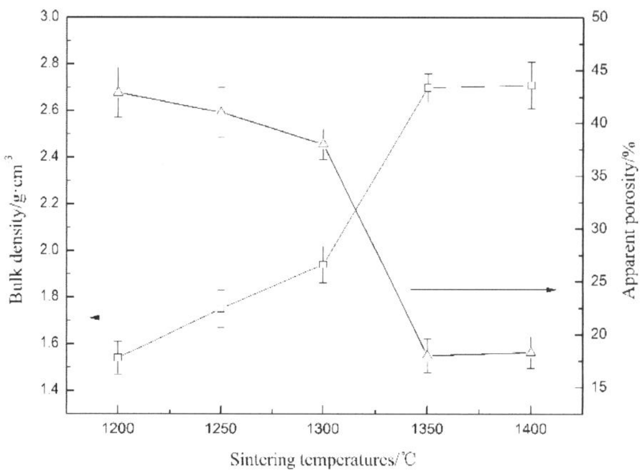

Figure 3. Bulk density and apparent porosity of ceramics sintered at different temperatures

Compared with the bauxite-sintered ceramics (as shown in Table 1), the samples obtained are low cost, and show a lower bulk density while similar bending strength and apparent porosity. Hereby, the gangue was very probably selected for the preparation of proppants that involved in the exploitation of coalbed methane. The low cost results mainly from the cheap starting materials of gangue and from the lower sintering temperatures.

Table 1. Comparison of two kinds of proppant ceramics

Raw materials	Sintering temperature[℃]	Bending strength[MPa]	Bulk density [g.cm^{-3}]	Apparent porosity
Bauxite	>1450	96	3.8	15%
Gangue	1250	85	2.7	18%

Figure 4. shows the XRD patterns of the ceramics sintered at different temperatures. At 1200°C, the proppants had a few peaks corresponding to the quartz phase (SiO_2) and mullite crystals ($3Al_2O_3 \cdot 2SiO_2$); with the increase of sintering temperature, the peaks at 25.7° and 26.0° associated with the mullite crystals were generally enhanced and sharper, whereas the quartz-related peak located at 21.8° weakened.

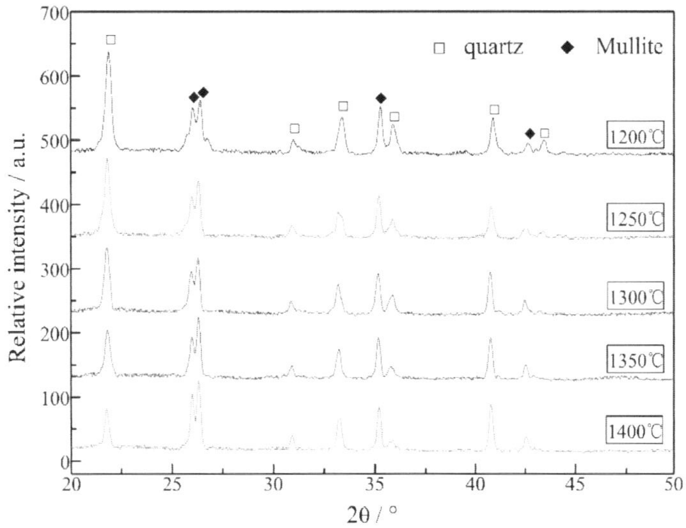

Figure 4. XRD patterns of ceramics at different sintering temperatures

Figure 5. shows SEM photos of fractured ceramics sintered at different temperatures. With the increase in sintering temperature, the porosity of ceramics decreases drastically. When the temperature reaches 1350℃, the bond bridge between crystal phase and glass phase becomes obviously improved and smooth.

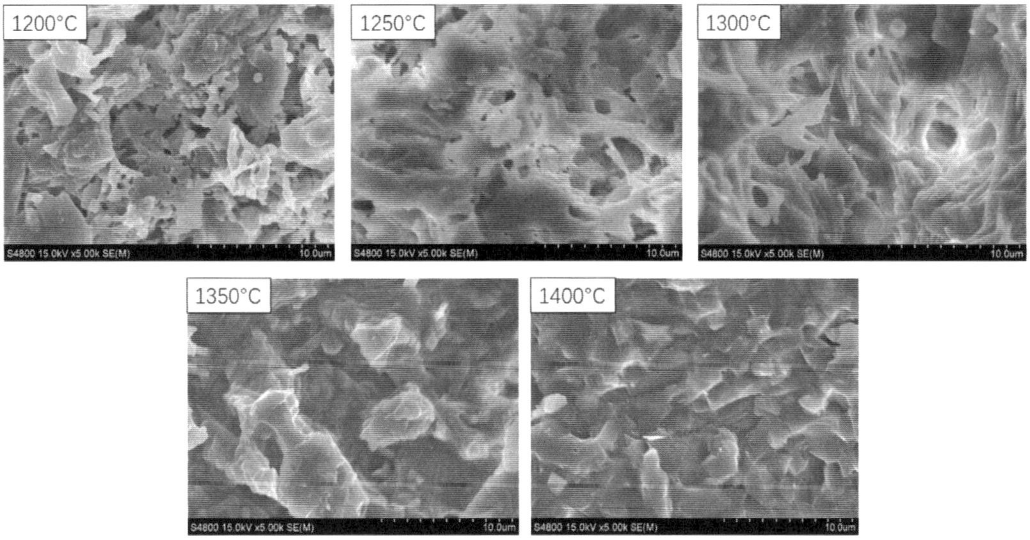

Figure 5. SEM images of the fractured ceramics at different sintering temperatures.

4. Discussion

The main crystalline phases of proppant ceramics sintered at 1200℃ with coal gangue as the raw material are the mullite and quartz (see in figure 4). However, only the flake-like structures were observed in the 1200℃ SEM photographs, and no regular geometric crystals appeared. When the temperature raised up to 1250℃, the materials began to melt and generally form amorphous glass phase (liquid phase). With the increase of liquid content, the speeds of diffusion and flow-mass transfer increased, and the liquid filled the gap among the particles, whereas, the particles flowed in the liquid phase, and the rearrangement occurred among the particles, and hereby, the porosity of ceramics was greatly reduced. The diffraction peak of quartz phase went weaker as the temperature increased, which indicated that the quartz and other oxides formed into the amorphous state.

Although the XRD results showed the existence of mullite crystals below 1250℃, it was not observed in the SEM photograph until 1300℃. According to the TGA-DSC results, the mullite embryo formed below 1100℃. The presence of liquid phase facilitates the formation of acicular mullite.As the temperature increase up to 1300℃, These mullite crystal grows in size within the liquid phase through a dissolution and reprecipitation process from small-scale (fine acicular-like) shapes to thicker, large-scale, rod-like shapes. The viscous flow mechanism of liquid phase promoted the intense arrangement of the rod-like mullite grains [17] and the decrease of grain boundary between pores and gaps, which can lead to the densification of ceramic. Porosity can greatly determine the strength of the ceramics, and the source of cracks usually started at position of the pores[18],so, the stress at the pores of ceramics was larger than other parts of matrix.Glassphase can flow to the crack, the crack are thus covered with glassy phase. More glassy phase is consumed by alumina to form mullite at higher temperature and enhanced the strength of ceramics. Moreove, just as we know that the fiber-toughening effects improved the toughness and the strength of ceramics, the rod-like mullite also hindered the growth of internal micro-cracks.

5. Conclusions

In this work, the low-cost proppant ceramics sintered at 1350°C presented the best performance with a bending strength of 85MPa, an apparent porosity of 18% and a bulk density of 2.70g / cm3, which were very close to these of the bauxite-sintered, and thus the gangue were very probably selected for the proppants preparation in the exploitation of coalbed methane. During the sintering process, the oxides in the raw materials melted into amorphous liquid phase, and the diffusion mass transfer accelerated, promoted the growth of mullite embryo, the arrangement of the rod-like mullite grains.

6. References

[1] Zhao X, Jian B, Xu Q, Liu J, Zhao Y and Duan P 2000 *Petroleum scienc*. 43 89-96.
[2] Zou C 2013 *Unconventional Petroleum Geology*. 111–147.
[3] Xu Z, Liu Q, Zheng Q, Cheng H and Wu Y 2016 *J. geochem explor*. 165 94-101.
[4] Lu S, Cheng Y and Qin L 2015 *J china u min techno*. 25 819-825.
[5] Song, D, Liu, Z, Wang, E, Qiu L, Gao Q and Xu, Z 2015 I*nt rock mech min*.78, 230-239.
[6] Zhao J Z, Liu Z L and Li Y M 2015 *Mater. Lett.* 152 74.
[7] Dong Z, Xia J, Fan C and Cao J 2015 *Constr build mater.* 100 63-69.
[8] Zhao P F, Tian Y M and Liu A 2015 *China ' s Refractories*. 24 36-41.
[9] Ma X X, Tian Y M, Zhou Y, Wang K Y, Chai Y S and Li Z G 2016 *Mater. Lett.* 180 127
[10] Zhang D, Fan G, Ma L and Wang X 2011 *Int j coal geol.* 86 190-196.
[11] Gu M, Dao E and K Mohanty 2015 *Fuel*. 150 191-201.
[12] Yang N andYue W 2000 *The Handbook of Inorganic Metalloid Materials Atlas(*Wuhan: Wuhan University of Technology Press)
[13] Kakali G, Perraki T, Tsivilis S and Badogiannis E 2001 *Appl clay sci.*20 73-80.
[14] Xie Z and Xi Y 2001 *Cement concrete res.* 31 1245–49.
[15] Wang C L, Ni W, Zhang Q S, Wang S, Gai G S and Wang W K 2016 *Constr build mater.* 104 109-115.
[16] Liu Z L, Zhao J Z and Li Y M 2016 *Mater. Lett.* 171 302.
[17] Chen C Y, Lan G S and Tuan W H 2000 *Jeur ceram soc.* 20 2519-2525.
[18] Kong X C, Tian Y M, Chai Y S, Zhao P F, Wang K Y and Li Z G 2015 *Ceram. Int.* 4294–4300

Acknowledgments

This work was supported by the Scientific and Technological Innovation Programs of Higher Education Institutions in Shanxi (Grant 2016-59), the Talents Training Project of Shanxi Graduate

IOP Conf. Series: Materials Science and Engineering **230** (2017) 012022 doi:10.1088/1757-899X/230/1/012022

Joint Training Base (Grant 2016-37), and the Shanxi Province Science Foundation (Grant 201601D102019).

Preparation of Grinding Aid Using Waste Acid Residue from Plasticizer Plant

Lingxiao Li [1, a *], Yanchao Feng [1,b], Manchao Liu[1,c] and Fengqing Zhao[1,2]

[1] Department of Chemical Engineering, Hebei University of Science & Technology, Shijiazhuang, 050018, China
[2] Hebei Engineering Research Center of Solid Waste Utilization, Shijiazhuang, 050000, China
E-mail: Zhaofq3366@126.com

Abstract. The grinding aid for granulated blast-furnace slag were prepared from waste acid residue from plasticizer plant through neutralization, de-methanol and granulation process. In this process, sulfuric acid was transformed into gypsum which has much contribution for grinding effect by combined use with the glycerol and poly glycerin in the waste. Fly ash was used for granulation for the composite grinding aid. Methanol can be recycled in the process. The result showed that the suitable addition of grinding aid is 0.03 % of granulated blast-furnace slag (mass). In this case, the specific surface area is 14% higher than that of the blank. Compared with the common grinding aids, it has excellent performance and low cost.

1. Introduction

With the rapid development of economy, the demand of building materials is increasing. In recent years, with the cement industry reducing excess production capacity to save energy and reduce carbon dioxide emission, the demand for concrete mineral admixture will increase year by year. Commonly used mineral admixtures are silica fume, fine grinding slag, fly ash, blast furnace slag and so on. In order to improve the activity of these materials, they are necessary to be ground to certain fineness. The addition of grinding aids is of great significance to improve the activity of raw materials and reduce energy consumption.

Grinding aids have found widely use in cement and concrete industry [1-3]. The research of grinding aids is mainly focused on the development of high efficiency grinding aids and composite grinding aids recently. As early as the beginning of twentieth century, many scholars used alcohol, alcohol amine, inorganic salts and polymers for slag grinding. The grinding mechanism was investigated, which mainly includes: strength weakening theory by X.J.Gao [4], film theory by Wu Xiusheng [5] and particle dispersion theory by A.T.Albayrak [6]. The development of composite grinding aids mainly adopts the mix design methods. Jiao Xiaofei et al [7] tested four kinds of common grinding aids (including glycerin, triethanolamine, sodium stearate, sodium lignosulfonate). The results showed that the grinding effect of glycerol is the best, sodium stearate gives optimal particle size distribution. Lin Hui et al [8] studied the influences of molecular size and functional groups of esters on the grinding effect. It turned out that the esters of large molecules and more polar possesses better grinding effect, and effect only when the powder fineness reaches certain fineness. At the same time, esters can improve the material density, increase the later strength. Lin Maosong et al [9] used polyol and sodium stearate to prepare composite grinding aids for steel slag and granulated blast-furnace slag. In this case, the grinding time is reduced by 20% with the target fineness of 425 m^2/kg compared with

IOP Conf. Series: Materials Science and Engineering **230** (2017) 012023 doi:10.1088/1757-899X/230/1/012023

the blank.

Considering the industrial waste residue contains glycerol, which is a commonly used grinding aid for cement, we utilized this material, through pre-treatment, to prepare grinding aid in order to improve the performance of the products and reduce cost. In the same time, we try to find a rational path of using waste acid residue from plasticizer plant.

2. Experimental

2.1. Materials

The waste acid residue which is highly corrosive and difficult to dispose is from fatty acid methyl ester plasticizer plant with the compositions: methanol 6~8 %, glycerin poly-glycerol 42~46 %, sulfuric acid 36~40 %, water 10~12 %. Lime (with the calcium oxide content of 90%), fly ash and granulated blast-furnace slag were from Hebei Jingye Group. Triethanolamine, thri-isopropanolamine and sodium stearate with CP grade are commercially available.

2.2. Methods

First of all, the concentrated sulfuric acid in the waste acid residue is neutralized by lime, the product with grinding effect, calcium sulfate, is obtained. Neutralization heat and a small amount of external heat were used to steam out harmful methanol. Then, add some fly ash in the mixture, which mainly includes calcium sulfate, glycerol and polyglycerol, mixed. The mixture is granulated to obtain grinding aid. See figure 1.

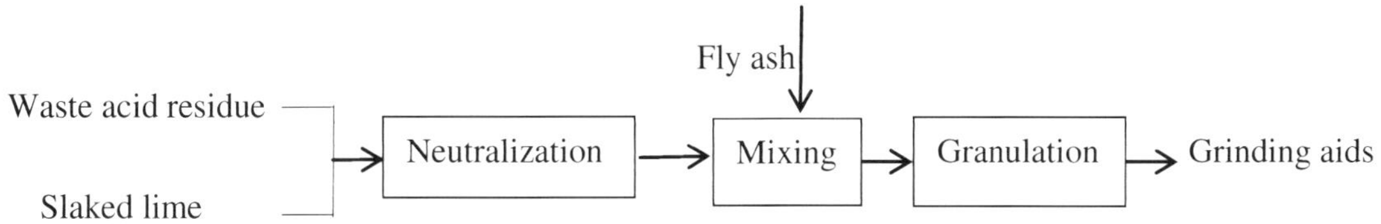

Figure 1. Waste acid residue treatment process.

3. Results and discussion

3.1. Determination of lime consumption

Waste acid residue contains much sulfuric acid, which can be neutralized with alkaline. Among the alkali materials, lime is relative cheap and available. More important, the neutralization product is gypsum which is an essential component in cementing materials. Less of lime will lead to excess in the waste acid residue. And it will lead to corrosion of equipment when used as grinding aid. As lime is add to the waste acid residue, it will react quickly with sulfuric acid to form dense calcium sulfate on the surface particle of the slake lime, resulting in incomplete of reaction in the core of the lime particle, and thus reducing the reaction rate. Therefore, appropriate addition of excessive amount of lime on theoretical basis is needed.

It can be seen from Table 1 that the theoretical dosage of lime is 100 %, which cannot ensured the system to be neutral after a week. Only part of the slaked lime participates in the reaction because of the formation of dense calcium sulfate. The proper dosage of lime should be 106 % of the theoretical amount to guarantee the pH value of the system to be 7 after a week.

Table 1. pH value of waste acid residue with various lime dosage

No	Lime dosage /% (Relative to theoretical value)	pH value after a week
1	100	4
2	101	4
3	102	5
4	103	5
5	104	5
6	105	6
7	106	7
8	107	8

3.2. Neutralization temperature control
On the one hand, we must control the temperature through the neutralization heat and external heating to guarantee the methanol completely removed. On the other hand, we should prevent the neutralization process from huge boiling or bubble entrainment. The temperature of removing methanol was determined to be 63~66℃. Keep the temperature until no liquid can be distillated off.

3.3. Grinding aids
The main component of the waste acid residue after neutralization and distillation are calcium sulfate, glycerin, polyglycerol and water, which can be used as grinding aids directly. However, it is hard for transport and storage. Therefore, fly ash is used to be solidifying material. The material proportioning of the final composition of the grinding aid (HB-3): waste acid residue 15 %, slaked lime 6 %, fly ash 75 % and water 4 %.

4. Application of HB-3 grinding aid
Granulated blast-furnace slag (4 kg) was mixed with the grinding aids, and then added into mill (SM Φ500mm×500mm), grinding for 70 min. The specific surface area of the power was determined. The results are shown in figure 2.

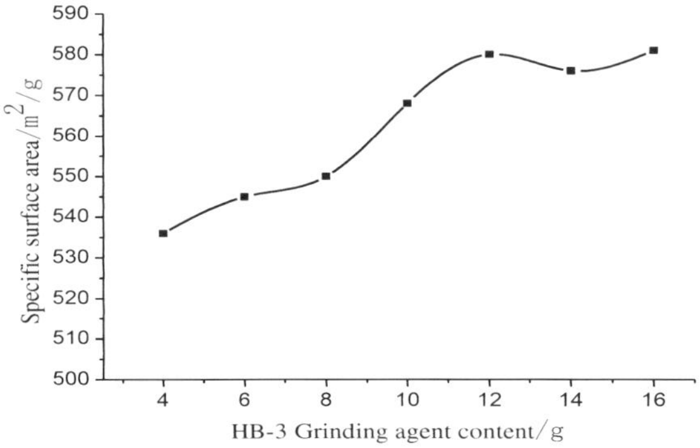

Figure 2. Grinding effect of different dosage of HB-3.

As is shown in the figure 2, the specific surface area is increased with the addition of HB-3 .When the dosage of HB-3 is 12 g, the specific surface area of the slag powder tends to be smooth. That is to say, the optimum dosage of HB-3 is 0.3 % of slag.

Table 2 gives the grinding effect of different grinding aids, HB-3 and commercially available grinding aids.

Table 2. Comparison of different grinding aids in application.

Grinding aids	Grinding aid addition /%	Specific surface area of powder (m2/kg)
Blank	0	514
HB-3	0.03	586
Tri-ethanolamine	0.012	572
Tri-isopropanolamine	0.012	590
Sodium stearate	0.012	600
Glycerol	0.012	596

As can be seen from Table 2, the effect of HB-3 grinding aid is obvious when used for slag grinding, and the specific surface area increased by 14% compared to the blank. HB-3 rivals commercial grinding aids in performance. However, the cost of HB-3 is much lower than the others.

5. Summary

HB-3 grinding aid with good performance was prepared using waste acid residue discharged from the plasticizer plant through neutralization, methanol removal and modification with fly ash. The grinding aid was applied to slag grinding, the specific surface area is 14 % higher than that of the blank, and the cost is much lower than commercial grinding aids.

This process provides a practical resource utilization method for waste acid residue from production of plasticizer. At the same time, a low cost grinding aid was obtained, which has much contribution for the energy saving and carbon emission in the production of granulated blast-furnace slag powder.

6. References

[1] Y L Sun, F Q Zhao and X M Chu 2014 Grinding Aids for Steel Slag and the Application *J. Advanced Materials Research.* **912-914** 158-61

[2] F Q Zhao, B Guo and H J Liu 2009 Utilization of High Carbon Fly Ash and Coal Gangue in Blended Cement: A Case Study *C. International Conference on Bioinformatics and Biomedical Engineering.* IEEE, 1-4

[3] F Q Zhao, W Ni and H J Wang 2007 Activated fly ash/slag blended cement *J. Resources Conservation & Recycling.* **52**(2) 303–13

[4] X. J. Gao, Y. Z. Yang and H. W. Deng 2011 Utilization of Beet Molasses as a Grinding Aid in Blended Cements *J. Construction and Building Materials.* **25**(9) 3782-89

[5] X S Wu 2008 Experimental study on slag grinding aids *J. Scientific information.* **23**(33): 450-51

[6] A. T. Albayrak, M. Yasar and M. A. Gurkaynak 2005 Investigation of the Effects of Fatty Acids on the Compressive Strength of the Concrete and the Grindability of the Cement *J. Cement and Concrete Research.* **35**(2): 400-4

[7] X F Jiao and T Han 2013 Experimental study on grinding aids for slag grinding *J. Building material technology and application.* **2** 21-2

[8] H Lin, X G Wang, D M Wang, D P Zheng, J G Xue, J H Zhao and X F Jian 2014 Effects of several small organic esters on slag grinding performance and strength of slag cement *J. Comprehensive utilization of fly ash.* **8**(25) 6-10

[9] M S Lin, L Y Zhang and Q Wang 2015 Experimental study on the application of powder composite grinding aids *J. Concrete.* **2** 63-5

Acknowledgement
We would like to show the best grateful to Yanchao Feng and Manchao Liu, as they had given much valuable help to the experiment.

IOP Conf. Series: Materials Science and Engineering **230** (2017) 012024 doi:10.1088/1757-899X/230/1/012024

Experimental Determination of Sulfur Partition Ratio in the $CaO-Al_2O_3-SiO_2-CaF_2-MgO$ Slag and Liquid Iron

Z Jinzhu[1], Y Yunsheng[1], F Wei[1], Z Jialiang[2] and Z Yi[2]

1 College of Materials and Metallurgy, Guizhou University, Guiyang 550025, Guizhou, China
2 Shougang Shuicheng Iron and Steel (Group) Co. Ltd, Liupanshui 553028, Guizhou, China
E-mail: zjz-yjx@163.com

Abstract. The external desulphurization of molten iron has become an important step in the production of steel and iron. The desulfurization degree of the high calcium slag, which was mainly taken from Shougang Shuicheng Iron and Steel (Group) Co. Limited, was investigated on basis of the fundamental theory of slag metal equilibrium reaction. The initial content of sulfur in the slag was adjusted to 2.60% mass perdent by adding analytical reagent CaS. The results show that the desulfurization degree of the high calcium slag increases obviously with the increase of temperature in the range 1593-1743K, and so the sulfur partition ratio. When the holding time of the hot metal and slag at controlled temperature was extended from 120 min to 180 min in the furnace, both the sulfur partition ratio and the desulfurization degree increased markedly.

1. Introduction

The external desulphurization of molten iron has become an important step in the production of steel and iron. It is meaningful for Shougang Shuicheng Iron and Steel (Group) Co. Ltd feeded with a wide and unstable iron ores to apply the ladle desulphurization. The average time of iron ladles from blast furnace to converter is about 4 hour in which the temperature drop of the liquid iron in 90 ton ladle is 130K or so (from about 1735K to 1605K averagely). Lower melting temperature is a basic and necessary characteristic of high calcium slag in the ladle desulfurization process. Acidic silicon dioxide can significantly lower the melting temperature of high calcium slag with a basicity decreased and against the desulfurization [1-2]. It is the reason that the weakly acidic aluminium sesquioxide is used to lower the melting temperature of high calcium slag [3,4]. Moreover, the CaF_2 with a lower melting point is a general selection to lower the melting temperature of high calcium slag [4,5]. The higher the sulfur partition ratio, the less the amount of the high calcium slag used in the ladle desulfurization process, and the lower the consumption of the process cost. Based on both the chemical composition of the blast furnace slag and the iron ladle situation of the ShouGang ShuiCheng Iron and Steel (Group) Co. Ltd, the $CaO-Al_2O_3-SiO_2-CaF_2-MgO$ slag with the lowest flowing temperature reported by literature [4-9] is selected for the sulfur partition ratio to be measured experimentally in the temperature range 1593-1743K and to discuss the effects on the sulfur partition ratio in high calcium desulfurization slag and liquid iron by temperature in this study.

2. Experimental

The high calcium desulfurization slag are mainly prepared by the blast furnace (BF) slag taken from Shougang Shuicheng Iron and Steel (Grope) Co. Limited. The main chemical composition of the BF slag after crushed, ball milled, magnetic separation and iron removal is shown in table 1. Based on the content requirements of the CaO-Al_2O_3-SiO_2-CaF_2-MgO slag with the lowest flowing temperature [4], analytical reagents $Ca(OH)_2$, Al_2O_3 and CaF_2 were added to the BF slag powder (see table 2). Mixed equally, the modified mixture was placed in a graphite crucible and heated in an electric resistance furnace to 1773K, maintained at the temperature for 30 min to ensure the mixture was fully melted, and then the high calcium slag was taken out and cooled by water.

The dried high calcium slag was ground to less than 200 mesh in an agate mortar, adding in an appropriate amount of analytical reagent CaS to make the sulphur content up to 2.60%Wt, and mixed equally, then the powder was pressed in a cylindrical sample mold to obtain testing samples, the size of which was φ3mm*3mm. The melting temperature of the slag was determined using a melting temperature measurer (model CQKJ-II). The temperature of the sample was directly measured by PtRh30-PtRh6 thermocouple. The accuracy of sample melting temperature was about ±3K. The temperature, at which the height of the sample is reduced to one half of its original height, is defined as hemisphere temperature; at a quarter of its original height is flowing temperature [3].

The main chemical composition of the pig iron pieces taken from Shougang Shuicheng Iron and Steel (Grope) Co. Limited is shown in table 3. Based on the mass of the iron about 210 grams, the high calcium slag powder with 2.60% sulphur was weighed according to a 12.0% mass ratio of iron (see table 4), placed in the graphite crucible with an inner size φ40mm*65mm under the iron pieces, and covered with a graphite cap; then, put the crucible into the well-type resistance furnace, heated up to 1773K and holding for 30 min, and lowered to the controlled temperature in 30 min as shown in the figure 1. The graphite crucible was taken out of the heating furnace according to the selected holding time 120 min or 180 min and poured out the hot metal and slag to cool rapidly. The samples were taken to analyze sulfur content.

Table 1. Main chemical composition of the BF slag (%Wt).

FeO	T.Fe	CaO	MgO	SiO_2	Al_2O_3	TiO_2	S	Ig
1.79	1.37	37.95	9.12	29.4	11.87	6.13	0.823	1.59

Table 2. Main chemical composition of the high calcium slag (%Wt).

CaO	Al_2O_3	SiO_2	CaF_2	FeO	MgO	TiO_2	S
47.0	20.0	13.48	10.0	0.82	4.18	2.75	0.38

Table 3. Main chemical composition of the pig iron (%Wt).

C	Si	Mn	Ti	P	S
4.87	0.18	0.36	0.24	0.087	0.026

3. Results and discussion

The analyzed sulfur contents in the hot metals and the slags are fully summarized in Table 4. The sulfur content in the iron decreased with temperature increase or holding time extension. But the sulfur content in the slag increased with temperature increase or holding time extension.

3.1. Hemisphere temperature and flowing temperature

The melting temperature of the high calcium slag was determined as 1606K for hemisphere temperature and 1613K for flowing temperature. The hemisphere temperature and the flowing temperature of the high calcium slag with 2.60% sulphur was determined as 1647K and 1670K respectively. The hemisphere temperature raises 41K with the sulphur content increased from 0.38% to 2.60%, and the flowing temperature raises 57K. It is evident that the melting temperature of the high calcium slag raises with the sulphur content increase.

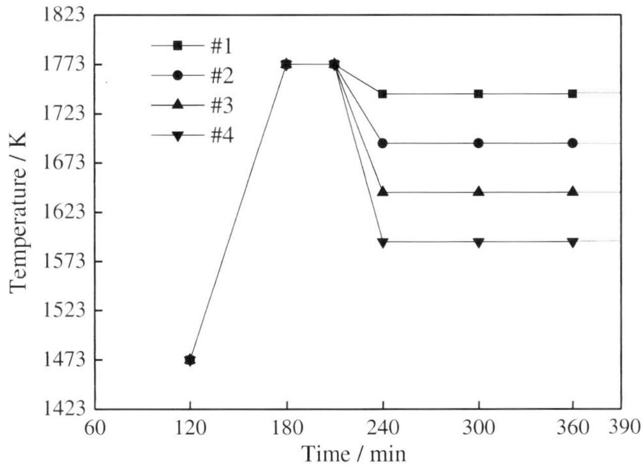

Figure 1. Heating and holding time of samples versus temperature.

Table 4. Sulfur content in samples and initial masses of both iron and slag.

Sample	1	2	3	4	5	6	7	8
Iron mass/g	206.19	225.24	208.63	204.24	201.95	204.22	240.32	225.24
Slag mass/g	24.74	27.03	25.04	24.51	24.23	24.51	28.84	27.03
holding temperature/K	1743	1693	1643	1593	1743	1693	1643	1593
holding time/min	120	120	120	120	180	180	180	180
sulfur in iron/ppm	90	130	190	220	60	80	130	170
sulfur in slag/%Wt	2.71	2.71	2.62	2.54	2.73	2.64	2.59	2.56

3.2. Effect of temperature on the desulfurization degree

When the holding time of the hot metal and slag at controlled temperature was 120 min in the furnace, the desulfurization degree of the high calcium slag increased correspondingly with temperature increase (see figure 2). Similarly, the desulfurization degree of the high calcium slag increased correspondingly with temperature increase in the range 1593-1743K when the holding time of the hot metal and slag at controlled temperature was extended from 120 min to 180 min in the furnace(see figure 2). Moreover, the desulfurization degrees of the samples increased correspondingly with the holding time extended from 120 min to 180 min at the same temperature (see figure 2). The higher desulfurization degree of the high calcium slag may be mainly correlated with the higher basicity or higher CaF_2 content [1,5].

3.3. Effect of temperature on sulphur partition ratio

When the holding time of the hot metal and slag at controlled temperature was 120 min in the furnace, the sulfur partition ratio of the sample increased correspondingly with temperature increase (see figure 3). Similarly, the sulfur partition ratio of the sample increased correspondingly with temperature increase when the holding time of the hot metal and slag at controlled temperature was extended from 120 min to 180 min in the furnace(see figure 3). Moreover, the sulfur partition ratio of the samples increased correspondingly with the holding time extended from 120 min to 180 min at the same temperature, and the increasing range of the sulfur partition ratio increased correspondingly with temperature increase in the range 1593-1743K (see figure 3). The higher sulfur partition ratio of the sample may be related to the multicomponent characteristics of the slag such as lower flowing temperature and better liquidity [1,3-5,8-10], and mainly correlated with the higher basicity and higher CaF_2 content[1,5-9].

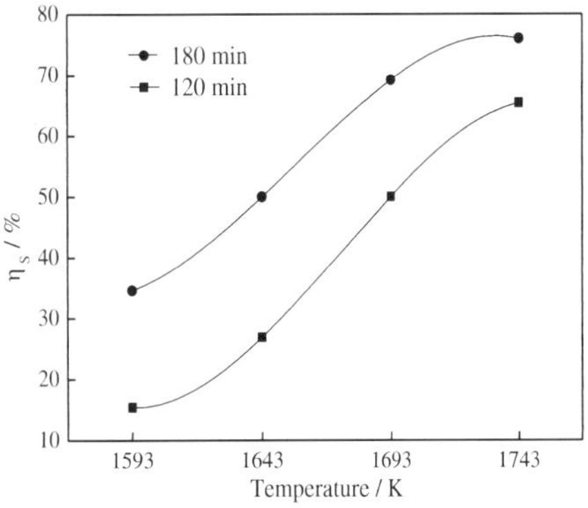

Figure 2. Effects of both temperature and holding time on the desul furization degree of the samples.

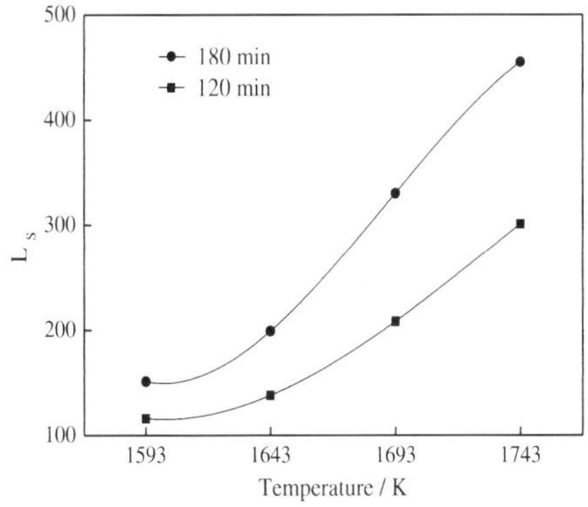

Figure 3. Effects of both temperature and holding time on the sulfur partition ratio of the samples.

4. Conclusions

The melting temperature of the high calcium slag raises with the sulphur content increase, the flowing temperature raises 57K from 1613K with the sulphur content increased from 0.38% to 2.60%.

The desulfurization degree of the high calcium slag increased correspondingly with both the temperature increase in the range 1593-1743K and the holding time extension from 120 min to 180 min. The sulfur partition ratio of the sample increased correspondingly with both the temperature increase in the range 1593-1743K and the holding time extension from 120 min to 180 min. The higher sulfur partition ratio of the sample may be related to the multicomponent characteristics of the slag, and mainly correlated with the higher basicity and higher CaF_2 conten.

5. References

[1] Xin T and Chushao X 1995 ISIJ International Sulphur *Distribution between CaO-SiO_2-TiO_2-Al_2O_3-MgO Slag and Carbon-saturated Iron at 1773K* chapter 35 pp 367-371

[2] Fengman S, Guo W, Yu S and Yibo H 2012 Northeastern University (Natural Science) *Mechanism analysis and Research on Desulphurization of molten iron with aluminium and calcium* chapter 12 pp 1737-1740

[3] Jinzhu Z, Li-li S and Wanzhong A 2010 Journal of iron and steel reseach *Melting Property of High-Alumina and Low-Titania BF Slag* chapter 22 pp 16-19

[4] Wei F, JinZhu Z, Wanzhang A and Jun H 2015 Advanced Materials Research *Effect of Al_2O_3 and CaF_2 on melting temperature of high calcium ladle desulfurization slag* chapter1094 pp 325-328

[5] Zhongqiang S, Maofa J, Lianke L and Yinchang C 2004 Journal of Iron And Steel Research *Determination of Sulfide Capacity, Sulfur Distribution Ratio and Desulfurization Percentage in Ladle Furnace Refining Process* chapter 16 pp 23-26

[6] Xia L, Yuehua D and Bangfu H 2012 Journal of Iron And Steel Research *A Thermodynamic Model for Calculation of Sulfur Distribution Ratio between CaO-SiO_2-Al_2O_3-MgO-TiO_2 BF Slag and Hot Metal* chapter 24 pp 19-23

[7] Xiaodong M, Mao C, Haifa X, Jinming Z, Geoff W and Baojun Z 2016 ISIJ International *Sulphide Capacity of CaO–SiO_2-Al_2O_3–MgO System Relevant to Low MgO Blast Furnace Slags* chapter 56 pp 2126–2131

[8] Xuemin Y, Jinsha J, Rucai D, Chengbin S and Hanjie G 2009 ISIJ International *A Thermodynamic Model for Calculating Sulphur Distribution Ratio between CaO–SiO_2–MgO–Al_2O_3 Ironmaking*

Slags and Carbon Saturated Hot Metal Based on the Ion and Molecule Coexistence Theory chapter 49 pp 1828–1837

[9] Zhongshan R, Xiaojun H and Kuochih C 2013 Journal of Iron And Steel *Research International Calculation and analysis of sulfide capacities for* CaO-Al_2O_3- SiO_2-MgO-TiO_2 *slags* chapter 20 pp 21-25

[10] Jinzhu Z ,LinW, Minghua L and Bineng Y 2011 Key Engineering Material *Study on viscosity of titania bearing blast furnace slag* chapter 469 pp 1742-1745

Acknowledgments

The authors gratefully acknowledge the financial support provided by National Natural Science Foundation of China (Grant No. 51274074).

IOP Conf. Series: Materials Science and Engineering **230** (2017) 012025 doi:10.1088/1757-899X/230/1/012025

Cellular Automaton Simulation for the Effects of Uneven Distribution of Dislocation Density and Small-Sized Precipitated Particles on Dynamic Recrystallization

X J Guan[1, 2, 3] and B J Yu[1, 2]

[1] Key Laboratory for Liquid-Solid Structural Evolution and Processing of Materials (Ministry of Education), xjguan@sdu.edu.cn, Jinan, 250061, P R China
[2] School of Material Science and Engineering, Shandong University, Jinan, 250061, P R China

Abstract. Based on the fundamental metallurgical principles, a modified cellular automaton model which considers the influences of uneven distribution of dislocation density and small-sized precipitated particles was proposed to simulate the dynamic recrystallization (DRX) performances of steels with the three kinds of equivalent particle`s size. The results show that the model can truly simulate the nucleation order and multi-round feature of DRX as well as the law of small-sized precipitated particle`s retarding DRX; The finer are the particles, the stronger gets the action of their pinning grain boundary.

1. Introduction

Dynamic recrystallization (DRX) plays a significant role in thermo-mechanical processing due to its improving mechanical properties by refining grain[1]. In the process of DRX, grain boundary migration decides the mean grain size of microstructure and is retarded by micro-alloying elements in metallic alloys as pinning effect of precipitated particles (Zener pinning)[2].

Many researches have been focused on DRX in recent decades. In these researches, if only the kinetics is of interest, the Johnson-Mehl-Avrami-Kolmogorov (JMAK) approach[3,4] can present a reasonable description. If information on microstructure development during DRX is simultaneously required, the models with adequate temporal and spatial resolution are needed. Therefore, numerical models, such as Q state Monte Carlo (MC)[5,6], phase field (PF)[7] and cellular automaton (CA)[8-15] have been proposed to simulate DRX. Compared with MC and PF method, the CA method is used more because of its more flexible and adaptable to temporal and spatial scale. Goetz and Seetharaman[8] first applied CA method to simulate the DRX features such as the necklace type of microstructures at high strain rates and its kinetics obeying the JMAK equation. Then, many researchers have been focused on DRX simulation based on the CA method[9-15]. However, the investigation with regard to the effects of precipitated particles on DRX is little. This work thoroughly simulates the DRX microstructure evolution in three steels with the different mean sizes and same volume fraction of precipitated particles by using a new CA model, respectively, and discusses the related effects.

2. The CA model for DRX

For the sake of simplifying the model, two assumptions are proposed as follows.

(1) The precipitated circular particles randomly distribute in the deformed matrix, and their shape and size do not change with deformation.

IOP Conf. Series: Materials Science and Engineering **230** (2017) 012025 doi:10.1088/1757-899X/230/1/012025

(2) For all primary grains and newly formed DRX grains, the dislocation density, ρ_0, is uniform and identical, and is set as $10^9/m^2$.

Except for the two equations for dislocation density distribution and R-grain growth kinetics, the rest content of the CA-model used in this paper are same as Reference [12].

2.1. Dislocation density distribution

If the effects of grain size and its boundary on dislocation density are considered, the mesoscopic dislocation density increment of a cell (i, j), $\Delta\rho_{i,j}$, can be expressed as[13]:

$$\Delta\rho_{i,j} = C\cdot\Delta\rho\left(\frac{d_{\mathrm{m}}}{d_{i,j}}\right)^2\cdot\left[\frac{1}{k}\sum_{n=1}^{k}\left(L_n / d_{\mathrm{CA}}\right)^{-2}\right] \tag{1}$$

where C is a constant to keep the total dislocation density of the simulated region to be invariable; $\Delta\rho$ is the average dislocation density increment of the simulating region; d_{m} is the average grain size of the simulating region; $d_{i,j}$ is the size of grain cell (i, j) exists in; k is the number of grain boundaries of surrounded cell (i, j); L_n is the distance between cell (i, j) and its nth grain boundary; d_{CA} is lattice constant.

2.2. R-grain growth kinetics

After nucleation, the newly formed R-grains will be driven to the deformed matrix of surrounded them by the difference of dislocation density. The velocity of grain boundary moving relates to driving force. Except for the changes of stored energy and grain boundary energy associated with the growth of R-grain, the effects of precipitated particle`s pinning grain boundary on the driving force should also be considered in the material with precipitates. So, the driving force, $\Delta F_{i,j}$, can be expressed as[13]:

$$\Delta F_{i,j} = \tau\Delta\rho_{i,j} - 2\frac{\gamma_{i,j}}{r_{i,j}} - p_z = \tau\Delta\rho_{i,j} - 2\frac{\gamma_{i,j}}{r_{i,j}} - \frac{3\gamma_{\mathrm{m}} f_{\mathrm{sec}}}{2r_{\mathrm{sec}}} \tag{2}$$

where τ is the dislocation line energy; $\rho_{i,j}$, represents the dislocation density difference between the R-grain and its surrounding matrix; $\gamma_{i,j}$ and $r_{i,j}$ are the boundary energy and radius of the R-grain, respectively; γ_{m} is the energy of the grain boundary with a high-angle (taken as 15°); f_{sec} and r_{sec} are the volume fraction and equivalent radius of the precipitated particles, respectively. It is obvious that the third term at right hand of Equation (2) reflects the pinning effect.

2.3. Simulation procedure

A 2-D square lattice consisting of 1500×1500 cells, corresponding to an area of 300 μm ×300 μm in the real sample were employed in this study. The presented simulation results were implemented in a CA code written in the Visual Fortran 6 programming language. The initial microstructure with the average grain diameter of some 48 μm was set by a normal grain growth program[14], and the maximum grain orientation number is 180.

A series of simulations for the DRX behaviors of an HPS485wf steel deformed at T=1373 K, $\dot{\varepsilon}$=0.1 s^{-1} and ε=0.8 were carried out. And the steel`s experimental data obtained at the same deforming conditions[15] were compared with the simulated one. The states of precipitated particles for the simulations are presented in Table 1, and the corresponding materials are marked as the serial numbers. The other parameters in the simulations are listed in Table 2.

Table 1. States of precipitated particles.

Serial number	Volume fraction, f_{sec} (%)	Equivalent radius, r_{sec} (nm)
No. 1	0	0
No. 2	1.5	200
No. 3	1.5	100

Table 2. Parameters of material used in the simulations[12].

T_m (K)	k_B (JK^{-1})	Q_{act} ($kJmol^{-1}$)	b (nm)	M (MPa)	δD_b (m^3s^{-1})	Q_b ($kJmol^{-1}$)	γ_m (Jm^{-2})
1810	1.38×10^{-23}	396.84	0.256	8.1×10^4	7.5×10^{-14}	159	0.76

3. Results and discussion

It is shown in figure 1 that with increasing strain, R-grains continuously nucleate at the grain boundaries of small-sized grains, then at that of large-sized grains, and gradually grow into surrounding deformed matrix to replace the initial microstructures. Nine circles of DRX occur at the deformation end (ε=0.8). It is seen that the improved CA model more truly simulates the DRX performances compared with that of conventional CA simulations.

The simulated and experimental stress-strain curves for No.1, No.2 and No.3 are compared in figure 2(a). The simulated curves show the obvious DRX characteristic with single peak pattern and the values of steady stress corresponding to the some 75% of peak stress ones, which is same as the experimental one. Compared with No.1, the simulated stress-strain curves of No.2 and No.3 are better agreement with the experimental results. It is seen in figure 2(b) that there is a common incubation stage of DRX in the materials before the strain reaches 0.1 and their volume fractions of DRX increase with raising strain, which is also same as the experiment. In addition, the DRX rate of No.1 is faster than that of No.2, and even faster than that of No.3, which is due to the effects of precipitated particles on resisting grain boundary migration. When the volume fraction of precipitated particles is fixed, the smaller is the particle size and the much intensive is their resistance effect. Therefore, the model of considered the small-sized particle`s effect is more necessary to the DRX simulation of micro-alloying steel, such as the HPS485wf steel.

In addition, compared with the idea Avrami exponent of JMAK equation, n_{Avrami}, the values from simulation results also are reasonable, and the effect of precipitated particles on it is little.

In summary, the CA model and its program of this paper not only better simulate the nucleation order and multi-round feature of DRX, but also the more truly reflect the effect of small-sized precipitated particles on DRX performances in the steel.

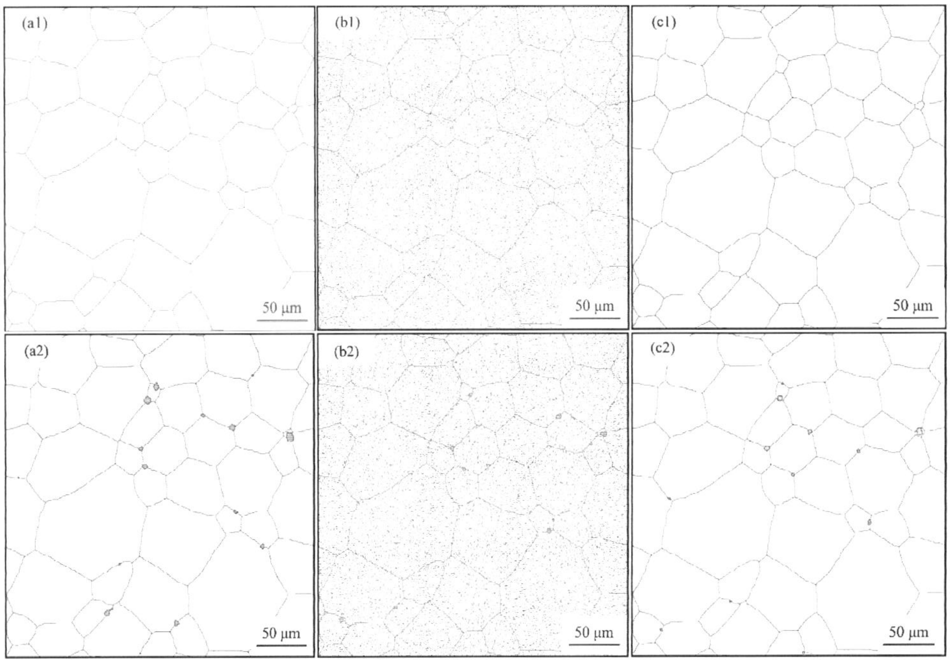

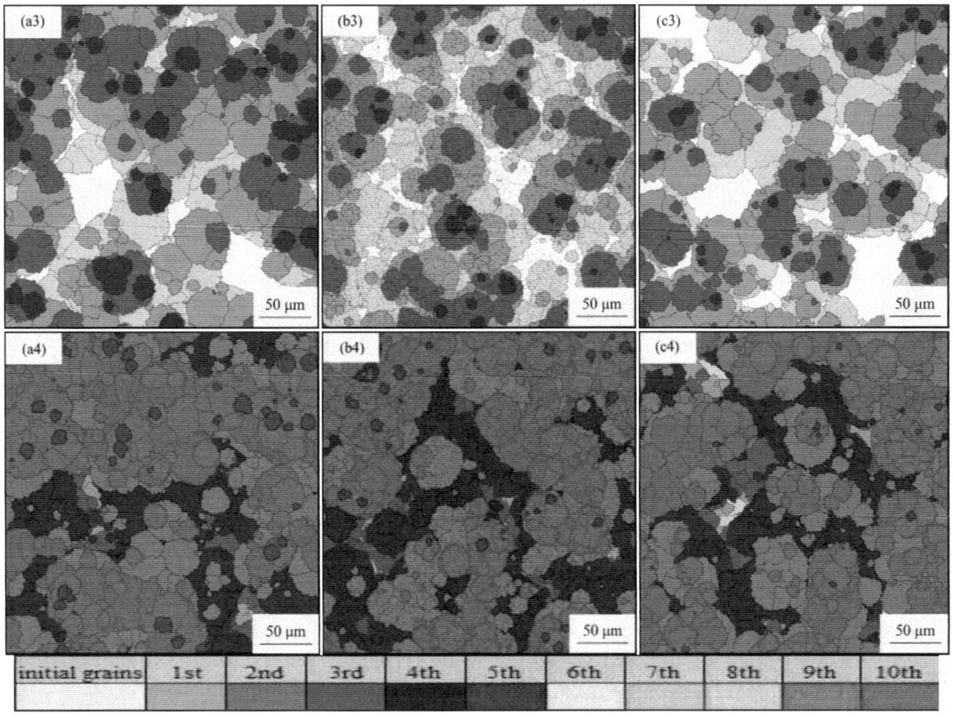

Figure 1. Microstructure evolutions of dynamic recrystallization for the materials with different precipitated particles. (a1~a4) r_{sec}=0 nm, f_{sec}=0; (b1~b4) r_{sec}=200 nm, f_{sec}=0.015; (c1~c4) r_{sec}=100 nm, f_{sec}= 0.015; (a1, b1, c1) ε=0; (a2, b2, c2) ε=0.1; (a3, b3, c3) ε=0.4; (a4, b4, c4) ε=0.8.

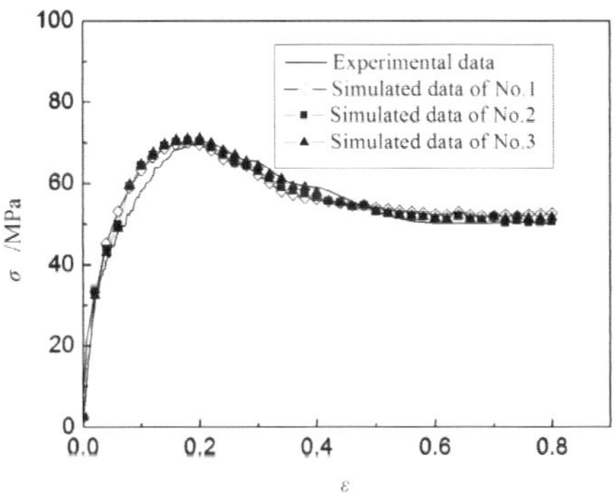

(a)

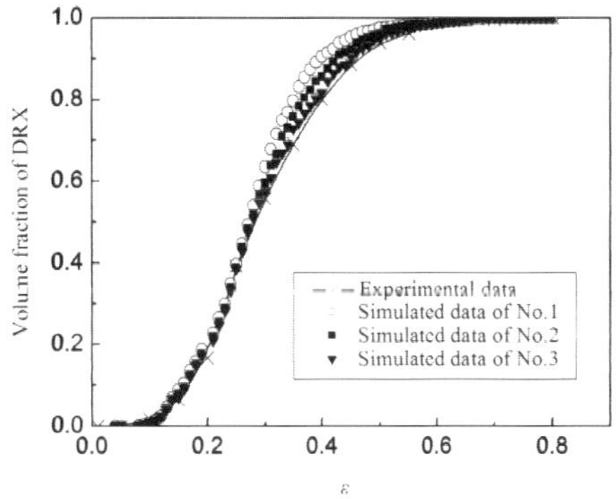

(b)

Figure 2. Stress-strain curves and kinetics of dynamic recrystallization for the materials with different precipitated particles (a) Stress-strain curves (b) kinetics.

4. Conclusions

A modified CA model is successfully used for simulating the two effects of finer precipitated particles on DRX, that is, the dislocation accumulating and the grain boundary migration resisting, which expresses the more real nucleation order and multi-round feature of DRX as well as the correct law of small-sized particle`s pinning R-grain boundary. The resisting action strengthens with the decrease of the small-sized particle`s average size at the same volume fraction, but do not affect the nucleation behaviors of DRX.

5. References

[1] Montheillet F, Lurdos O and Damamme G 2009 A grain scale approach for modeling steady-state discontinuous dynamic recrystallization *Acta. Mater.* **57** 1602-12

[2] Semiatin S L, Weaver D S, Fagin P N, Glavicic M G, Goetz R L, Frey N D, Kramb R C and. Anthony M M 2004 Deformation and recrystallization behavior during hot working of a coarse-grain nickel-based superalloy ingot material *Metall. Mater. Trans.* A **35** 679-93

[3] Avrami M 1939 Kinetics of phase change I: general theory, *J. Chem. Phys.* **7** 1103-12

[4] Jonas J J, Quelennec X and Jiang L 2009 The Avrami kinetics of dynamic recrystallization, *Acta. Mater.* **57** 2748-56

[5] Rollett A D, Luton M J and Srolovitz D J 1992 Microstructural simulation of dynamic recrystallization *Acta. Metall. Mater.* **40** 43-55

[6] Peczak P 1995 A Monte Carlo study of influence of deformation temperature on dynamic recrystallization *Acta. Metall. Mater.* **43** 1279-91

[7] Takaki T, Hisakuni Y and Hirouchi T 2009 Multi-phase-field simulations for dynamic recrystallization *Comput. Mater. Sci.* **45** 881-88

[8] Goetz R L and Seetharaman V 1998 Modeling dynamic recrystallization using cellular automata *Scrip. Mater.* **38** 405-13

[9] Ding R and Guo Z X 2001 Coupled quantitative simulation of microstructural evolution and plastic flow during dynamic recrystallization *Acta. Mater.* **49** 3163-75

[10] Lee H W and Im YT 2010 Cellular automata modeling of grain coarsening and refinement during the dynamic recrystallization of pure copper *Mater. Trans.* **51** 1614-20

[11] Guan X J and Yu B J 2013 Simulation of the dynamic recrystallization in HPS485wf steel deformed at different conditions *J. Xuzhou Inst. Techno.* **28** 1-4

[12] Guan X J, Yu B J, Wang L J, Zhou L J and FU J 2014 Simulation of dynamic recrystallization for advanced steel HPS485wf using cellular automata *Trans. Mater. Heat Treat.* **35** 218-24

[13] Yu B J 2012 Modeling and simulation of dynamic recrystallization behavior for material with second phase particles using cellular automata method *Doctoral Dissertation* (Jinan: Shandong University) pp 69-72

[14] Ma X F, Guan X J, Liu Y T, Shen X M, Wang L J, Song S T and Zeng Q K 2008 Cellular automaton model for grain growth based on modified transition rule *Chin. J. Nonferrous Met.* **189** 138-44

[15] Wang L J, Guan X J, Zhao J, Yu B J, Zeng Q K, Sun Q and Liu Q Q 2010 Dynamic recrystallization and mathematic model of HPS482wf steel *Trans. Mater. Heat Treat.* **31** 147-51

Acknowledgements

We sincerely acknowledged the support from the Specialized Research Fund for Doctoral Program of Higher Education of China (No.200804220021).

IOP Conf. Series: Materials Science and Engineering **230** (2017) 012026 doi:10.1088/1757-899X/230/1/012026

Optimization Method of a Low Cost, High Performance Ceramic Proppant by Orthogonal Experimental Design

Y Zhou[1], Y M Tian[2], K Y Wang[1], G Li[1], X W Zou[1], Y S Chai[1]

[1]Collaborative Innovation Center of Shanxi Key Basic Materials, School of Materials Science and Engineering, Taiyuan University of Science and Technology, Taiyuan, Shanxi, PR China

[2]Shanxi Engineering Vocational College, Taiyuan, Shanxi, PR China

E-mail: zhouyi478w@163.com

Abstract. This study focused on optimization method of a ceramic proppant material with both low cost and high performance that met the requirements of Chinese Petroleum and Gas Industry Standard (SY/T 5108-2006). The orthogonal experimental design of $L_9(3^4)$ was employed to study the significance sequence of three factors, including weight ratio of white clay to bauxite, dolomite content and sintering temperature. For the crush resistance, both the range analysis and variance analysis reflected the optimally experimental condition was weight ratio of white clay to bauxite=3/7, dolomite content=3 wt.%, temperature=1350℃. For the bulk density, the most important factor was the sintering temperature, followed by the dolomite content, and then the ratio of white clay to bauxite.

1. Introduction

Proppant material is a key factor widely employed in the hydraulic fracturing process which is an effective method for exploitation of unconventional oil and gas resources.[1] Choosing the suitable proppant during the mining operation can improve service years of the gas well, increase its production rate and reduce the mining cost.[2,3] In the early mining process, the quartz sands was selected as the proppant. Although the quartz sand is of low cost and easy to obtain, it is usually limited by the unfavorable factors, such as low breakage resistance, low level of sphericity, poor diversion ability, etc. Ceramic proppant material, by contrast, possesses high breakage resistance, good diversion ability, high level of sphericity, and so on.[4] At present, manufacturing ceramic proppant with high performances mainly depends on the high grade bauxite as raw material. Consequently, the following two questions would be caused. First, reserves of the high grade bauxite resource would be drastically reduced. Second, adoptively high grade bauxite with high cost would enhance the cost of mining the unconventional oil and gas.[5,6] Selecting raw materials with low cost to replace the high grade bauxite is deemed as future trends for both the proppant industry and the unconventional oil and gas industry. At present, the methods of additives, such as adding iron ore, manganese ore, etc. are generally adopted to improve the performance of the ceramic proppants.[7,8] However, limited research on the low cost and low grade bauxite with dolomite ore as additive is available.

For the present study, the low-cost 3rd grade bauxite and white clay were chosen as the starting materials and dolomite as the additive to prepare the ceramic proppants. The mechanical properties of the proppants were thoroughly investigated using the orthogonal experiment design.

IOP Conf. Series: Materials Science and Engineering **230** (2017) 012026 doi:10.1088/1757-899X/230/1/012026

2. Experimental

2.1. Materials preparation
The low-cost raw materials of 3rd-grade bauxite and white clay, both of which are widely produced from Yangquan in Shanxi Province, were ground to an average size of 0.05 mm (300 mesh) and then calcined at 300℃ to remove the adsorbed water and other volatile impurities. The chemical compositions of the raw materials were listed in Table 1. The mixture of bauxite and white clay was moved to a powerful ball machine (R02, Eirich Co. Ltd., Germany) to form spherical green bodies. Then they were sieved to remain the spheres with 0.5-0.9 mm in size. These spheres were placed single layer in a rectangular corundum crucible and then sintered in a box-type electrical furnace (KF1700, Nanjing Boyuntong Instrument Technology Co. Ltd., China) in air atmosphere at a temperature range of 1250℃ and 1350℃ for 2 h with the heating rate of 5℃ /min and a furnace cooling. The as-sintered proppants were passed through a set of 20/40-mesh sieves (aperture size of 0.85-0.425 mm) to finally obtain the particular ceramic proppants.

2.2. Measurements
The bulk density and crush ratio under a constant pressure of 52 MPa were measured according to the Chinese Petroleum and Gas Industry Standard (SY/T 5108-2006). The bulk density is calculated from the formula

$$\rho_b = \frac{M}{V_s}. \qquad (1)$$

where M is the weight of the ceramic proppant [g] and V_s is the naturally stacked volume [cm^3]. The crush ratio of the proppant is calculated by the formula

$$\eta = \frac{w_c}{w_0} \times 100\%. \qquad (2)$$

where w_c is the weight of crushed specimens [g] after testing and w_0 is the weight of initial specimens [g]. The crush resistance of the proppant sample is expressed by

$$R_c = 1 - \eta = (1 - \frac{w_c}{w_0}) \times 100\%. \qquad (3)$$

Table 1. The chemical compositions of raw materials. [wt.%]

components	Al_2O_3	SiO_2	Fe_2O_3	CaO	MgO	LOI
Bauxite	59.7	13.9	6.5	0.15	0.24	19.51
White clay	29.6	40.6	5.7	--	--	24.1
Dolomite	0.23	0.46	0.2	34.7	19.24	45.17

3. Orthogonal experimental design
This paper aims to study the effect of the following three factors: (A) white clay/bauxite ratio (W/B), (B) dolomite content, (C) sintering temperature on the crush resistance and bulk density of the ceramic proppants. Since the full factorial design requires 27 groups of experiments, using the orthogonal experimental method can effectively reduce the experiment number to 9 groups. Each factor has three levels (Table 2). A standard orthogonal experimental design table of $L_9(3^4)$ was adopted. Since there are only three factors studied, a blank column was set as the error column (Table 3).

Table 2. The experiment variables design.

Levels	Factors		
	W/B (A)	Dolomite content (%) (B)	Temperature (℃) (C)
1	1/9	1	1250
2	2/8	2	1300
3	3/7	3	1350

Table 3 The orthogonal experimental design table.

Exp. No.	A	B	C	Blank(error)	Crush resistance (%)	Bulk density (g/cm^3)
1	1/9	1	1250	1	86.3	1.29
2	1/9	2	1300	2	89.9	1.32
3	1/9	3	1350	3	90.6	1.47
4	2/8	1	1300	3	88.7	1.35
5	2/8	2	1350	1	89.8	1.44
6	2/8	3	1250	2	88.2	1.33
7	3/7	1	1350	2	90.7	1.38
8	3/7	2	1250	3	91.2	1.41
9	3/7	3	1300	1	92.7	1.36

4. Results and discussion

4.1. Range analysis of the crush resistance

Since the optimal ceramic proppant sample or the best experimental condition could not be obtained directly from the experimental data in Table 3, a further orthogonal analysis was required. In the range analysis of orthogonal experiment (Table 4), I_j, II_j, III_j, are defined as the sum of the values of three levels in the specific factor *j* (*j*=A, B, C, D). R_j represents the range between the maximum and minimum value of I_j, II_j, III_j, in a specific factor *j*. The dominant degree of each factor can be evaluated by comparing R_j. As seen from Table 4, the influence of the three factors on the crush resistance follows the sequence: A>B>C. This result indicates the most important factor for the crush resistance is the white clay/bauxite ratio (W/B), followed by the dolomite content, and then the sintering temperature. The optimal sample was obtained as $A_3B_3C_3$, which means the best experimental condition is: W/B=3/7, dolomite content=3 wt.%, temperature=1350℃.

Table 4 The range analysis on crush resistance.

	A	B	C	error
I_j	266.8	265.7	265.7	268.8
II_j	266.7	270.9	271.1	268.8
III_j	274.6	271.5	271.3	270.5
R_j	7.9	5.8	5.6	1.7

4.2. Variance analysis of the crush resistance

The variance analysis was introduced to investigate whether the effect of factors on the crush resistance was significant or could be ignored. The sum of squares (SS), degree of freedom (DF), F value (F) for all factors were shown in Table 5. The most important parameter is F value, which is obtained from the following equation:

$$F_j = \frac{SS_j / DF_j}{SS_e / DF_e} . \tag{4}$$

where SS_j and SS_e are the sums of squares in the *j* column and the error column, DF_j and DF_e are the degree of freedom in the *j* column and the error column. The *F* value is used to indicate the magnitudes of the effects of factors and the prominence of data can be verified by *F*–test. To a specific F_α, obtained from the distribution table of *F* value, if $F_j > F_\alpha$, the effect of factor *j* is prominent. On the other hand, if $F_j \leq F_\alpha$, the effect of factor *j* for the results is not prominent. The critical values of F_α

(α=0.01, 0.05, 0.1, 0.25) obtained from the distribution table of *F* value were also listed in Table 5. For example, as the *F* value of factor A (21) is greater than $F_{0.05}$ (19), the effect of weight ratio of white clay to bauxite on the crush resistance is prominent in α=0.05. Obviously effect of the three factors follows the sequence: A>B>C, as is consistent with the result of range analysis. It indicates that the most important factor of the crush resistance is the weight ratio of white clay to bauxite, followed by the dolomite content, and then the sintering temperature.

Table 5 The variance analysis on crush resistance.

	A	B	C	error
SS	13.68	6.76	6.7	0.65
DF	2	2	2	2
V	6.84	3.38	3.35	0.325
F_j	21	10.4	10.3	
$F_{0.01}$	99	99	99	
$F_{0.05}$	19	19	19	
$F_{0.1}$	9	9	9	
$F_{0.25}$	3	3	3	

4.3. Range analysis of the bulk density

Listed in the last column of Table 3 is the bulk density of the ceramic proppants. The range analysis on the bulk density is shown in Table 6. The influence of the three factors on the bulk density follows the sequence: C>B>A, according to the value of R_j. It means the most important factor for the bulk density is the sintering temperature, followed by the dolomite content, and then the ratio of white clay to 3rd bauxite (W/B). This may be resulted from the facts that enhancing the sintering temperature and increasing the content of dolomite could be easier to promote the generation of liquid phase, thus increasing the bulk density of the ceramic proppants.[9,10] Further studies on effect of several factors on the structure and mechanism of liquid phase generation have been in progress.

Table 6 The range analysis on bulk density.

	A	B	C	error
I_j	4.08	4.02	4.03	4.09
II_j	4.12	4.17	4.03	4.03
III_j	4.15	4.16	4.29	4.23
R_j	0.07	0.15	0.26	0.2

5. Conclusion

(1) Low-cost, high performance ceramic proppants made from locally produced 3rd-grade bauxite, white clay and the addition of dolomite were successfully prepared.

(2) The experimental condition of ceramic proppants with the best crush resistance was determined to be ratio of white clay to bauxite=3/7, dolomite content=3 wt.%, temperature=1350℃, by the orthogonal analyses.

(3) According to the range analysis of bulk density, the most important factor for the bulk density lies in the sintering temperature, followed by the dolomite content (%), and then the ratio of white clay to the 3rd bauxite.

(4) In order to satisfy the ceramic proppant industry standard, the optimally experimental condition is $A_3B_3C_3$ according to the orthogonal analyses.

6. References

[1] Lee D, Elsworth D, Yasuhara H, Weaver J, and Rickman R 2010 *J. Pet. Sci. Eng.* **74** 67

[2] Zhao J, Liu Z, and Li Y 2015 *Mater. Letters* **152** 72
[3] Wen Q, Zhang S, Wang L, Liu Y, and Li X *J. Pet. Sci. Eng.* **55** 221
[4] Yu S 1987 *Oil Drill. Prod. Tech.* **9** 35
[5] Luscher W, Hellmann J, Scheetz B, and Wilson B 2006 *Int. J. Appl. Ceram. Tec.* **3** 157
[6] Wray P 2008 *Am. Ceram. Soc. Bul.* **87** 26
[7] Gao H, You T, Wu H, and Lou Y 2006 *Panz. Sci. Tech. Info.* **10** 43
[8] Ma X, Yao X, Hua S, and Chen Y 2009 *J. Chin. Ceram. Soc.* **37** 280
[9] Niesyt M, and Psiuk B 2017 *Ceram. Int.* **43** 51
[10] Ma X, Tian Y, Zhou Y, Wang K, Chai Y., Li Z 2016 *Mater. Letters* **180** 127

Acknowledgement
The authors would like to thank the funding from Talents Training Project of Shanxi Graduate Joint Training Base (2016-37), Shanxi Province Science Foundation (201601D102019), Scientific and Technological Innovation Programs of Higher Education Institutions in Shanxi (2016-59), Fund for Shanxi Key Subjects Construction(FSKSC), and Doctoral Research Start-up Fund of Taiyuan University of Science and Technology (20142025).

IOP Conf. Series: Materials Science and Engineering **230** (2017) 012027 doi:10.1088/1757-899X/230/1/012027

Molten Slag Bath Reduction: Carbon-Thermal Reduction of Blast Furnace Dust in Molten Blast Furnace Slag

X Yin[1,2], C M Zhang[1,2], Y H Cai[1,3], C M Zhao[1,2], J Yang[1] and B Y Li[1]

[1] University of Science and Technology Liaoning, Anshan 114051, Liaoning, China
[2] Key Laboratory of Chemical Metallurgy Engineering of Liaoning Province, Anshan 114051, Liaoning, China
[3] Pangang Group Chengdu Iron & Steel Co., Ltd. Panzhihua 617000, Sichuan, China
E-mail: q6513057@aliyun.com

Abstract: By adding BF dust into molten BF slag, it was conducted to explore the feasibility of carbon-thermal reduction of BF dust in molten BF slag bath. The experimental results indicate that, BF dust can achieve self-reduction in the bath. It is convenient for the reduction products that are iron-based metal particles separating from slag. Besides, the reduction of BF dust is endothermic. When the mass ratio between BF slag and BF dust is 10:1, the temperature of BF slag after reduction can meet the requirement of dry processing treatment. According to the comparison by X-ray diffraction, the main mineral phases of BF slag are relatively stable before and after reduction. Undoubtedly, it is a novel method to cope with the metallurgical solid waste in molten BF slag bath.

1. Introduction

In China, the disposal of metallurgical solid waste has always been the hot topic in ironmaking and steelmaking industry. With the overproduction of steel in recent years, it is inevitable that the generation of metallurgical solid waste has been growing significantly as well. In ironmaking, the major solid waste are blast furnace (BF) slag and BF dust. For BF slag, it is estimated that the average production of BF slag is up to 346kg per tonne hot metal[1]. Currently, these slags are either water granulated or air cooled. In general, water granulation is the most common treatment used to produce glassy materials that can be widely used in concrete and cement[2-4]. But before water granulation, the high quality sensible heat of molten slag is always wasted in vain without reasonable utilization[5]. It has been estimated that the waste sensible heat of the molten BF slag at 1773K reached 18.5kg CE (kilogram of coal equivalent)[6]. But at present, not only the efficiency of heat recovery is still at a low level, but also the quality of this kind of energy is poor[5]. Therefore, from the view of energy saving, it is important and valuable to use the sensible heat in a high-efficiency way.

In addition to BF slag, BF dust is also a kind of by-product generated in the ironmaking process. It is estimated that the amount of BF dust is 15-50 kg per ton hot metal[7]. Since the various ingredients of BF dust are stemmed from the raw material added in BF, the dust contains high values of carbon and iron accompanied with harmful elements of Zn, K and Na. Based on this, there has been several successful attempts to separate and recycle valuable part of the BF dust, such as carbon and iron as well as zinc[8-10]. But the residual part that is regarded as the so-called useless waste is still a tough problem that has to be faced. Moreover, the disposal of BF dust on the landfill can obviously and inevitably cause environmental pollution as the leaching of some toxic metals (such as Pb and As)[11]. Thus, there needs more appropriate ways to recycle BF dust. Shaheer A. Mikhail et al.[12] reported that carbon-rich BF dust could be regarded as reductant, not only realizing self-reduction, but also making the iron oxides in basic oxygen furnace dust reduced at the same time. This research provides an idea that with the help of the high temperature platform, it is feasible to realize the total recycling of BF dust.

Therefore, the aim of this paper is to explore the feasibility of both recycling BF dust and the direct use of high-quality sensible heat of molten BF slag. For the sake of achieving the goal, the experiments were carried out with the addition of BF dust into molten BF slag bath. Both the products and mineral phases of slag before and after reduction were studied by several methods, such as X-ray diffraction (XRD) and scanning electron microscopy (SEM) coupled with energy dispersive spectrometer (EDS). Meanwhile, the heat use of molten BF bath was evaluated as well.

2. Experimental

2.1. Experimental materials

Based on practical composition from AnSteel, BF slag used in the experiment was synthesized by analytical reagents. BF dust was sampled from the spot in AnSteel. The composition of BF slag and BF dust was listed in table 1. The binary basicity of BF slag and BF dust were 1.11 and 1.04, respectively.

Table 1. The main chemical compositions of BF slag and BF dust (w_t/%).

Sample	CaO	SiO_2	Al_2O_3	MgO	C	Fe_2O_3	Fe_3O_4	NiO	CoO
BF slag	40	36	14	10	--	--	--	--	--
BF dust	4.93	8.92	3.41	1.65	25.2	29.6	7.53	14.18	4.58

2.2. Sample and method

In the present work, analytical reagents were adopted to synthesize BF slag. Firstly, all the ingredients, including SiO_2, CaO, MgO and Al_2O_3, were mixed homogeneously together as the designed mass ratio shown in table 1. Then the mixture was put into the graphite crucible and heated in a high temperature furnace for 30mins at 1500℃ in pure nitrogen, where the graphite crucible can be protected from oxidation. After synthesis, the BF slag was cooled in the air and crushed for further use.

BF dust were made into pellets mixed with 2% bentonite binder and appropriate water by a pelletizer. Then, the pellets were put in a drying oven for 2 hours at 120℃. The diameter of cooled pellets is about 10.6mm shown in figure 1.

Figure 1. The scale of BF dust pellets.

Generally, combing with the output ratio between BF slag and BF dust as well as the volume of corundum crucible, the usage of BF slag was confirmed as 80g, while the usage of BF dust was decided to be 8g, 12g and 16g at each experiment respectively.

Figure 2 shows the experimental apparatus used for thermal reduction of BF dust in molten BF slag bath. The experiments were conducted at 1500℃ with the temperature duration of 30mins, so that the solid BF slag could fully melt into liquid. Then, couples of BF dust pellets were added into the molten slag bath to make carbon-thermal reduction. 30mins later, the corundum crucible was taken out and cooled in the air. The magnet was used to separate the products after reduction. X-ray diffraction (XRD, Cu Kα, λ=1.54178Å, 40kV and 40mA, 2θ =10°-90°, stepsize=0.013, interval=5s) was used for measuring the mineral phases of slag, and scanning electron microscope (SEM, type: JSM6480LV) coupled with energy dispersive spectrometer (EDS) was employed to make microstructure observation of the product. Actually, though there was obvious reduction weight loss at each experiment, this work is mainly emphasized on the effect of recycling BF dust and the sensible heat use of molten BF slag. The kinetics of reduction of BF dust will be discussed later in another paper specially.

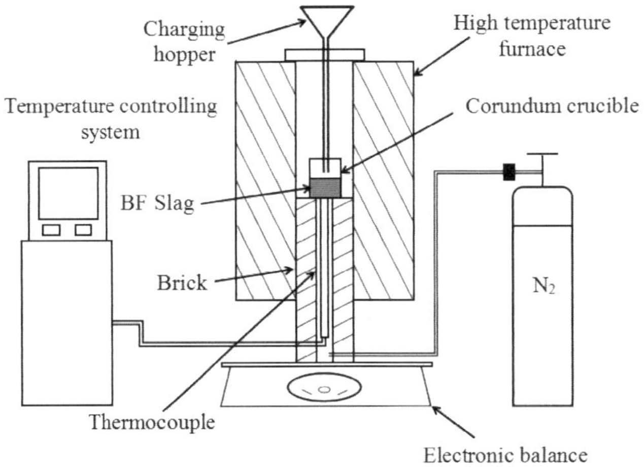

Figure 2. Experimental apparatus sketch at high temperature.

2.3. Thermodynamics of carbon-thermal reduction in BF slag bath

As shown in table 1, the mole radio of carbon and oxygen is calculated as 1:0.45, which means that the amount of carbon in BF dust is excess. figure 3 shows the standard Gibbs free energy changes of different reactions at high temperature[13]. It can be seen that the standard Gibbs free energy changes are all negative under the temperature from 1100℃ to 1500℃. And it is obvious that the carbon-thermal reduction reactions of metal oxides, especially NiO, CoO, FeO, Fe_3O_4, Fe_2O_3, are able to happen in molten slag bath.

Figure 3 also shows the diagram of oxygen potential of part of oxides in BF dust[13]. Clearly, carbon is prior to reduce the oxide of Ni and Co instead of Fe. Meanwhile, iron oxides are doomed to be reduced into Fe step by step at 1500℃. Thus, from the view of thermodynamics, molten BF slag bath reduction is able to achieve the recycling of BF dust.

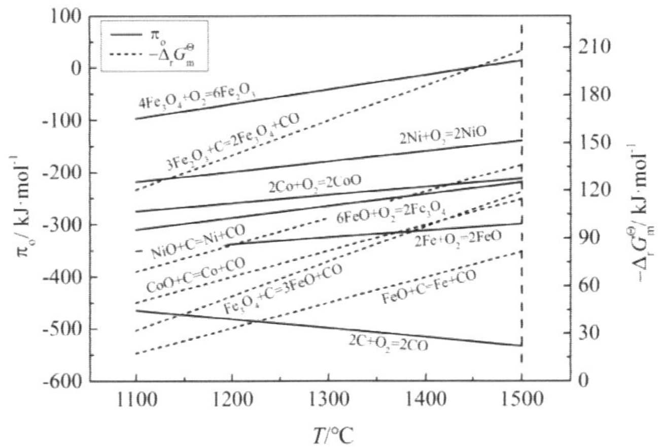

Figure 3. The oxygen potential of oxides and standard Gibbs free energy changes of reactions in slag bath at high temperatures.

2.4. Mechanism on carbon-thermal reduction of BF dust in slag bath

Since the BF dust pellets were added into the slag bath, solid-solid reactions firstly took place at the reaction point. On account of BF dust containing both carbon and metal oxides, the slag bath reduction of BF dust pellets is similar to the reduction of iron ore-coal composite pellets[14]. The direct reduction reactions can be seen in figure 3. Then, with CO gradually generated, the indirect reduction are simplified as equation (1):

$$MeO + CO = Me + CO_2 \tag{1}$$

Where Me represents metals reduced by CO. In addition to the direct and indirect reduction, the gasification of carbon is supposed to be taken into account for the excess carbon existed in the pellets. Boudouard reaction (gasification of carbon) is expressed as equation (2):

$$C + CO_2 = 2CO \tag{2}$$

In common, the direct reduction and the gasification of carbon are endothermic process, which means that the enthalpy change of the total reduction reactions is positive ($\Delta H > 0$). So, when the temperature is increasing, it is beneficial for the generation of CO. As the bond of direct reduction and the gasification of carbon, CO plays a key role on the rate of the whole reduction. Due to the limited contact area of solid, the rate of reduction reactions is relatively slow at initial period. However, when the partial pressure of CO rises to some value, the interface of reactions turns from solid-solid area to gas-solid area, causing the reduction rate growing fast. The reduction process of BF dust pellet in slag bath is shown in figure 4. Thus, based on the analysis above, the carbon-thermal reduction of BF dust in slag bath belongs to solid-solid phase reaction and gas-solid phase reaction.

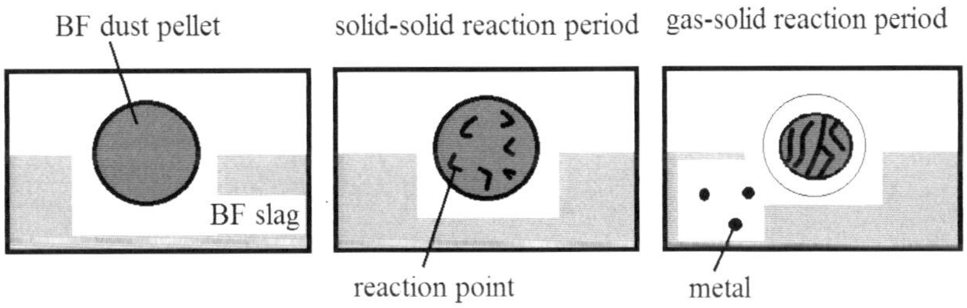

Figure 4. Reaction periods of BF dust reduction in BF slag bath.

3 Results and discussion

3.1. Metal particles separated from BF slag

When the pellets were added into the molten BF slag, the reduction reactions were violent in the crucible with weight loss significantly, and amounts of smoke generated accompanied with burst-out sound of pellets. Figure 5 shows that the metal particles gather and grow up with different sizes after reduction. And it is clear that most of the particles were attracted by magnet. It indicates that the metal particles are easy to be separated from the slag.

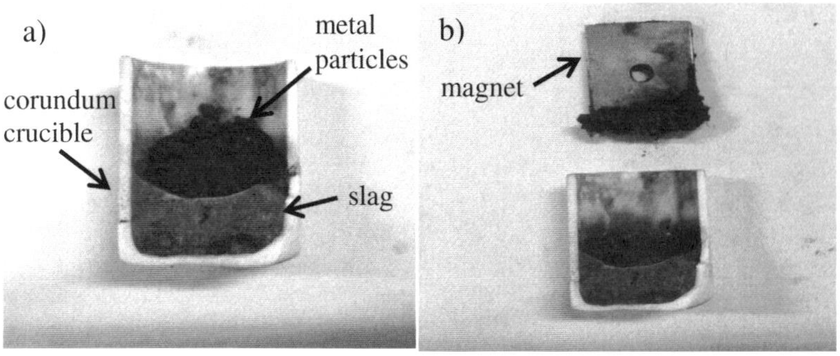

Figure 5. Metal particles after cooling (a) and their magnetic (b).

SEM images in figure 6 shows that the main elements of metal particle contains Fe, Ni, Co and Si. It means that most metal oxides in the BF dust are reduced in BF slag bath, and the reduced metals are enriched in the particle where finally iron-based alloy is generated. Moreover, the physical interface of metal particle with the surrounding slag is clear, which indicates that it is easy to separate the metal particles from the slag after cooling. In figure 6(ii), it shows that there are dose of carbon inside the crack. It is probably because during the process of reduction, the metal particles were graphitized by the excess carbon in BF dust. This kind of graphitization is common to be found in the carbon-saturated iron[15].

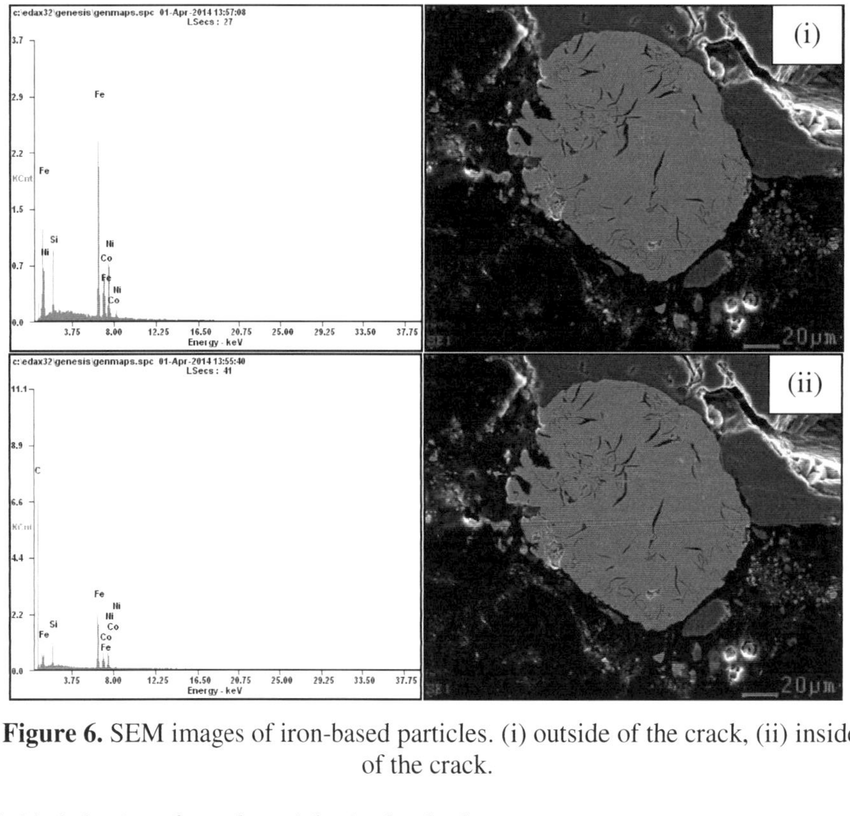

Figure 6. SEM images of iron-based particles. (i) outside of the crack, (ii) inside of the crack.

3.2. Unsinkable behavior of metal particles in slag bath

In this work, it is very interesting to find that most of metal particles are floated on the slag surface after cooling, and it is difficult to find metal particles sinking into the slag or on the bottom of the crucible as shown in figure 5. Though it is seemingly convenient for separating metal products from the surface of slag, still the unsinkable metal particles are disadvantageous for the following treatment of BF slag. Because when it is slagging, molten slag and the surface metal products are unavoidable to be mixed together to some extend, which brings unnecessary load to the normal treatment of BF slag. In usual, the density of iron is 7.9 $g\cdot cm^{-3}$, approximately 2.5 times higher than that of BF slag. From this viewpoint, the gathering metal particles are supposed to sink in the slag, even on the bottom of the crucible. But, for one thing, as the stirring capacity of CO is limited in the carbon-thermal reduction, it is hard to make amounts of metal droplets gather and grow up. For another thing, by adding the pellets into the bath, there needs energy supplement for the carbon-thermal reduction of BF dust. As a result, it is inevitable to impact the system temperature that directly affects the viscosity of slag. With the viscosity increasing, it leads to the fluidity of slag turning to bad, which is neither advantageous for the metal particles to sink nor good for the following treatment of BF slag. Therefore, considering the reasons above, it is an alternative way to properly improve the slag bath temperature coupled with the stirring operation, so that not only the metal products are able to sink on the bottom of slag, but also the fluidity of slag can be ensured to meet the requirements of following treatment.

3.3. Heat use of molten BF slag

In China, the slagging temperature of BF slag is usually above 1450℃, containing heat about 1770 MJ per ton hot metal[16]. Based on this, it can be calculated that 80g BF slag contains heat nearly 146.48kJ at 1500℃. In order to estimate the heat use of molten BF slag, the enthalpy changes are calculated by the equation of Kirchhoff as follow:

$$\Delta_r H_m^\Theta = \Delta_r H_{m,298}^\Theta + \int_{298}^{T} \Delta c_p \mathrm{d}T \tag{3}$$

Where, Δc_p: heat capacity from 298K to the certain temperature, $kJ\cdot mol^{-1}\cdot K^{-1}$; $\Delta_r H_{m,298}^\Theta$: standard enthalpy change of reactions at 298K, $kJ\cdot mol^{-1}$; $\Delta_r H_m^\Theta$: standard enthalpy change of reactions from 298K to the certain temperature, $kJ\cdot mol^{-1}$. Here, it is assumed that ferrous oxides are reduced to Fe by one step. According to the the chemical composition of BF dust in Table 1, the calculated results of $\Delta_r H_m^\Theta$ for each reduction reaction are shown in table 2.

Table 2. The standard enthalpy changes of main reduction reactions in BF dust.

No.	Reactions	$\Delta_r H_m^\Theta$ /($kJ\cdot mol^{-1}$)
1	$Fe_2O_3+3C=2Fe+3CO$	494.00
2	$Fe_3O_4+4C=3Fe+4CO$	676.38
3	$NiO+C=Ni+CO$	138.08
4	$CoO+C=Co+CO$	128.41

Clearly, the results of $\Delta_r H_m^\Theta$ are all positive in table 2, it means that the absorption of heat is apparent. Besides, due to not participating in the reduction, these ingredients such as SiO_2, CaO, MgO and Al_2O_3 in BF dust are heated in the slag bath, so the physical absorption of heat is taken into account as well. Therefore, the total heat consumption of pellets is calculated by the income and outcome of heat caused by physical absorption and chemical reactions. The calculated results are that when the reactions were complete, the system temperature was predicted to be 1358.3℃, 1288.1 ℃ and 1217.8 ℃ with the added pellets of 8g, 12g and 16g, respectively. The temperatures above were almost in accord with the results showing in the experimental controlling system. Particularly, when the amount of BF dust pellets is 8g, namely the mass ratio between BF slag and BF dust is 10:1, the temperature after carbon-thermal reduction is just fit for the dry processing at 1348 ℃ to 1361 ℃ for BF slag[17]. Hence, in order to meet relevant requirement of various slag treatment, the amounts ratio between BF dust and BF slag should correspondingly make some adjustment.

3.4. Mineral phases of BF slag

The mineral phases of BF slag before and after slag bath measured by XRD is shown in figure 7. It can be found that, there are no Fe diffraction peaks in the BF slag after slag bath. Based on this, it is in accordance with the mentioned results of few metal particles sinking in the BF slag. Moreover, compared with the original BF slag, the main mineral phases of BF slag after slag bath are basically the same. Besides, the mineral phase gehlenite ($Ca_2Al(AlSiO_7)$) disappeared in the BF slag, while grossularite ($Ca_3Al_2Si_3O_{12}$) appeared. This transformation is probably caused by the added ingredients of CaO and SiO_2 in BF dust as shown in equation 4.

$$2CaO \cdot Al_2O_3 \cdot SiO_2 + CaO + 2SiO_2 = 3CaO \cdot Al_2O_3 \cdot 3SiO_2 \tag{4}$$

Generally speaking, it is still common to find grossularite in $CaO\text{-}SiO_2\text{-}Al_2O_3$ system. The XRD results show that the main mineral phases of BF slag is not changed violently by the addition of BF dust. Also, it means that BF slag successfully contains the residual part of BF dust that can not be reduced during the slag bath. Undoubtedly, the method of BF slag bath is an alternative choice to be taken seriously in the field of energy saving and solid waste recycling in the future.

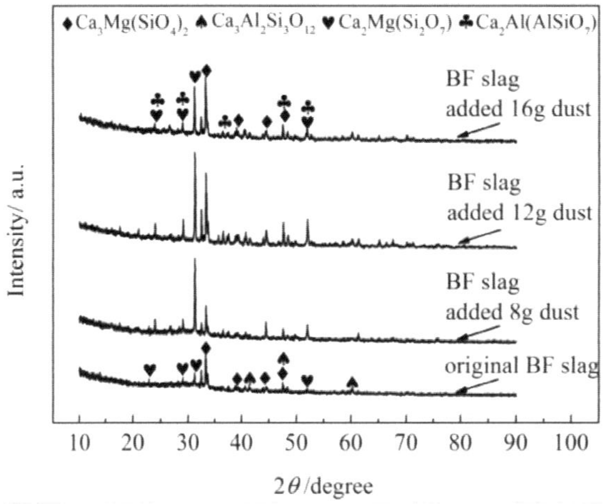

Figure 7. XRD images of BF slag before and after slag bath.

4. Conclusions

(1)With the sensible heat of slag bath at high temperature, self-reduction of BF dust is able to be realized. The reduction products are iron-based alloy particles, which remain on the surface of slag and are easy to be separated and recycled by magnet.
(2)The process of BF dust reduction in molten slag bath is endothermic. In this work, when the mass ratio between BF slag and BF dust is 10:1, the temperature of BF slag after reduction is able to meet the requirement of dry processing treatment.
(3)The residual part of BF dust after reduction is fully dissolved in the BF slag, and the main mineral phases of BF slag are relatively stable before and after reduction. Meanwhile, XRD results show that there are not metal particles in the BF slag after reduction.

5. References

[1] Zhang Z S, Lian F, Liao H Q, Xu L H, Li X Y and Zhao H N 2011 *Adv. Mater. Res.* 299 206-10
[2] Özbay E, Erdemir M, Durmuş H İ 2016 *Constr. Build. Mater.* 105 423-34
[3] Dippenaar R 2005 *Ironmak. Steelmak.* 32 35-46
[4] Siddique R 2014 *J. Mater. Cycles Waste Manag.* 16 460-7
[5] Li G, Guo M 2014 *Waste Biomass Valor.* 5 317-25
[6] Wang J J, Cai J J, Chen C X, Li G S, Zhang Q 2007 *Ind. Heat.* 36 1-3 (in Chinese)
[7] Yu L C 2003 *Baosteel Techn.* 22-5 (in Chinese)
[8] Yehia A, Elrahiem F H 2005 *Eur. J. Mineral Process. Environ. Prot.* 5 184-9
[9] Saleh H I, Hassan K M 2004 *J. Chem. Technol. Biotechnol.* 79 397-402.

IOP Conf. Series: Materials Science and Engineering **230** (2017) 012027 doi:10.1088/1757-899X/230/1/012027

[10] Zeydabadi B A, Mowla D, Shariat M H, Kalajahi J F 1997 *Hydrometallurgy* 47 113-25
[11] Das B, Prakash S, Reddy P S R Biswal S K Mohapatra B K Misra V N 2002 *Eur. J. Mineral Process. Environ. Prot.* 2 61-8
[12] Mikhail S A, Turcotte A M 1998 *Thermochim. Acta.* 311 113-9
[13] Huang X H 2013 *Iron and steel metallurgy principle* (Beijing: Metallurgical Industry Press) P 352 (in Chinese)
[14] Man Y, Feng J X, Li F J, Ge Q, Chen Y M Zhou J Z 2014 *Powder Technol.* 256 361-6
[15] Qi K, Yu F Y, Bai F D, Yan Z M, Wang Z X, Li T J 2009 *Mater. Design* 30 4511-15
[16] Shun L I, Zhang G, Xie G 2013 *Ind. Heat.* 42 58-9 (in Chinese)
[17] Bin D U, Zhang Y G 2013 *Ind. Furnace* 35 55-7 (in Chinese)

Acknowledgments
Thanks for the financial support from the National Natural Science Foundation of China by the grant of No.51174111. Still thanks a lot for AnSteel for providing the BF dust needed in this work.

Analyzing of Nonuniform Hardness for Anchorage Clamp Plate after Heat Treatment

Liu Zhenmin[1,2,3,a], Wang Lijun[2,3,b], Ruan Shipeng[1,2,3,c], Wang Ningtao[2,3]

[1]Material Science and Engineering Institme, Beijing University of Science and Technology, Beijing 100083, China
[2]Xingtai Iron and Steel Corp. , Ltd. , Xingtai 054027, China
[3]Hebei Wire Engineering Technology Research Center, Xingtai 054027, China
Email: [a] Liuzm9530@xtsteel.com, [b]xcyyyj@xtsteel.com, [c]ruansp@xtsteel.com

Abstract. 20CrMnTi wire rod can be used to manufacture the anchorage clamp plate. In the normal production, the test results show that the hardness of anchorage clamp plate is nonuniform. In this paper, the chemical composition， microstructure, grain size of clamp plate billet and the microstructure, hardness of the finished clamp plate product were studied and analyzed. The results show that improper heat treatment technics causes nonuniform hardness for anchorage clamp plate.

1. Introduction

Prestressed anchor working clamp is one of the most critical parts in the prestressed anchorage system. The structure shape of work clamp is the outline for the cone, the internal structure is non-standard thread. After more than 10 production processes, the part was cutted along longitudinal incision, and finally processed by heat treatment. The specific production process is With 24mm specifications of 20CrMnTi hot rolled wire, annealing, pickling, phosphating, cold forming (clamp blank), blank hole tapping, cut into two piece, carbonitriding, quenching, tempering, and finished product inspection. According to the hardness requirements design of clamp, China's current production control range is HRC58-64 or HRA78-84. The hardness of clamp must be achieved by heat treatment, otherwise it is difficult to anchor. On the other hand, the hardness of clamp can not be too high, otherwise the clamp will crack or even break[1].

In the process of normal production inspection, it is found that the hardness of the clamp is uneven. In order to find out the cause of uneven hardness of clamp, the chemical composition, microstructure, grain size of clamp blank and the hardness and microstructure of finished product were analyzed. It is useful for the improvement of heat treatment process.

IOP Conf. Series: Materials Science and Engineering **230** (2017) 012028 doi:10.1088/1757-899X/230/1/012028

2. Experimental material and methods

The experimental materials are 5 pieces of clamp blank, 5 pieces of qualified clip and more than 20 pieces of clamp with uneven hardness.

Firstly, the chemical composition, microstructure and grain size of clamp blank and hardness of the finished product were tested. Then the characteristics of hardness distribution were analyzed. Finally, the microstructure of the samples with different hardness was analyzed in order to determine the causes of uneven hardness. The grain size of the clamp blank is tested according to GB/T 6394. The chemical composition of blank samples need to be heading into an appropriate shape and then testing. The hardness of the finished clip is measured by the automatic hardness tester with AT200DR-TM model. The microstructure was analyzed by means of metallographic microscope and scanning electron microscope.

3. Test result

3.1. Analysis of the clamping piece blank

3.1.1 Analysis of the composition of the blank

The chemical composition of clamp plate were listed in table 1.

Table 1 The chemical composition of clamp plate（mass fraction, %）.

grade	chemical composition								
	C	S	Si	Mn	P	Ni	Cu	Cr	Ti
20CrMnTi	0.18	0.005	0.25	1.00	0.011	0.017	0.05	1.20	0.10

According to the results of the component analysis, the content of each alloy element is within the national standard.

3.1.2. Microstructure and grain size of blank

There is fiber zonal texture in the clip blank after cold working(Caused by hot rod itself and cold working). The microstructure and the grain size of clamp blank are shown in figure 1 which are fine and uniform, and the grain size is 9 grade.

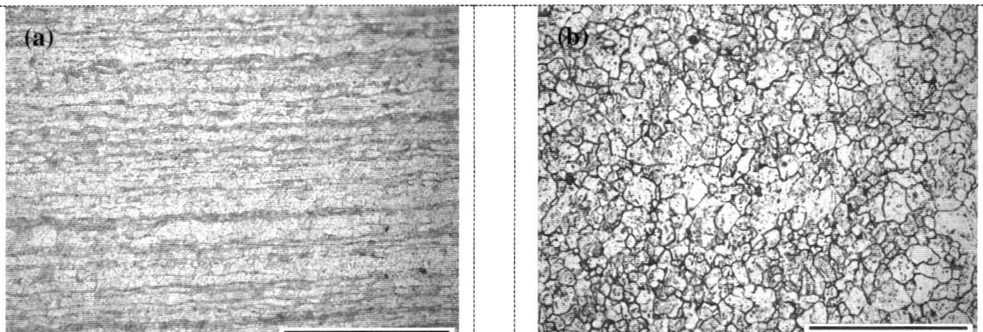

Figure 1. The microstructure（a）and the grain size（b）of clamp plate billet.

3.2. Performance and microstructure analysis of the finished clamp

3.2.1 Analysis of the hardness of the finished clamp

The hardness test value of qualified samples and unqualified samples are shown in figure 2. The so-called nonuniform hardness is that the hardness value of some samples is higher or lower than the hardness value of qualified samples.

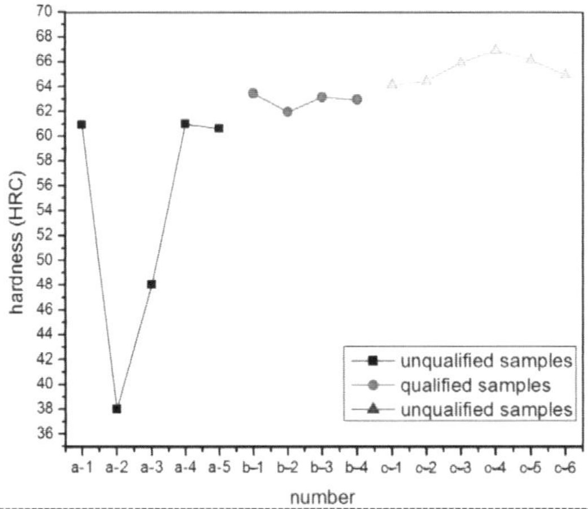

Figure 2. The hardness result of clamp plate.

3.2.2. Analysis of microstructure

As you know, the performance is determined by the microstructure. Due to the nonuniform hardness of the surface of unqualified products, the microstructure of the unqualified products surface must be different from the qualified products. So the analysis of the surface microstructure is very important [2,3].

The polished surface of qualified samples and unqualified samples were observed in optical microscope. But there is no difference between them. Moreover, we did not find any empty in the surface microstructure.

1) samples with qualified surface hardness

From figure 3(a) we can know that the macrostructure of cross section of qualified samples can be divided into four regions: external area of carburized layer, fine grain zone and inner tooth of carburized layer.

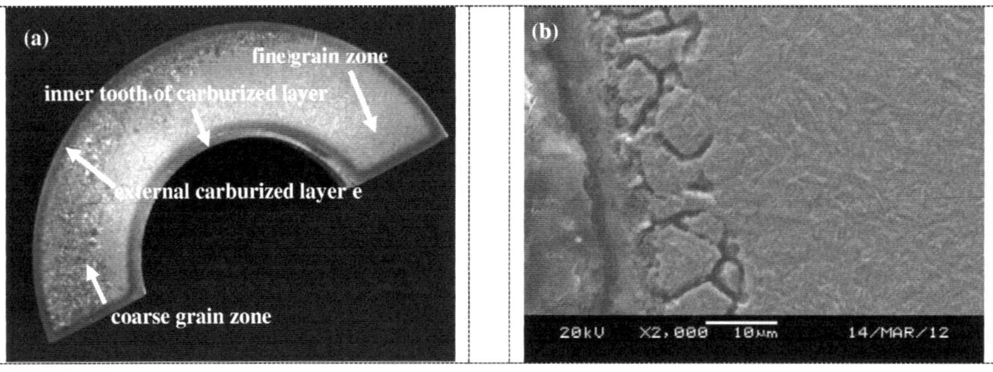

(a) The surface Texture Distribution, 10X.

(b) The thickness of surface troostite is 15μm.

Figure 3. The macrostructure (a) and microstructure (b) of clamp plate with competent surface hardness.

figure 3(b) is the surface microstructure of the samples. The surface layer microstructure was martensite and troostite which along the crystal boundary. The microstructure in the core of the sample was lath martensite. No carbides were observed in the carburized layer.

2) samples with lower surface hardness

The macrostructure of cross section of samples with lower surface hardness can be also divided into four regions: external area of carburized layer, coarse grain zone, fine grain zone and inner tooth of carburized layer. Its coarse grain zone is lath martensite.

The surface microstructure of samples with lower surface hardness was observed. The surface microstructure is a large number of troostite and a small quantity of martensite in the surface carburized layer. There is little distribution of carbide particles in the surface carburized layer(figure 4(a)). A large number of ferrite was found in the core of the samples, see figure 4(b).

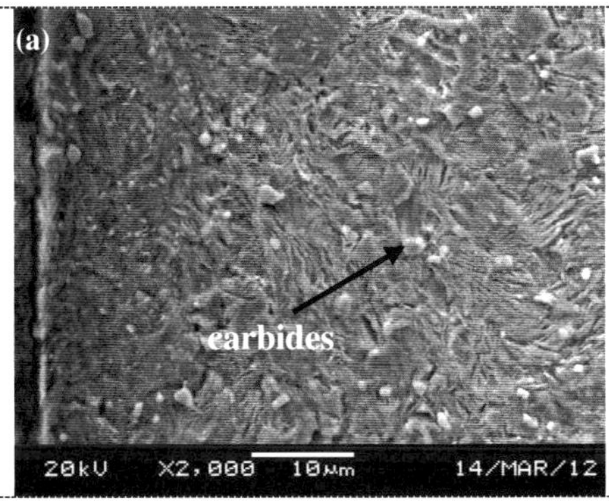

(a) carbon particles in layer zoon.

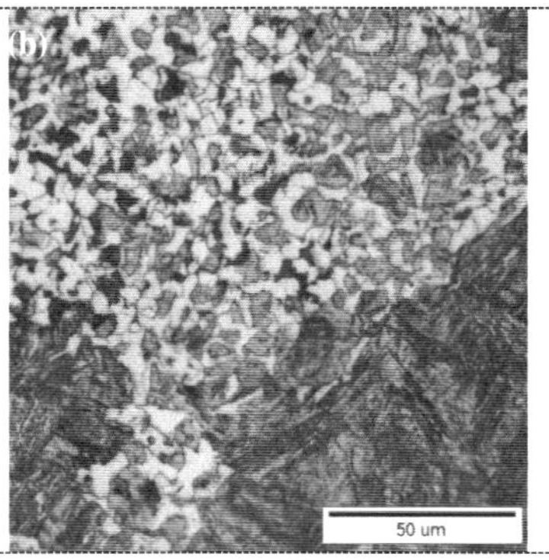

(b) Large numbers of ferrite and part of martensite and bainite.

Figure 4. The microstructure of lower hardness sample (a, b).

The microstructure in the core of the sample is a large number of ferrite, a small quantity of martensite and bainite, which shows that the effect of quenching is not good so that completely martensite structure is not obtained.

3) samples with higher surface hardness

The macrostructure of cross section of samples with higher surface hardness can be also divided into four regions: external area of carburized layer, coarse grain zone, fine grain zone and inner tooth of carburized layer. Its coarse grain zone is lath martensite.

The microstructure of the surface layer and the heart of the samples with higher surface hardness was observed, see figure 5. The surface microstructure is martensite and intergranular troostite. A large number of larger carbides distributed at carburized layer were found in figure 5(b). The center of sample is lath martensite.

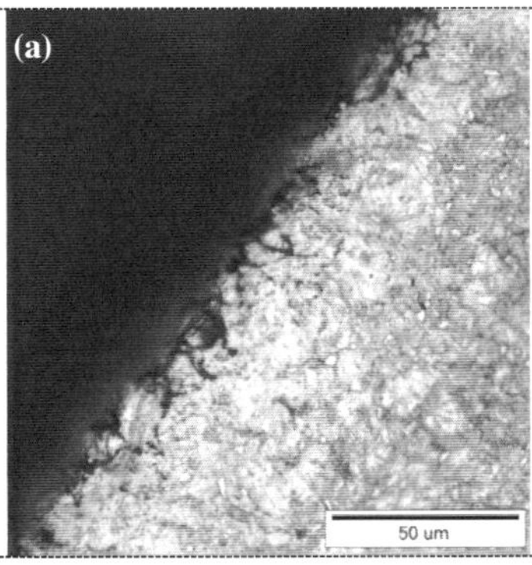

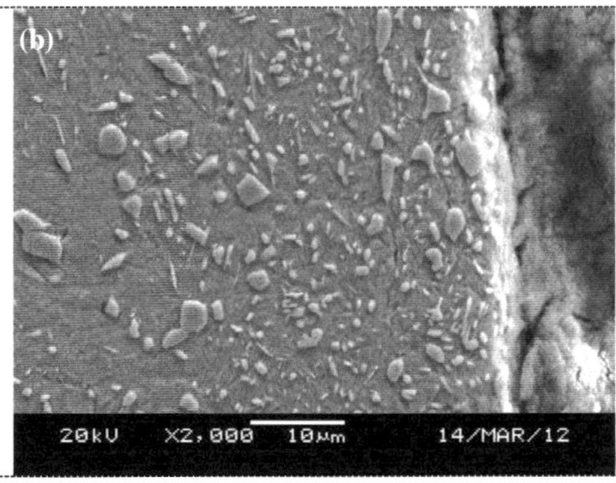

(a) Martensite and troostite.

(b) Large numbers of carbides in layer zoon.

Figure 5. The microstructure of higher hardness sample（a, b）.

4. Test results discussion

The surface hardness of products depends on the surface microstructure. Troostite, martensite and carbonitriding degree in surface layer microstructure determine the surface hardness. The carbonitriding degree can be characterized by the quantity of carbide in carburized layer. Carbonitriding degree is mainly depended on carbonitriding temperature and atmosphere under the premise of chemical composition meet the requirements.

4.1. Analysis of unqualified samples with lower surface hardness

Causes for the lower surface hardness after carbonitriding are low concentration of surface carbon and nitrogen, decarburization, nitrogen removal, too much residual austenite; black organization formed in the surface layer. There is no obvious decarburization and excessive residual austenite but major black organization in the surface layer with microstructure test. Therefore, the main reason for lower surface hardness is too much black organization. According to the different morphology, the black organization can be divided into 3 categories: black, black belt and black net [4,5]. Black organization structure was punctate appeared in the surface of gear member. Black organization structure is the micro morphology of surface holes. There is a very thin oxide on surface of black belt and inside are troostite, bainite, and containing alloy oxide, carbon nitride and a little martensite, residual austenite as well. Black net is mainly trooslite and bainite.

The main reason for lower surface hardness is a large number of Troostite formed in the surface layer. Troostite more, hardness value lower. There are two reasons for the lower hardness. One possibility is low temperature of carbonitriding and poor activity of furnace gas. Austenite is not stable enough because of low surface carbon content and then transform into troostite. May also be decarburization caused by slow cooling after carbonitriding and surface decarburization during quenching heating process. In addition, The content of alloy elements and C, N in the austenite decreased because of Cr, Mn and other alloy elements dissolve in carbon nitride in the surface layer. Then, the stability of austenite decreased when quenching and finally leads to reticular troostite formation. We have observed the troostite at the surface layer in all the samples, and just how many. Moreover, on the basis of the above factors, more generation of troostite was achieved because of cooling rate of quenching is insufficient[6].

4.2. Analysis of unqualified samples with higher surface hardness

Although the surface layer of unqualified samples with higher surface hardness is martensite and a small amount of troostite. But high surface hardness was achieved eventually because of the high degree of carbonitriding (more carbide in the carburized layer).

5. Conclusion

- Coarse grains both exist in the surface hardness qualified samples and unqualified samples. The coarse grain was formed in the carbonitriding process and subsequent heat treatment process which has nothing to do with raw materials.
- The main reasons for the unqualified products with lower surface hardness are due to a large number of troostite at surface layer generated after heat treatment.
- The main reasons for unqualified products with higher surface hardness are due to over carburizing at the surface after carbonitriding.

6. Reference

[1] Zhou Minghua. Factors Influencing the Performance of the Anchorage with Wedged Jaws[J]. BRIDGE CONSTRUCTION, 2001(2), p 58-61

[2] Ma Yongjie. Analysis and Prevent Measures on 20CrMnTi Carbonitriding Abnormal Microstructures[J]. FOUNDRY TECHNOLOGY, 2010, 31 (2) , p 245-247

[3] Cui Zhongqi. Physical Metallurgy and Heat Treatment[M]. Beijing: China Machine Press.1997

[4] Cui Zhanquan, Wang Kunlin, Wu Run. metallography and heat treatment[M]. Beijing: Peking University Press.2010.10

[5] Tang Xuezhe, Ji Jianlong. Causes and measures in low carbon steel carbonitriding black structure[J]. metal working (hot working).2011, 11

[6] Yang Bingbinb. Effect of Carbonitrided Microstructure on Stress and Properties[J]. Hot working Technology, 2011, 40(4) , p 174-175

Acknowledgments

I would like to express my gratitude to all those who helped me during the writing of this paper. First of all, I appreciate my firm who give me a comfortable working atmosphere. Secondly, I want to thanks my workmates for their help. I want to let them know that you are my mentor of my working life. I am also deeply indebted to my leader for his direct and indirect help to me during the project research. Also I have a word to editors, thank you very much for your reviewing of this paper. Finally, I am indebted to my family members for their continuous support and encouragement.

IOP Conf. Series: Materials Science and Engineering **230** (2017) 012029 doi:10.1088/1757-899X/230/1/012029

Preparation and Saturable Absorption Property of Graphene on the Optic Fiber Side by Transferring CVD- Graphene Grown on Ni

Liyun Ding*, Bing Xu, Chuang Xu, Jun Huang, Donglai Guo

National Engineering Laboratory for Fiber Optic Sensing Technology, Wuhan University of Technology, Wuhan, 430070, China
E-mail: adlyw@whut.edu.cn

Abstract. A key enabler in the mode-locked fiber lasers has been the discovery of novel saturable absorbers (SA) such as carbon nanotubes (CNT) and graphene. We demonstrate chemical vapor deposition (CVD)-graphene on nickel (Ni) transferred onto optical fibers preserves its nonlinearity. The influence of the temperature on the growth of graphene was explored by Raman spectra, scanning electron microscope (SEM) and atomic force microscope (AFM), respectively. The results show the temperature will affect the thickness and surface uniformity of graphene on Ni film. Besides, the mechanism of CVD process and the saturable absorption property of the graphene are illustrated, respectively.

1. Introduction

Graphene is a single layer of carbon atoms arranged in two-dimensional honeycomb crystal lattice, and has been attracting significant attention due to its distinct mechanical, structural, electrical, and chemical properties, which make it promising in fields of electronics, sensing, and optoelectronics devices. One application of graphene is acting as a saturable absorber, which was firstly demonstrated in 2009. [1] When the light irradiates on graphene, graphene shows a nonlinear absorption to it, which is attributed to the SA mechanism. Graphene is believed to be an excellent SA material, due to the unique energy band gap structure with advantages of low absorption intensity, deep modulation depth, high damage threshold, fast recovery time and wide range of saturable absorption (300-2500 nm). Typically, there are three different structures of saturable absorber device, sandwich structure, evanescent wave coupling structure and waveguide-incorporated structure, realized by various methods. For example, Sun et al. [2] and Bao et al. [3] applied evaporated graphene-PVA composite and CVD-graphene in mode-locked ultrafast laser of sandwich structure, respectively. In 2015, Azooz et al. used the graphene oxide in mode-locked thulium ytterbium co-doped fiber laser [4]; and Zhao et al. use graphene film as saturable absorber for an Optical Modulator in Pulsed Laser Generation [5] in 2015.

In this paper, we demonstrate the influence of temperature on CVD-graphene growth on Ni film. The influence of the temperature on the growth of graphene was explored by Raman spectra, scanning electron microscope (SEM) and atomic force microscope (AFM), respectively. Besides, the transmission curve of the CVD-grown graphene is measured, and the mechanism of CVD process and the saturable absorption property of the graphene are illustrated.

2. Experimental

IOP Conf. Series: Materials Science and Engineering **230** (2017) 012029 doi:10.1088/1757-899X/230/1/012029

2.1. Materials and apparatus.
Ethyl alcohol, acetone, nitric acid, Ferric chloride, phosphoric acid and poly (methyl methacrylate) (PMMA) were purchased from Sigma-Aldrich Shanghai (Shanghai, China). Vacuum pump used in this experiment was from Woosung Vacuum Co. Gas Mixing System used in this experiment was from MTI Corporation. Mini tube furnace was from Thermcraft.

2.2. Preparation of Graphene.
The graphene was synthesized by a laboratory-built ultralow-pressure thermal CVD system with a quartz tube furnace. CH_4 served as the carbon source and Ni film as the catalyst. The Ni film was placed on a quartz plate and then loaded into the center of the quartz tube furnace. After the pressure decreased below 20 mTorr by air pump, then Ar, methane, hydrogen was injected until the gas detection circuit became normal. The preparation process included heating, annealing, growth and cooling with different temperatures shown in figure 1. Firstly, the reactor was heated to the pretreatment temperature. Annealing is the first chemical reaction process of the whole process, to clean and chemically modify the surface of Ni, and optimize its crystallinity, roughness and grain size as much as possible for the subsequent steps of the growth of graphene.

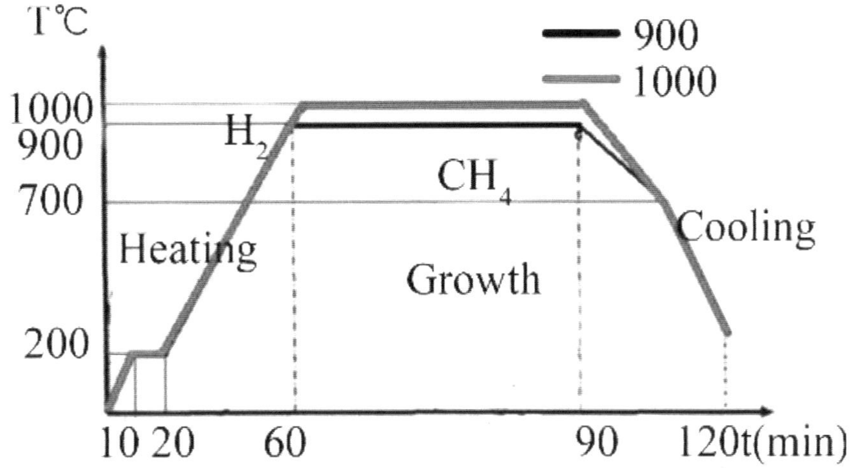

Figure 1. Temperature changes through the preparation of graphene.

2.3. Transfer of graphene to the optical fiber side.
PMMA was spin coated onto the graphene on the Ni film, followed by heating at 150 °C for 10 min. Then the sample was immersed in ferric chloride and deionized water to dissolve the Ni film and remove the residual ferric chloride, respectively. After transferred onto the optical fiber side, the PMMA/graphene was heated at 150 °C for 10 min to gain a good contact between graphene and the optical fiber. Finally, the fiber side was immersed in acetone to dissolve the PMMA, and thoroughly rinsed.

2.4. Measurement of the graphene absorber.
We tested the saturable absorption property of the graphene SA with a laser diode at 473 nm as the input light source. The measured transmission experimental apparatus of the prepared graphene SA is shown in figure 2. The light intensity at both ends of the optical fiber is measured by optical power meter. With the increase of light intensity, the increase of transmittance rate is decreased gradually, even invariant.

IOP Conf. Series: Materials Science and Engineering **230** (2017) 012029 doi:10.1088/1757-899X/230/1/012029

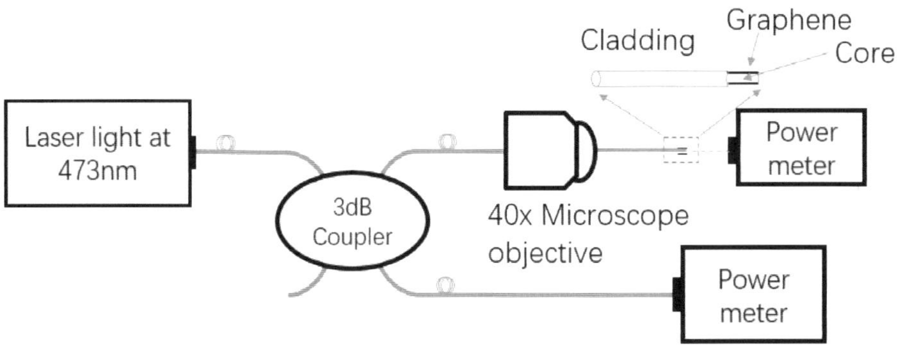

Figure 2. The measured transmission schematic diagram of the prepared graphene SA.

3. Results and discussions

3.1. Raman spectrum graphene.
The Raman spectra of the samples synthesized on Ni film at temperature of 900 °C and 1000°C are shown in figure 3. The two most intense features are the G peak at ~1580 cm-1 and a band at ~2680 cm-1, historically named 2D, since it is the second most prominent peak in graphene samples. The G peak is due to the doubly degenerate zone center E_{2g} mode. And the 2D peak is caused by two reverse momentum of the carbon atoms in double phonon resonance transition, its movement and shape is closely related to graphene's layers. The G peak and 2D peak indicate the existence of graphene, and the absence of D and G' peaks prove that there is no disorder of graphene. [6] The ratio between the intensity of 2D peak and G peak (I_{2D}/I_G) is relative to the number of graphene layers. It is clear that I_{2D}/I_G is about 1.2 at 900 °C and I_{2D}/I_G is about 0.64 at 1000 °C in figure 3. Therefore, that the thickness of CVD-graphene film at 900 °C is thinner than that at 1000 °C.

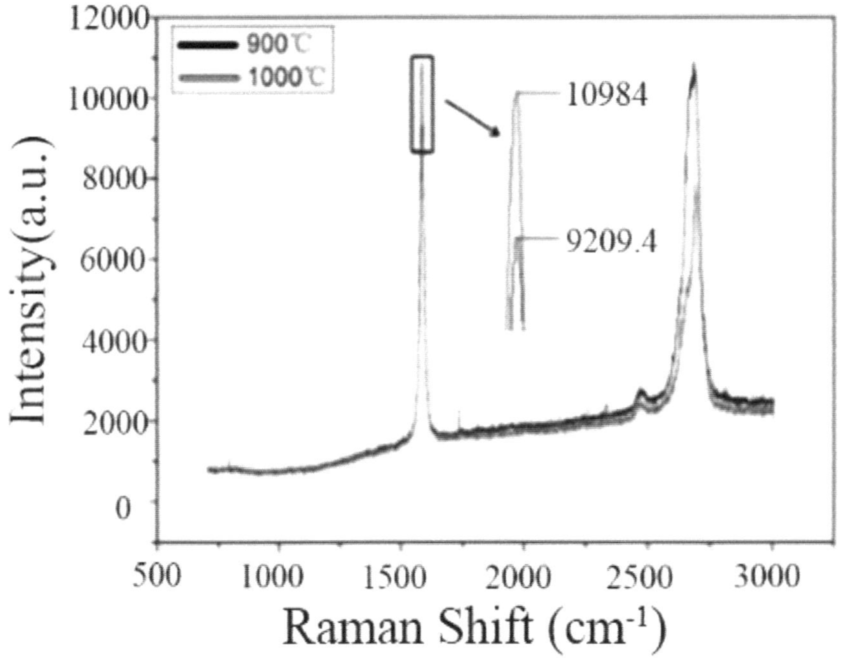

Figure 3. Raman spectra of graphene grown at 1000 °C (a) and 900 °C (b), respectively.

3.2. SEM images of graphene.

The SEM images of CVD-grown graphene at 900 °C and 1000 °C are shown in figure 4(a) and (b), respectively. The graphene film grown at 900 °C has good continuity and uniformity, while the graphene film grown at 1000 °C has many folds and some nanoplatelets graphene agglomeration. It is because the graphene growth on Ni is a deposition-segregation process (figure 4.c) [7]; ^{12}C and ^{13}C graphene pattern grown on the Cu film are separated, while on the Ni, they are evenly distributed. This indicated that graphene growth on Ni is deposition-segregation mechanism, while growth on Cu is a surface process. And the carbon solubility of nickel increases with the temperature increasing, so the carbon atoms agglomerate into graphene folds.

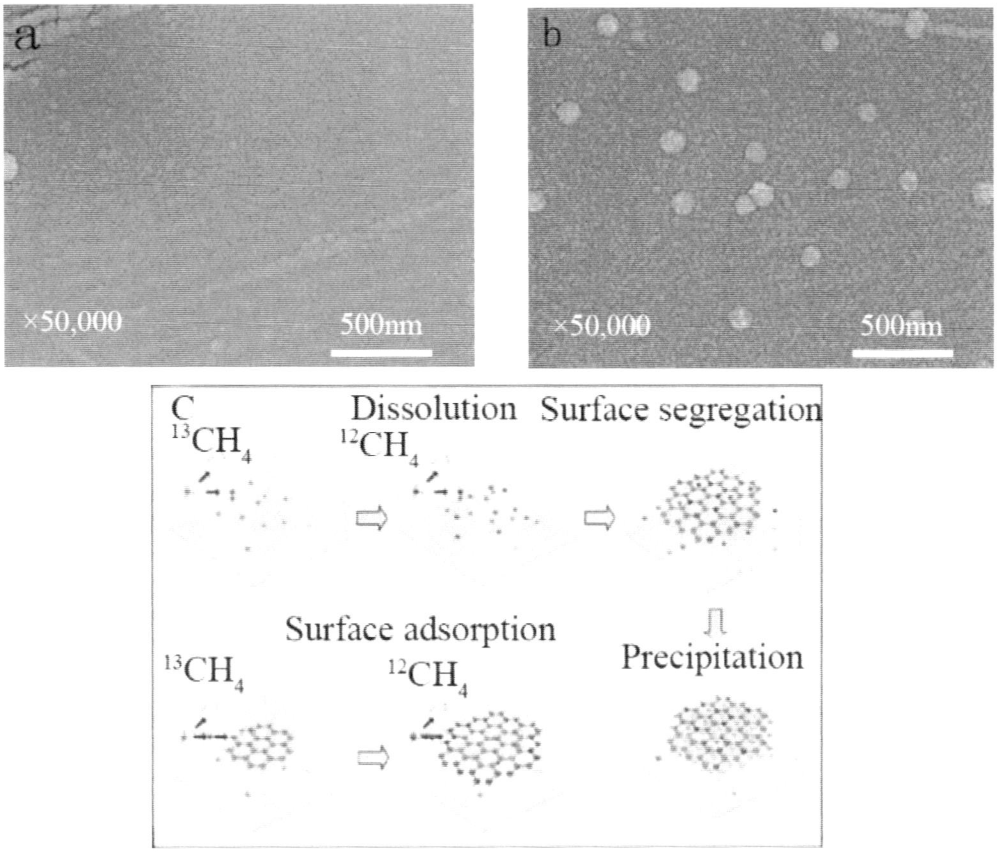

Figure 4. SEM of graphene prepared at 900 °C (a) and 1000 °C (b), respectively. (c) The mechanism schematic diagram of deposition and surface precipitation in the process of growth of graphene.

3.3. AFM images of graphene.

It is evident that the graphene film is roughly uniform in the AFM images as shown in figure.5. It can be seen that the thickness distribution of the graphene film is around 36 nm, except for a few folds of 174 nm. The thickness of single layer graphene is theoretically 0.334 nm [8], but experimentally 0.7~1.2 nm due to the surface absorption-caused change of carbon atom's van der Waals radius. Therefore, the number of graphene layers is about 30 layers. From the results by SEM and AFM, we can see that the CVD-graphene film grown at 900 °C has better continuity and uniformity than that at 1000°C.

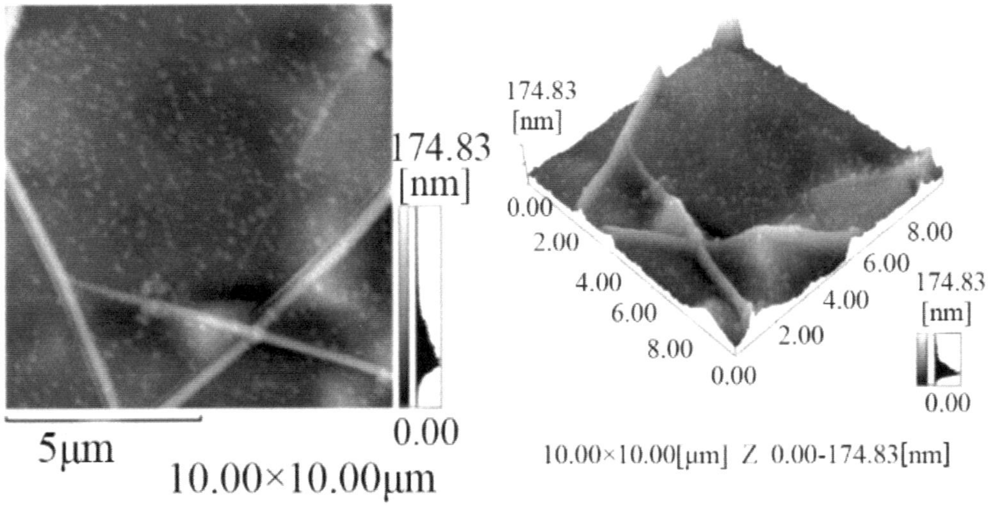

Figure 5. AFM images of graphene growth on 900°C

3.4. Graphene absorber experiment.

The measured transmission curve of the optical decorated graphene in side SA is shown in figure 6. The modulation depth is ~9.5% and the nonsaturable loss is ~50%. Note that the modulation depth is dependent on the amount of graphene layer on the fiber side [9]. The nonsaturable loss is less than the other article (~63.1%) [10].

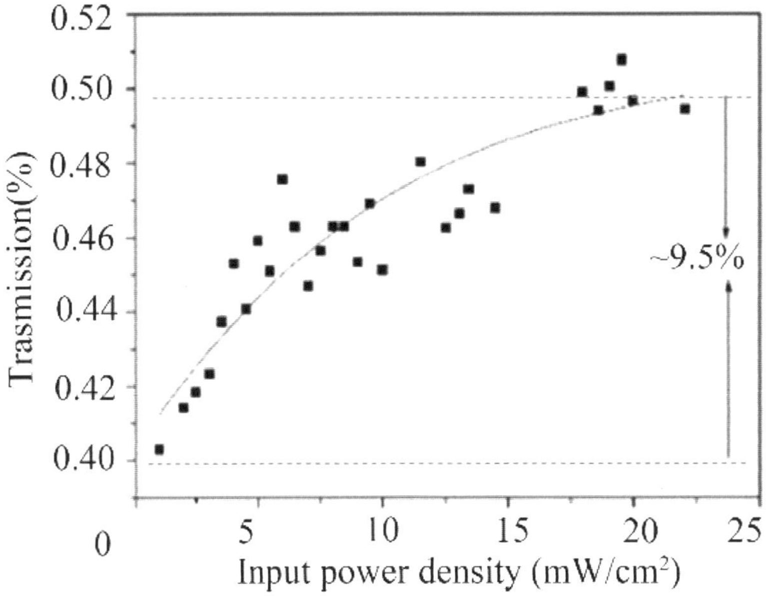

Figure 6. Transmission curve of graphene SA.

4. Conclusion

In this paper, it was found that the number of graphene layers is controllable via modulating the reaction temperature of the CVD process on Ni film. The results show that the CVD-graphene at 900 °C on Ni film has better uniformity than that at 1000 °C. The modulation depth of the CVD-graphene SA is 8% and the nonculturable loss is 50%, which indicating a promising prospect in mode-locked leaser.

References

[1] Hasan T, Sun Z, Wang F, et al. Nanotube–polymer composites for ultrafast photonics[J]. *Adv. Mater,* 2010, 21(38-39):3874-3899.

[2] Z. Sun, T. Hasan, F. Torrisi, D. Popa, G. Privitera, et al., Graphene mode-locked ultrafast laser, *Acs Nano,* 4(2009) 803-810.

[3] Q. Bao, H. Zhang, Y. Wang, Ni, Z. et al., Atomic-layer graphene as a saturable absorber for ultrafast pulsed lasers, *Adv. Funct. Mater*, 19(2009) 3077–3083.

[4] Azooz, Harun, Ahmad, et al. Mode-locked thulium ytterbium Co-doped fiber laser with graphene oxide paper saturable absorber, *Chin Phys Lett,* 32(2015) 59-62.

[5] G. Zhao, J. Hou, Y. Wu, J. He, et al., Preparation of 2D MoS 2 /graphene heterostructure through a monolayer intercalation method and its application as an optical modulator in pulsed laser generation, *Adv Optical Mater,* 3(2015) 937–942.

[6] Ferrari, A. C.; Basko, D. M. Nat. *Nanotechnol.* 2013, 8, 235.

[7] Park H J, Meyer J, Roth S, et al. Growth and properties of few-layer graphene prepared by chemical vapor deposition[J]. *Carbon,* 2010, 48(4):1088-1094.

[8] P. Blake, E. W. Hill, A. H. Castro Neto, et al. Making graphene visible[J]. *Appl Phys Lett,* 91(2007) 666-47.

[9] F. Bonaccorso, Z. Sun, T. Hasan, et al. Graphene photonics and optoelectronics, *Nat Photonics,* 4(2010) 611-622.

[10] Z. C. Luo, W. J. Cao, A. P. Luo, and W. C. Xu, Optical deposition of graphene saturable absorber integrated in a fiber laser using a slot collimator for passive mode-locking, *Appl Phys Express,* 5 (2012) 055103-055103-3.

Acknowledgments

This work is supported by Financial Support from the National Natural Science Foundation of China (NO. 61575150, NO. 61377092).

Control of Low Melting Point Mno-Sio$_2$-Al$_2$o$_3$ Inclusions in Low Carbon Thin-Strip Continuous Casting Steel

Jing Chen, Qing Zhu, Di Huang, Shaobo Zheng, Jieyu Zhang, Huigai Li

State Key Laboratory of Advanced Special Steel & Shanghai Key Laboratory of Advanced Ferrometallurgy & School of Materials Science and Engineering, Shanghai University, Shanghai 200072, China.
E-mail:chen.jing.1217@hotmail.com

Abstract. There is a significant difference in the demand for molten steel quality between thin-strip continuous casting and traditional continuous casting. In order to make sure the better surface quality of the thin strips, to generate an oxidation film on the surface of cooling roller is required. This will require that the higher oxygen potential in molten steel and inclusions with low melting point. In this article, the possibility of producing low-melting inclusions which is mainly consisted of SiO_2 and MnO is studied by controlling the initial oxygen potential and addition order of deoxidizing alloys. The interaction activity between each component in the ternary system of Al_2O_3-SiO_2-MnO is obtained by Action Concentration model. The equal [Mn], [Si], [O], [Al] curve under the temperature of 1823K and equilibrium condition in ternary system of Al_2O_3-SiO_2-MnO is obtained by relative thermodynamic calculation as well. The control method for getting the low-melting point inclusion is as below. While the weight percentage of Si is 0.35% and the one of Mn is 0.90%, in order to maintain the melting point of inclusion around 1200°C, the free oxygen potential in melted steel F[O] should be maintained between 0.002% ~ 0.004%. On the contrary, the requirement for acid dissolved [Al] content in melted steel is as low as 0.0001% ~ 0.0005%.

1. Introduction

Thin-strip continuous casting is a new technology which connects the continuous casting and continuous rolling by using a twin roll caster to produce the strip products directly from molten metal[1]. It makes the steel making process more compact, more successive, more efficient and more environment-friendly. The molten steel could be casted into a thin-strip steel with thickness less than 5mm, which achieves the integration of casting and rolling. In a twin roll caster, molten metal is introduced between a pair of counter-rotated horizontal casting rolls, which are colled so that metal sheels solidify on the moving roll surfaces and brought together at the nip between the rolls to produce a solidified thin-strip product delivered downwardly from the nip[2].

When casting thin-strip in a twin roll caster, the molten metal in the casting pool will be generally at a very high temperature of 1500℃ and above, and the high cooling rates are required during solidification, which is as high as 1000K/s. It is significant to achieve a high heat flux and extensive nucleation at the initial solidification process on the strip surfaces to form the metal shells. As Lazar S[3] described that the heat flux on the initial solidification can be increased by adjusting the steel melt chemistry. So that a substantial proportion of the oxides formed as deoxidation products are liquid state at the initial solidification temperature so as to form a substantially liquid layer at the interface between the molten metal and the casting surface for the role of covering slags in casting. Obviously, it is not advantageous to cast with clean steel in thin-strip continous casting in which the

number of inclusions in steel is required. This will require that the higher oxygen potential in molten steel and inclusions with low melting point. The molten steel contains a distribution of oxide inclusions (typically MnO, SiO_2, CaO and Al_2O_3) sufficient to provide an adequate density of nucleation sites on the roll surfaces for initial and continued solidification and the resulting strip product exhibits a characteristic distribution of solidified inclusions and surface characteristics.
According to Suito and Inoue [4], inclusion composition can be estimated from thermodynamic as a function of steel composition and the reaction temperature, provided that the equilibrium between steel and inclusion is established. Okuyama [5] proved that the reactions between inclusions and steel is fast enough to establish the local equilibrium between inclusions and molten steel.

2. Experimental

2.1. Preparation of Metal Samples

A Medium-frequency induction furnace was used for experiments in the present work. The experimental procedures were as follows:

1. Melt 32Kg steel in furnace at 1873K (1600℃).
2. Add 385g MnFe alloys into the molten steel for deoxidation.
3. After 2minutes, add SiFe alloys into the molten steel for continuous deoxidation to generate MnO-SiO_2 inclusions.
4. After 2, 5, 10, 15, 25, 30minutes, take steel samples No.1 to No.5 using copper-formed samplers.

2.2. Chemical Analysis and SEM Investigation of Inclusions

The chemical compositions of in the steel samples were analyzed by Inductively Coupled Plasma-atomic Emission Spectroscopy (C, Si, Mn), Nitrogen-oxygen determinator (T[O], N), Chemistry Analysis (S), respectively. And the inclusions were analyzed by scanning Electron Microscopy (SEM)-Energy Dispersive X-ray Spectroscopy (EDX). The compositions of the steel before and after the experiments were shown in Table 1.
The inclusion particles were extracted from metal sample by a potentiostatic electrolysis using nonaqueous organic electrolytic. The dissolved sample weight during electrolysis was in the range of 0.2 to 0.3 g. The inclusion particles extracted were separated using a polycarbonate film filter with an open pore size of 0.8 μm, and were estimated three- dimensionally (3-D) by using SEM-EDX at magnifications of 5000 and 10000. The size of each inclusion was evaluated as the equivalent spherical diameter, dV, which was measured on the SEM photographs using a semi-automatic image analyzer.

Table 1. Chemical Composition of the Thin-Strip

samples	T[O]	[N]	Fe	C	Si	Mn	S
1	0.0167	0.0043	98.4	0.101	0.354	0.898	0.0209
2	0.0206	0.0056	98.5	0.0988	0.335	0.87	0.0229
3	0.0262	0.0039	98.5	0.102	0.323	0.855	0.0234
4	0.0114	0.0087	98.6	0.0927	0.295	0.817	0.0207
5	0.0066	0.0043	98.6	0.0908	0.255	0.782	0.0201

3. Thermodynamic Fundamentals

In this work, reactions between liquid steel and inclusions are listed in Table 2.Where, $\Delta G_i^\theta$is the standard Gibbs free energy, K is the equilibrium constant, $a_i$is the activity of element i. Here, the activity of element i (a_i) in the steel melts can be estimated from Eq.(1) and Eq.(2).

$$\lg f_i = \sum_{j=1}^{n} e_i^j [\%j] \quad (1)$$
$$a_i = f_i \cdot [\%i]. \quad (2)$$

Table 2. Chemical reactions occuring between molten steel and inclusions and their standard Gibbs free energy[6,7]

No.	Chemical Reaction	ΔG^0 [J·mol^{-1}]	K
1	$2(MnO)+[Si]=(SiO_2)+2[Mn]$	$-5700-34.8T$	$K_1=\dfrac{a_{SiO_2}\cdot a_{[Mn]}^2}{a_{MnO}^2\cdot a_{[Si]}}$
2	$[Si]+2[O]=(SiO_2)$	$-580550+221.03T$	$K_2=\dfrac{a_{SiO_2}^3\cdot a_{[Al]}^4}{a_{Al_2O_3}^2\cdot a_{[Si]}^3}$
3	$2(Al_2O_3)+3[Si]=3(SiO_2)+4[Al]$	$658200-107.1T$	$K_3=\dfrac{a_{SiO_2}}{a_{[Si]}\cdot a_{[O]}^2}$

Where, e_i^j is the first-order interaction coefficient of element j to i relative to dilute solution whose value can be obtained by Eq.(3) and Table 3, $[\%j]$ is the mass percentage content in the liquid steel of element i, f_i is the activity coefficient.

$$e_i^j(T)=\left(\frac{2557}{T}-0.365\right)e_i^j(1873K) \quad (3)$$

Table 3. First-order activity interaction coefficients e_i^j of solutes at 1873K[9]

i	j				
	Al	C	O	Mn	Si
Al	63/T+0.011	0.091	34.740/T+11.95	-	0.056
C	0.043	158/T+0.0581	-0.34	-0.12	162/T-0.008
O	-20600/T+7.15	-0.043	-1750/T+0.734	-0.021	-0.131
Mn	-	-0.07	-0.083	0	0
Si	0.058	380/T-0.023	-0.23	0.002	34.5/T+0.089

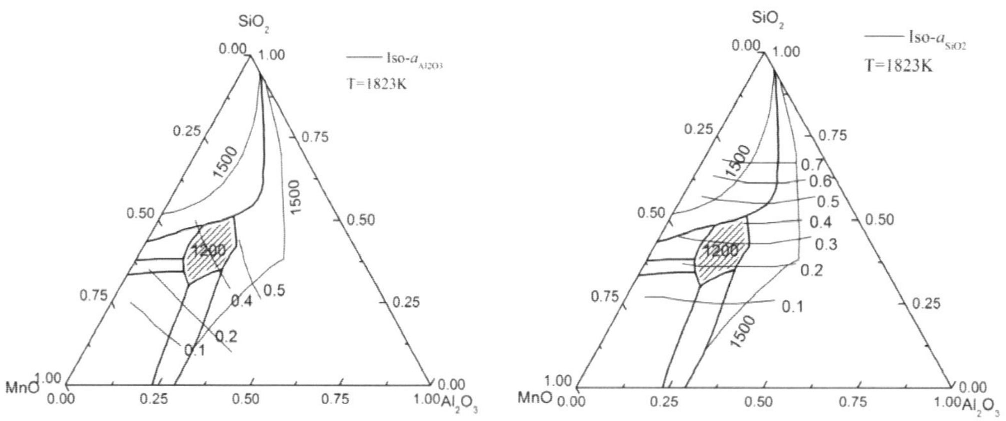

(a) Calculated iso-activity curves of Al_2O_3 (b) Calculated iso-activity curves of SiO_2

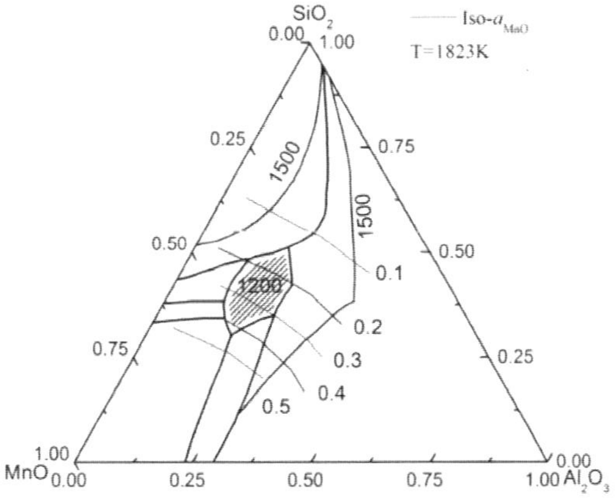

(c) Calculated iso-activity curves of MnO

Figure 1. Calculated iso-activity curves of Al_2O_3/SiO_2/MnO in Al_2O_3-SiO_2-MnO melts at 1823K

Several thermodynamic models for liquid slag systems have been develop over several decades by many researchers and some of these have shown good reproducibility to experimental data and are convenient to use with a computer. In this work, the Action Concentration[8] model is used to calculate the activity coefficient of the oxidation in the melts. Figure 1 shows the calculated iso-activity curves of Al_2O_3/SiO_2/MnO in Al_2O_3-SiO_2-MnO melts at 1823K.

4. Results and discussion

4.1. Characteristics and Variation of Inclusions in Mn/Si Deoxidezed Steel

It's desired to keep inclusions in a liquid state in molten steel during the casting stage and to make the inclusions soft enough during the rolling and plastic deformation stage. Low melting point and soft inclusions are in general easily elongated during plastic deformation of the steel. The formation of inclusions is depended on both steel chemistry and prevailing temperature. Even though FeSi alloys and FeMn alloys are mainly added to deoxidize molten steel, the alloys containing small amounts of Al (0.02~3.5wt%) enables on the formation behavior of inclusions was investigate. Mn/Si deoxidized liquid steel with a small amount of Al and O may produce inclusions of MnO-SiO_2-Al_2O_3 type.
Figure 2 is a diagram of the liquidus surface of the MnO-SiO_2-Al_2O_3 system optimized by FactSage. The low melting point region of the MnO SiO_2 Al_2O_3 ternary oxide system is good target for the inclusions composition.

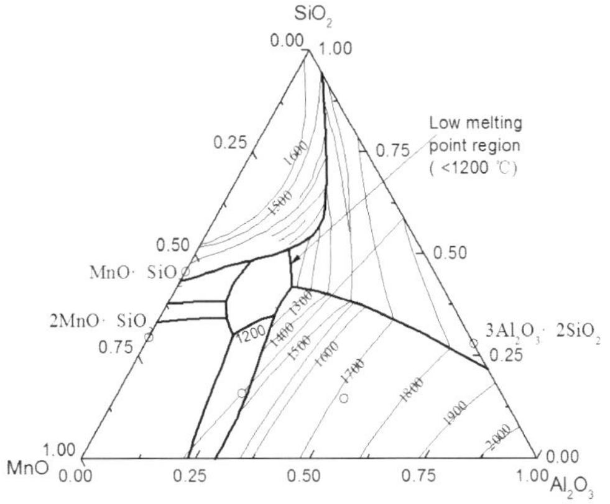

Figure 2. Calculated liquidus surface of the MnO-SiO_2-Al_2O_3 system

4.2. SEM Investigation of Inclusions Deoxidating by Mn/Si

It was found by the observation using SEM-EDS that MnO-SiO_2-Al_2O_3 inclusions widely dispersed in steel samples, globular, the typical inclusion detected was shown in figure 3. The change of inclusions compositions observed by the reaction time are showed in figure 4 and the 3-D morphology and the corresponding compositions of MnO-SiO_2-Al_2O_3 type inclusions are showed in figure 5. In particularly, they were small (about 2~5 μ m), spherical, and their composition is as below Al_2O_3: 30%~35%, SiO_2: 30%~35%, MnO: 30%~35%. The melting point is within the range between 1300°C~1500°C, which is lower than the liquidus temperature of 1522°C.

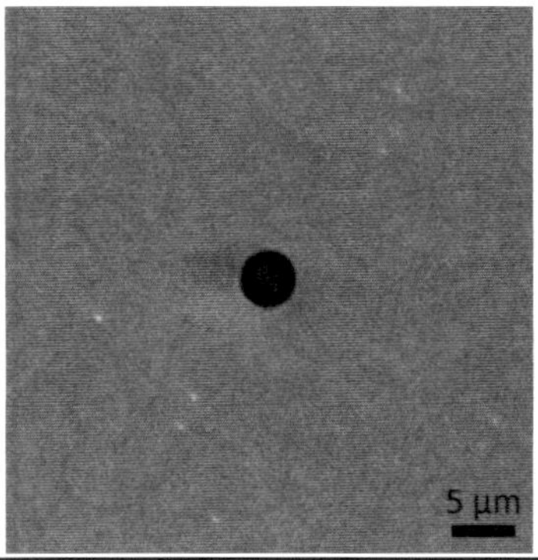

Element	O	Al	Si	Mn
Atom（%）	64.37	13.70	12.80	9.13

Figure 3. Typical inclusion detected in steel sample No.1.

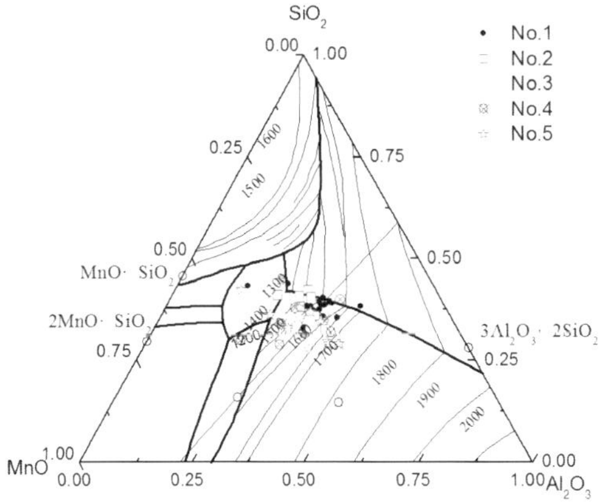

Figure 4. Changement pattern of inclusion in thin-strip along with reaction time

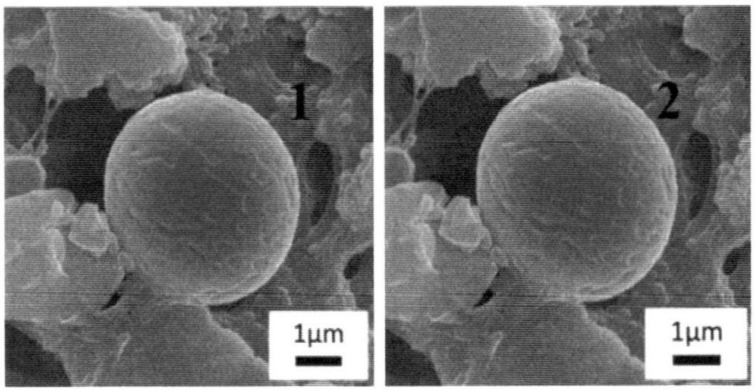

Element	C	O	F	Al	Si	Mn	Pt
1	17.04	53.26	4.10	10.04	9.48	4.97	0.74
2	17.1	50.43	3.54	11.27	10.26	6.67	0.73

Figure 5. The ternary morphology of inclusions in the sample No.1

4.3. Equilibrium between the Observed MnO-SiO_2-Al2O3 inclusions and molten steel

Theoretically, The [Mn] content in liquid Fe with C=0.10 and Si=0.35 mass%, which is in equilibrium with the inclusions of MnO-SiO_2-Al_2O_3 system, it can be deduced from Reaction (1) that the inclusion composition is expressed as a function of the activities of Mn, Si, activity of MnO and SiO_2 and temperature.

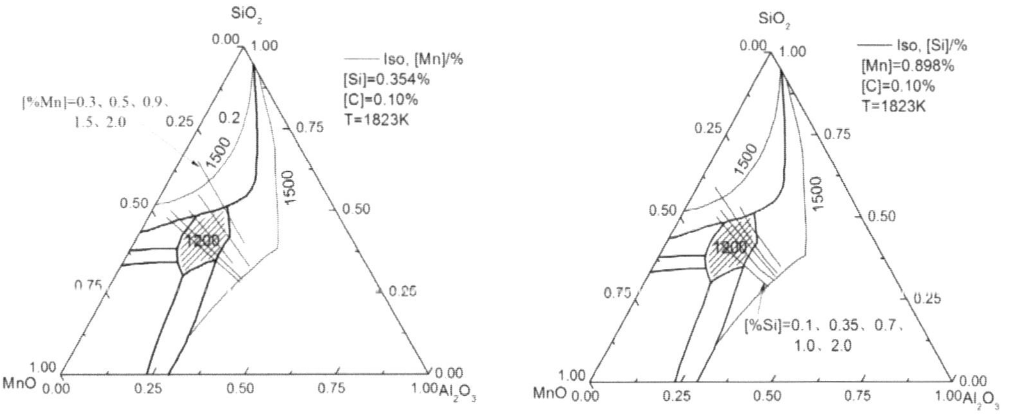

(a) Iso-[Mn] (b) Iso-[Si]

Figure 6. Iso-[Mn] and Iso-[Si] lines on the equilibrium between the molten steel and Al_2O_3-SiO_2-MnO inclusions at 1823K

The iso-Mn(mass%) lines at C=0.10 and Si=0.35 mass% were obtained at 1823K by the iterative method, using the ΔG_1^{θ} value, the activities of MnO(s) and SiO_2(s) and the respective interaction coefficients for the activity coefficients of $f_{Si}$and f_{Mn}. The results are shown in figure 6(a). The iso-Si(mass%) lines were also calculated from Reaction (1) at C=0.10 and Mn=0.90 mass% in a similar manner and the results are shown in figure 6(b). The hatched areas in figure 6 correspond to the inclusions compositions in equilibrium with the steels containing [Mn]=0.3~2.0 mass% and [Si]=0.1~2.0 mass%.

The iso-[Al] (mass ppm) lines were calculated from Reaction (2) at C=0.10 and Si=0.35 mass%, by using the ΔG_2^{θ} value, the activities of Al_2O_3(s) and SiO_2(s) and the respective interaction coefficients for the activity coefficients of f_{Si} and f_{Al}. The results are shown in figure 7(a).

The iso-[O] (mass ppm) lines can be calculated from Reaction (3) at C=0.10 and Si=0.35 mass% in a similar manner and the results are shown in figure 7(b).

The solid lines indicated in figure 7 (a) and 7(b) have the same meaning as those in figure 6(a) and 6(b). And the dash-dotted line indicated correspond to the iso-[Mn]=0.9 (mass%) in figure 6(a). It can be seen in figure 7(a) and 7(b) that the [Al] and [O] contents in the hatched areas are 1-5 ppm and 20-40ppm, respectively, a certain inclusion composition in the hatched areas can be fixed by controlling either the content of [Al] or [O]. When the Al level exceeds 5 ppm, the inclusions composition with C=0.10, Si=0.35 and Mn=0.9 mass% don't located in the low melting point areas. Thus, the Al content must be controlled below 5 ppm by using a low Al-containing alloys. Therefore, the control method for getting the low-melting point inclusion is as below. With the increasing of Mn, Si and Al contents in melted steel, the corresponding contents of MnO, SiO_2 or Al_2O_3 in complex composite inclusion is increasing relevantly. While the mass percentage of Si is 0.35% and Mn is 0.90%, in order to maintain the melting point of inclusion around 1200°C, the free oxygen potential in melted steel F[O] should be maintained between 0.002% ~ 0.004%. On the contrary, the requirement for acid dissolved Al content in melted steel is as low as 0.0001% ~ 0.0005%.

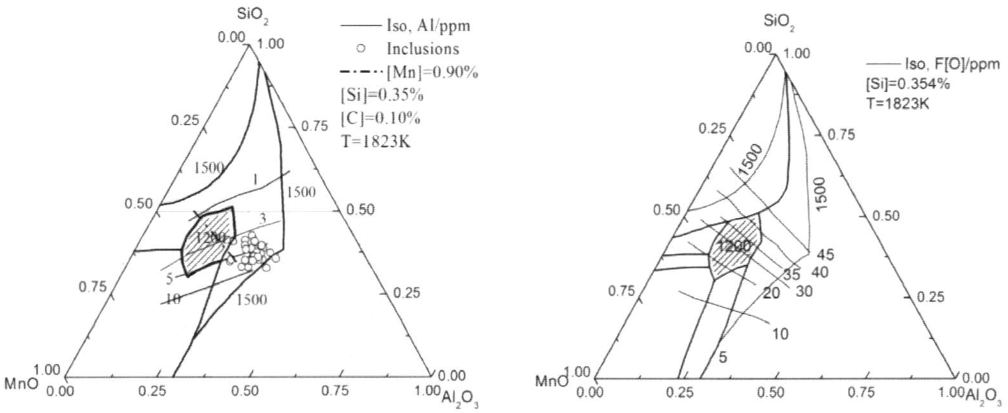

(a) Iso-[Al] (b) Iso-[O]

Figure 7. Iso-[Al] and Iso-[O] lines on the equilibrium between the molten steel and Al_2O_3-SiO_2-MnO inclusions at 1823K

5. Conclusions

Inclusions chemistry of Mn/Si deoxidized steel was studied through both thermodynamic calculation and laboratory experimental confirmation. It has proved that the computational thermodynamics provides a powerful tool for engineering inclusions and precipitates in steel. The following conclusions were obtained:

(1) For metal samples with total content of Al<0.002%, Si:0.25~0.35% and Mn:0.77~0.90%, MnO-SiO_2-Al_2O_3 type inclusions were generated in liquid steel. The particles were small (about 2~5μm), spherical, and their composition is as below Al_2O_3: 30%~35%, SiO_2: 30%~35%, MnO: 30%~35%. The melting point is within the range between 1300°C~1500°C, which is lower than the liquidus temperature of 1522°C.

(2) It is possible to predict the inclusions composition from steel chemistry. For instance, While the mass percentage of Si is 0.35% and the Mn is 0.90%, in order to maintain the melting point of inclusion around 1200°C, the free oxygen potential in melted steel F[O] should be maintained between 0.002% ~ 0.004%. On the contrary, the requirement for acid dissolved Al content in melted steel is as low as 0.0001% ~ 0.0005%.

IOP Conf. Series: Materials Science and Engineering **230** (2017) 012030 doi:10.1088/1757-899X/230/1/012030

6. References

[1] Clay Cross, Rama Ballav Mahapatra, Walter W.Blejde, Steve Leonard Wigman. LADLE REFINING OF STEEL[P]. US:6547849B2, Apr.15, 2003.

[2] Kiroyuki Otsuka, Koshiro Yamane, Satoshi Terasaki, Mark Schlichting, Rama Ballav Mahapatra. THIN CAST STEEL STRIP WITH REDUCED MICROCRACKING[P]. US:7975754B2, Jul.12, 2011.

[3] Lazar Strezov. CASTING OF METAL[P]. US:5720336, Feb. 24, 1998.

[4] Suito H, Inoue R. Thermodynamics on control of inclusions composition in ultraclean steels[J]. ISIJ international 1996; 36: 528-36.

[5] Ohta H, Suito H. Activities in CaO-MgO-Al2O3 Slags and Deoxidation Equilibria of Al, Mg, and Ca[J]. ISIJ international 1996; 36: 983-90.

[6] Gaye H, Gatellier C, Nadif M, et al. Slag metal reactions and control of the residual inclusions composition in secondary steel making[J]. Rev Metall Cah Inf Tech. 1987, 84(11):759.

[7] Ohta H, Sutto H. Activities in MnO-SiO2-Al2O3 Slags and Deoxidation Equilibria of Mn and Si[J]. Metall Mater Trans B. 1996, 27(B):263.

[8] Xiaochun Ma, Guoguang Cheng. Calculating Model of Action Concentration For MnO-SiO2-Al2O3 Slag System[J]. SHANGHAI METALS, 2011, 33(2):40-44.

[9] Fujisawa T, Sakao H. Equilibrium Relations Between The Liquid Iron Components and The Deoxidation Products Resulting From Mn-Si-Al alloys[J]. Tetsu-to-Hagane. 1977, 63(9):1494.

IOP Conf. Series: Materials Science and Engineering **230** (2017) 012031 doi:10.1088/1757-899X/230/1/012031

Study on Carbon Reduction of Guizhou Oolitic Hematite by Graphite in Muffle Furnace

Zheng Zhong, Jinzhu Zhang *, Run Huang, Mingming Li and Jie Wu

School of Materials and Metallurgy, Guizhou University, Guiyang 550025, China
E-mail: zjz-yjx@163.com
*Corresponding author

Abstract. Oolitic hematite is an important iron ore resource in China. The mixed powder of the Guizhou oolitic hematite, the graphite, and the calcium hydroxide was pressed in a cylindrical sample mold by a 16 Mpa pressure. The dried samples at 378K for 12h were reduced in a muffle furnace at the temperature between 1323K and 1473K. The microstructure and the metallization degree of the sample were studied by means of optical microscope and chemical analysis. The results show that the higher metallization degree of the Hezhang oolitic hematite was up to 80.2% at 1323K with carbon oxgen mole ratio 1.3, the metallization degree of the sample in a muffle furnace is lower than that in a microwave furrnace, the lower metallization degree of the sample should be concerned with that the redued metallic iron can be oxidized in the muffle furnace with the temperature cooling down very slowly.

1. Introduction

Oolitic hematite is widely distributed in China and mainly distributed in Hubei, Hunan, Jiangxi, Guangxi, Yunnan, Guizhou, Chongqing, and other 11 provinces. Oolitic hematite is an important iron ore resource in China. Oolitic hematite is a deposit sediment by hematite, limonite and gangue mineral such as quartz, clay, etc. generally shows irregular shell layers of parcels in the oolitic particle center outward, forms a multilayer structure. There are not obvious boundaries between iron rich ores and gangue layer, coated or tightly intereconnected with finely laminated.

The depth reduction of oolite hematite and direct metallization has become a new research hot spot. The growth of the metallic iron particles is affected by both interface reaction velocity and diffusion speed in the deep reduction process of the oolitic hematite. The metallic iron particles is developed, continually grew up, gathered by the smaller ones, and gradually tured into the larger ones at enough high temperature with the reduction time prolonged. Generally speaking, raising the reduction temperature and prolonging the reduction time are all beneficial to the iron grains growth, but the over-high binary basicity is bad for the iron grain growth [1]. It is reported that the microstructure of oolitic hematite has been changed and the oolitic structure has been damaged seriously after heating by microwave. The microwave heating is selectivity. The iron oxides in the oolitic hematite has a microwave absorptive capacity, otherwise, the apatite and silicate in the oolitic hematite. It is beneficial for the reduction of iron oxides in a microwave furnace [2].

In this study, the binary basicity was selected as 1.5, the holding time at reduction temperature was taken as 4h, the atomic ratio of Carbon in the graphite to Oxygen bonded with iron atoms in the oolitic hematite from1.1 to 1.4, and the reaction temperature between 1323K and 1473K in a muffle furnace were all taken into consideration to understand the effects on the iron metallization degree of Hezhang oolitic hematite.

2. Experiments

2.1. Oolitic hematite and reagent

The oolitic hematite ores taken from HeZhang of Guizhou province in China were crushed and ground into powder [3]. The main chemical composition of the oolitic hematite was shown in Table 1. The hematite is the main form of the component Iron. The main chemical composition of the graphite taken as reducing agent is as follows: 98.0% (mass percent) fixed carbon, 1.2% ash, 0.5% volatilc, and 0.24% moisture. The analytically pure calcium hydroxide ($Ca(OH)_2$) was used as a basic oxide to adjust the binary basicity of the samples.

Table 1. Main chemical composition of the oolitic hematite [mass percent,%]

composition	FeO	Fe_2O_3	SiO_2	Al_2O_3	CaO	P	MgO
content	9.44	62.65	9.78	8.71	1.56	0.69	0.71

Table 2. The size distribution of the oolitic hematite [mass percent,%]

Particle size [mm]	-0.074	0.074-0.098	0.098-0.125	0.125-0.2	0.2+
percentage	18.83	31.32	18.18	31.16	0.51

2.2. Experimental scheme

The binary basicity was selected as 1.5, the holding time at reduction temperature was taken as 4 hour, the atomic ratio of Carbon in the graphite to Oxygen bonded with iron atoms in the oolitic hematite were set as 1.1, 1.2, 1.3 and 1.4 respectively, and the reaction temperature was selected as 1323K, 1373K, 1423K and 1473K respectively, in a muffle furnace were all taken into consideration to understand the effects on the iron metallization degree of Hezhang oolitic hematite.

2.3. Experimental processes

Oolitic hematite was crushed and ball milled for 15 minutes. The powder particle distribution of the oolitic hematite in the experiment was shown in Table 2. The powders of the oolitic hematite, the graphite, and the calcium hydroxide were weighed according to the experimental parameters and mixed uniformly and completely. The mixture was extruded into the cylindrical pieces with a 30mm diameter and about 12mm height in the mold by a 16MPa pressure and dried at 378K for 12h. Then, the cylindrical sample was placed into the alumina crucible with a outer graphite crucible, covered with a graphite plate and an invert Φ91×85 mm graphite crucible, heated in a muffle furnace up to setting temperature, and held for the designed time. Four samples were placed closely to the right and left sides of the heating electrode rod in the chamber of muffle furnace [4].

The samples cooled to room temperature in the furnace were sawed off radially into four equal parts by hand. One part was used for microstructure determination, two parts were crushed and milled to 0.075mm or below for X-ray diffraction and chemical composition analysis.

3. Results and analysis

3.1. Effects of temperature on metallization degree of the sample

The metallization degree of the sample was decreased with temperature increase from 1323K to 1423K, increased from 1423K to 1473K, on the condition that the binary basicity was selected as 1.5, the holding time at reduction temperature was taken as 4 hour, the atomic ratio of Carbon in the graphite to Oxygen bonded with iron atoms in the oolitic hematite were set as 1.3 (see figure 1). The lowest metallization degree of the sample was gotten at 1423K, and the largest gained at 1323K.

Generally speaking, in the early stage of the sample reduction, the chemical reaction plays a dominant role for the reaction speed control; in the middle stage of the sample reduction, both the chemical

reaction and the mutual diffusion of carbon atoms or iron atoms become the dominant speed control step; in the late stage of the sample reduction, the diffusion of carbon atoms is the dominant speed control step. Meanwhile, the diffusion of iron atoms plays a dominant role for the metallic iron grains growth. Based on the reduction of the oolitic hematite and the diffusion of iron atoms in the sample, the metallic iron particles were developed, continually grew up, gathered by the smaller ones, and gradually turned into the larger ones at enough high temperature with the reduction time prolonged.
The metallization degree of the sample in this study is lower than that the carbon reduction [5] is executed in a microwave furnace. That the slag formation reaction can hinder the mutual diffusion of carbon atoms or iron atoms may be a main reason for the lower metallization degree, and the higher the temperature, the larger the effect. Moreover, the lower metallization degree of the sample should be concerned with that the reduced metallic iron can be oxidized in the muffle furnace with the temperature cooling down very slowly.

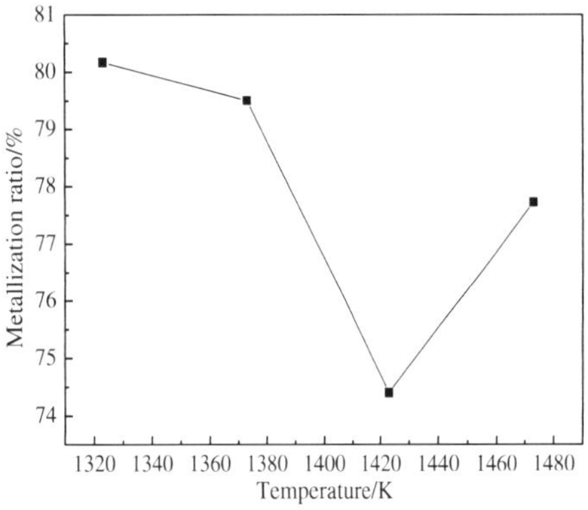

Figure 1. Effect of temperature on the metallization degree of the sample at Carbon Oxgen mole ratio1.3

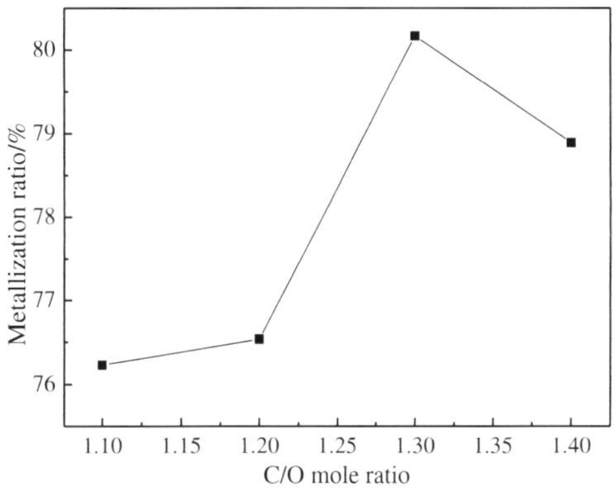

Figure 2. Effect of Carbon Oxgen mole ratio on the metallization degree of the sample at 1323K

3.2. Effects of Carbon Oxgen mole ratio on metallization degree

The metallization degree of the sample was increased with Carbon Oxgen mole ratio increase from 1.1 to 1.3, decreased from 1.3 to 1.4, on the condition that the binary basicity was selected as 1.5, the holding time at reduction temperature was taken as 4 hour, the reduction temperature was set as 1323K(see figure 2). The lowest metallization degree of the sample was gotten at 1.1, and the largest gained at 1.3.
At the early stage of the reduction, because the carbon content is not much, metallization rate curve smooth. With the increase of the ratio of carbon, increase the surface area per unit volume of iron oxide with a reducing agent [6], thereby increasing the contact area between the two, help iron oxide with a reducing agent of solid solid reaction, promoted the reduction of oolitic hematite, and with more carbon, total carbon gasification rate the more conducive to the reduction. At the late stage of the sample reduction, metallization rate decline may be at the root of excessive carbon reaction with other image, formed the organization of low density porosity low melting point material, prevented the reduction of its body and the diffusion of gas products.This is different from that of the microwave heating process, and the metallization rate curve will continue to rise.This may be because the heating process of iron oxides and gangue minerals are not consistent, the characteristics of microwave selective heating [7], so the changes are not the same. Due to the temperature difference between material form a larger thermal stress, so the oolitic hematite in the heating condition of the muffle furnace under the structure of no damage under microwave condition is serious.

3.3. Effect of temperature on the microstructure of the sample

The microstructure of the sample under various reduction temperatures [8] is different, on the condition that the binary basicity was selected as 1.5, the holding time at reduction temperature was taken as 4 hour, the

atomic ratio of Carbon in the graphite to Oxygen bonded with iron atoms in the oolitic hematite were set as 1.3(see figure 3).
When the temperature reaches 1323K, as shown in figure 3 (a), light grey part is iron oolite and matrix interaction. The oolitic structure was destroyed, and the metal iron were fundamental to the small particles scattered. When the temperature reaches 1493K, from figure 3 (b) can be seen in the disappearance of ooide the structure, iron particles in gangue material as crystal nucleus [9] grew up together. More and less dispersed iron particles, metallic iron ring mostly grew up together. This is due to the increase of temperature, the nucleation driving force for iron particles, reaction of activated carbon increased significantly, the dynamic conditions have been improved, which is beneficial to the iron particles grew up together. However, at this temperature, the damage degree is less than the extent of microwave heating. Because microwave heating in the muffle furnace. It is through the internal heating mode [10] increases the temperature of a substance. Iron oxides and gangue minerals in the microwave processing conditions, the disorganization.

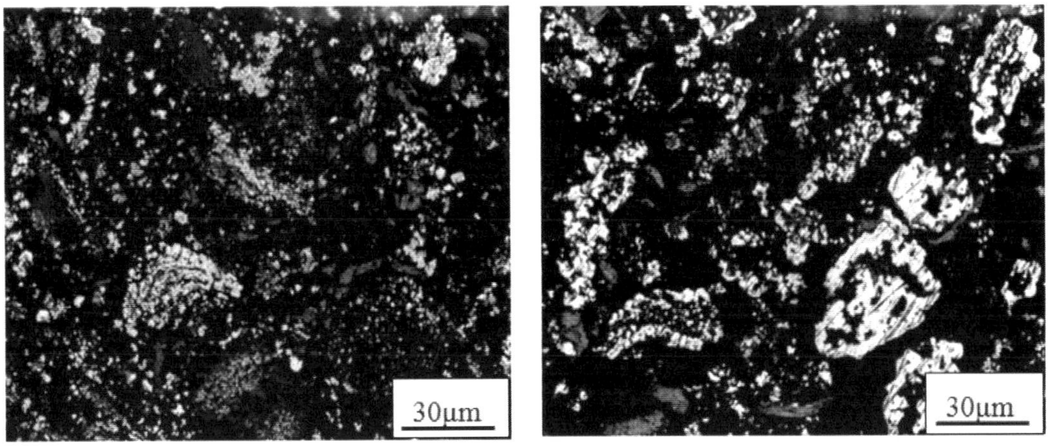

（a） 1323K （b） 1493K

Figure 3. Effect of reduction temperature on the microstructure of the sample at carbon oxgen mole ratio 1.3

4. Conclusions

(1) The metallization degree of the Hezhang oolitic hematite in a muffle furnace is lower than that in a microwave furrnace.
(2) The lower metallization degree of the sample should be concerned with that the redued metallic iron can be oxidized in the muffle furnace with the temperature cooling down very slowly.
(3) The higher metallization degree of the sample was up to 80.2% at 1323K withr carbon oxgen mole ratio 1.3.
(4) The metallic iron particles were grew up, gathered by the smaller ones, and gradually tured into the larger ones with the reduction temperature increase.

References

[1] Yongli L, Tichang S, Jue K, Chengyan X, Zhanhua L and Qian G 2012 Advanced materials research *Industry test on phosphorus removal and direct reduction of High-phosphorus oolitic hematite ore* chapter 32 pp 1538-1542
[2] Haonan C 2016 Guizhou university *The reduction of oolitic hematite and separation of metallic iron* chapter 26 pp 23-25
[3] Yushu Z, Yazhuo D and Wenqi G 2010 Metal mine *Progress of beneficiation on oolitic hematite ore* chapter 136 pp 92-96
[4] Shuwen C, Yongzhen Z and Qiping C 1997 Metal mine *Direct reduction of oolitic hematite ore based on magnetism produced in Guizhou* chapter 24 pp 14-17
[5] Guanglong J, Jinzhu Z, Benjun X and Song Y 2014 Advanced materials research *Study on the reduction of oolitic hematite ores in N_2-CO atmosphere* chapter 22 pp 3275-3280
[6] Wei W, Jianjun G, Jiaqing Z, Yuanhong Q, Jianchang W and Xidong Z 2016 Journal of iron and steel research international *Reduction of carbon-bearing pellets of oolitic hematite in a Shaft Furnace* chapter 26 pp 23-26

[7] Fengjiu L 2011 Applied mechanics and materials *Study on dispersion character of fine oolitic hematite ore particle* chapter 412 pp 1156-1159
[8] Jia Y, Wen N and Fei Z 2010 Metal Mine *The Growth Characteristics of Metallic Iron Grains in Depth Reduction of Oolitic Hematite* chapter 412 pp 52-56
[9] Dongbo H, Yanbing Z, Rufei W, Wei G and Xiaoming L 2016 Journal of iron and steel research international *Direct reduction of high-phosphorus oolitic hematite ore based on biomass pyrolysis* chapter 22 pp 23-32
[10] Yongsheng S, Shufei L, Guangquan S, Yuexin H and Yanjun L 2009 Metal mine The deep investigation of oolitic hematite ore reduction experiments chapter 21 pp 80-83

Acknowledgment

The authors are especially grateful to the Program of National Natural Science Foundation of China (Grant No. 51274074).

IOP Conf. Series: Materials Science and Engineering **230** (2017) 012032 doi:10.1088/1757-899X/230/1/012032

Laser Cladding Synthesis of Ta_2O_5 Coatings on a Ta Substrate

H Zhang[1], H Xie[1] and T M Yue[1,2]

[1] The Advanced Manufacturing Technology Research Centre Department of Industrial and Systems Engineering Hong Kong Polytechnic University Hong Kong
E-mail: tm.yue@polyu.edu.hk

Abstract. A five-layer overlaid Ta_2O_5 dense ceramic coating was successfully fabricated on a pure Ta substrate using laser cladding. The coating exhibits a graded microstructure with a single phase of α-Ta_2O_5 formed towards the top of the coating, and was metallurgically bonded to the Ta substrate. Both the coating and the coating-substrate interface region were free from porosity. The development of the coating's microstructure was discussed based on an existing Ta-O phase diagram.

1. Introduction

In recent years, the study of Ta_2O_5 ceramic coatings has received increasing attention because they exhibit many attractive properties, and these coatings can be used for a variety of engineering applications [1-2]. Its high dielectric constant makes it suitable for use in the area of microelectronics, such as for the production of capacitors. The metal pentoxide, Ta_2O_5, has a high melting point of 1872°C and can be used as a thermal barrier coating to protect turbine surfaces [3]. Ta_2O_5 is also commonly used as a high refractive index material due to its high transmittance, low optical loss and low self-emission at high temperatures [4]. Ta_2O_5 has long been considered an excellent biomaterial owing to its good corrosion and wear resistance, as well as excellent biocompatibility [5]. Many deposition methods have been experimented with to prepare Ta_2O_5 coatings on metal substrates, including hydrolysis condensation, cathodic arc deposition, radio frequency sputtering, DC reactive magnetron sputtering, and polymer-assisted deposition [6-10]. Nevertheless, these methods are mainly used for preparing thin Ta_2O_5 films in the micrometer range. The thickness of films prepared by these methods is often too thin to meet mechanical requirements for applications in the machinery and biomaterial industry. To prepare bulk or thick Ta_2O_5 coatings, vacuum hot-pressed sintering technology is the most favourable production route [11]. To prepare dense Ta_2O_5 material using solidification is not used because of its high melting point. This paper reports the successful fabrication of thick Ta_2O_5 coatings on Ta substrates using laser cladding, which can be considered as a form of rapid solidification process.

2. Experiment method

Powder of Ta_2O_5 (0.1-1 μm) was preplaced on a tantalum (99.9%) substrate to form a powder bed with a thickness of about 1.2 mm. A 5 kW continuous-wave CO_2-laser was used to melt this powder bed, and as a result, the first layer of Ta_2O_5 coating with a thickness of about 0.3-0.4 mm was formed. Subsequently, another four layers of Ta_2O_5 were built on top of the first layer using this powder bed laser melting method, and the final thickness of the coating became about 1.5 mm. The laser processing parameters that used were: a laser power 1.5 kW, laser scanning speed of 600 mm/min, overlap rate 40% with the diameter of the laser beam being 4.5 mm. Considering that the powder itself is a form of oxide, no inert shrouding gas was employed to shield the melt pool from the atmosphere.

The phases and microstructure were characterized using a Rigaku x-ray diffractometer with Cu-Kα radiation operated at 40 kV and 30 mA, and a Quanta 450 field emission scanning electron microscope equipped with an energy dispersive spectrometer.

3. Results and discussion

The XRD patterns in figure. 1 show that the original β-Ta_2O_5 powders are orthorhombic crystals (JCPDS card, No. 25-0922), while the coating was composed of a single α-Ta_2O_5 phase with tetragonal crystals (JCPDS card, No. 19-1300). For the latter, the presence of strong diffraction peaks of (0,0,12) and (1,0,1) suggests that the coating consists of highly crystallized phases. The difference in structures between the powder and the coating indicates that the powders had completely melted in the laser cladding process.

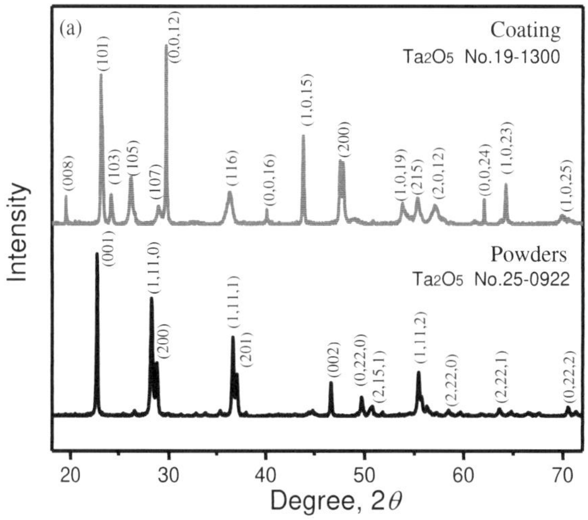

Figure 1. The XRD patterns obtained for the Ta_2O_5 powder and the Ta_2O_5 coating.

Figure 2a shows an overall image of the coating formed on a Ta substrate, no voids and un-melted Ta_2O_5 particles were observed in the coating and at the substrate-coating interface. This shows that laser cladding is a promising fabricating technique that can produce relatively thick and dense Ta_2O_5 coatings on pure Ta substrates. Moreover, due to the characteristic of high transmittance of Ta_2O_5, a large proportion of the laser light was transmitted through the powder bed and absorbed by the tantalum substrate. Thus, the Ta melted pool can be quickly heated up and caused the Ta_2O_5 powders to be melted.

An examination of the coating revealed that the coating can be broadly divided into two regions (figure 2b). According to results of the EDS analysis, the upper region was identified as tantalum pentoxide; whereas in the lower region, two phases were identified; they were tantalum pentoxide and a tantalum-rich solid solution, Ta(O). Close examination of the lower region revealed that it can be further sub-divided into four zones according to their distinct microstructures. Zone I composed of some unidirectional growth dendrites with elongated pockets of Ta_2O_5 trapped within the dendrites, Ta(O).

The EDS results revealed that the overall Ta content in Zone I varied from 49.3 at% to 35.9 at% along the dendrite growth direction. The results also show that the oxygen content in Ta(O) was as high as 21 at%, and Ta in Ta_2O_5 reached 31 at%. This is believed to be due to solute trapping caused by rapid solidification. Apparently, the high Ta content in Zone I was the result of laser melting of the Ta substrate and the Ta has mixed with melted Ta_2O_5 to form a liquid solution. According to the Ta-O phase diagram and the composition range of this zone (figure 2c), when the melt cooled down, it passed through an immiscibility gap and liquid-liquid phase separation occurs, i.e. $L \rightarrow L_1$(Ta-rich)+L_2(O-rich). The Ta-rich L_1 droplets will sink towards the substrate because of its

IOP Conf. Series: Materials Science and Engineering **230** (2017) 012032 doi:10.1088/1757-899X/230/1/012032

higher density as compared with L_2. As the melt temperature drops, a monotectic reaction, L_1(Ta-rich)$\rightarrow$Ta(O)+ L_2(O-rich) takes place. Once the nucleation of the primary phase, Ta(O), occurs on the surface of tantalum substrate, it grows in a columnar form along a positive thermal gradient, and at the same time, oxygen is rejected into the residual liquid. Finally, this residual liquid solidifies, and αTa_2O_5 forms on both sides of the Ta-rich columnar dendrites. It was noticed that most of the αTa_2O_5 phase appeared as elongated pockets and were trapped within the Ta(O) dendrites (arrows in figure 2d show some examples). This could happen if the Ta(O) advancing rate is faster than the buoyant velocity of L_2. As a consequence, droplets of L_2 are engulfed by Ta(O) crystals (figure 2e). The liquid in close proximity to the L_1-Ta(O) interface is enriched with oxygen because of solute rejection, and as a result, L_2 droplets nucleate on the Ta(O)-L_1 interface (figure 2e). Finally, according to the phase diagram, L_2 solidifies at a lower temperature through a eutectic reaction. However, a lamellar eutectic morphology was not observed. This could be caused by complete separation of the two eutectic products, Ta(O) being deposited onto the existing Ta(O) dendrites, while αTa_2O_5 solidifies and grows into elongated pockets. The creation of a rough interface with a skeleton of dendrites formed between the substrate and the coating is considered to be beneficial because, to some extent, a metallurgical bond is achieved and the graded microstructure would provide a gradual change in properties which can relieve stress concentration at the interface.

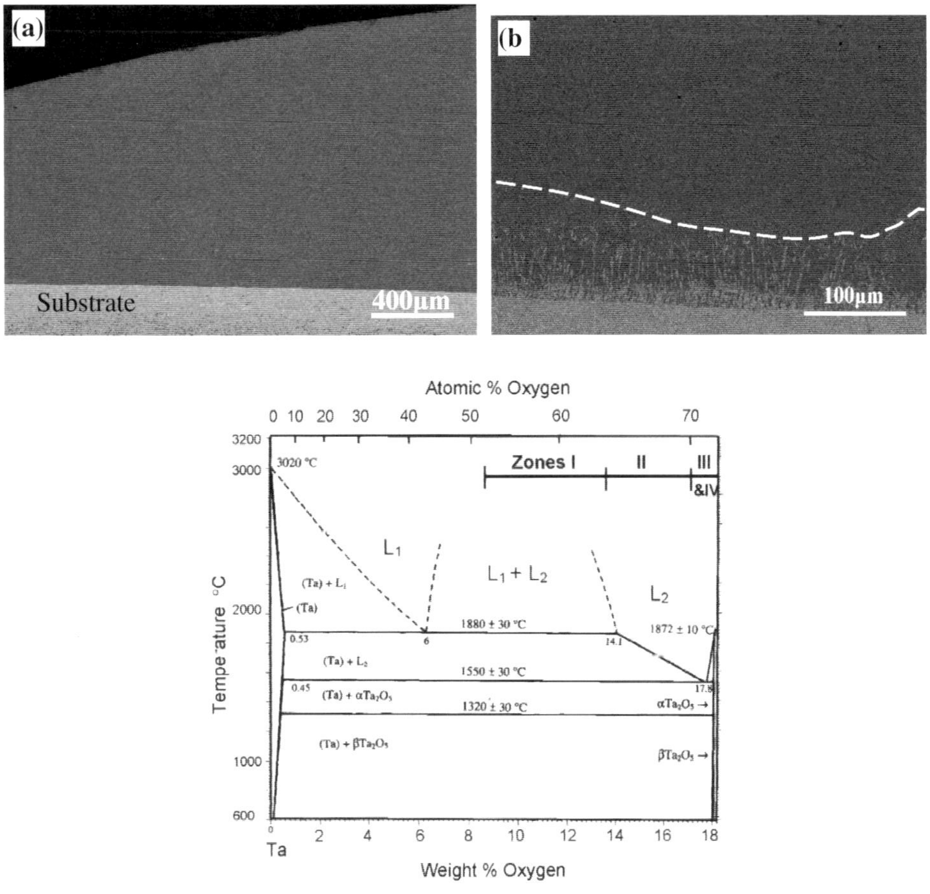

(c)

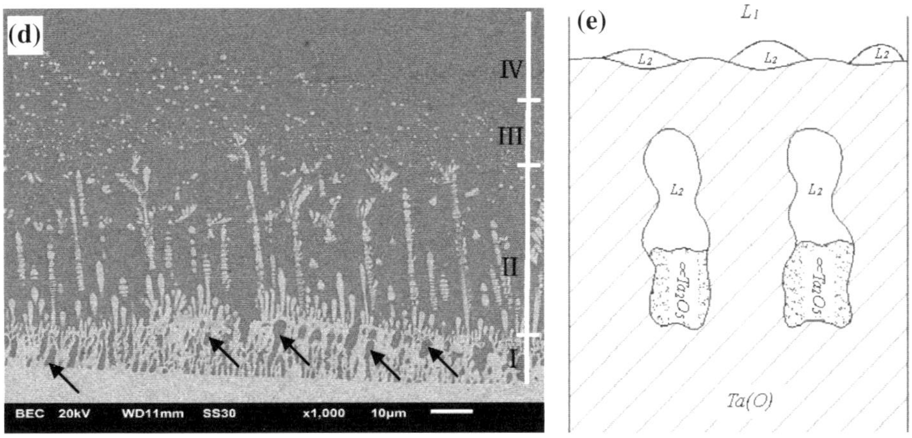

Figure 2. The whole laser-cladded coating on a Ta substrate (a); the microstructure near the interface (b); the Ta-O phase diagram [12] with the corresponding composition of different zones in the coating (c); a higher magnification of (b) arrows show examples of elongated pockets of Ta_2O_5 trapped in Ta(O) crystals (d); a schematic diagram showing growth morphology of the monotectic reaction products (e).

In Zone II, the microstructure was composed of some Ta-rich dendrites (light colour) and the Ta_2O_5 phase (dark). In this zone, the Ta content reached 30.5 at%; again, this was due to dilution of Ta of Zone I. When the temperature dropped, the Ta-O melt did not pass through the immiscible gap, but the primary Ta-rich phase nucleates directly onto the surface of the columnar crystals of Zone I, and resumes the columnar crystal growth. In Zone III, the overall Ta content was 28.7 at% and was close to the Ta_2O_5 phase of 28.57 at%. Some small spherical primary Ta(O) particles were found, the reason for this is not clear at this time, they may have originated from fragments of the columnar dendrites of Zone II, or nucleated as the primary phase. The remainder of the structure should be a eutectic phase with a very small fraction of Ta(O). However, in order to confirm this, a TEM study would be required. In Zone IV, the Ta content was found to be equal to that of Ta_2O_5, and dilution of Ta did not occur. As a result, a continuous Ta_2O_5 phase was formed. One of the problems encountered in interpreting the coating graded microstructure using existing phase Ta-O diagrams is that there are uncertainties with the Ta-O diagram as it was not completely established, especially towards the high oxygen regions.

4. Conclusions

(1) A dense graded Ta_2O_5 coating was successfully synthesized on a tantalum substrate by laser cladding using the pre-placed powder bed method with multi-layers. The quality of the coating is considered to be good, and is free from pores. Thus, this study reveals that laser cladding is a potentially effective method to produce relatively thick Ta_2O_5 coatings on Ta substrates.

(2) The graded microstructure was discussed using an existing Ta-O phase diagram. Despite the fact that the Ta-O diagram is incomplete, the experimental results of this study confirms the monotectic reaction of the diagram, and the morphology of the microstructure close to the coating-substrate interface can be explained in terms of the advancing rate of the solid-liquid interface.

5. References

[1] C Chaneliere, J L Autran and R A B Devine 1998 J. *Marer. Sci. Eng* 22 269
[2] Y Li, T T Zhao and S B Wei 2010 J. *Mater. Sci. Eng. C* 30 1227
[3] P Shang, S M Xiong and L H Li 2013 J. *Appl. Surf. Sci* 285 713
[4] F D Franco, M Santamaria and F D Quarto 2012 J. *Acta* 59 382
[5] T Lu, J Wen and S Qian 2015 J. *Biomater* 51 173

[6] YS Sun, J H Chang and H H Huang 2014 J. *Surf. Coat. Techn* 259 199
[7] B Díaz, J. S wiatowska and V Maurice 2013 J. *Electrochim. Acta* 90 232
[8] C L Tien 2008 J. Appl. *Surf. Sci* 255 2890
[9] C Mannequin, T Tsuruoka and T Hasegawa 2016 J. *Appl. Surf. Sci* 385 426
[10] W Hu, J Xu and X L Lu 2016 J. Appl. *Surf. Sci* 368 177
[11] M H Habibi, L Wang and J D Liang 2013 J. *Corros. Sci* 75 409
[12] S P Garg, N Krishnamurthy, A Awasthi and M Venkatraman 1996 J. *Phase Equilibria* 17 63

Acknowledgments

This paper was substantially supported by the research grants from the Hong Kong Polytechnic University (Project Nos. G-YN75 and G-YBE5). Zhang H. ould also thank the support from the Joint Fund of Iron and Steel Research by National Natural Science Foundation of China (NSFC) Grant No. U1560105.

IOP Conf. Series: Materials Science and Engineering **230** (2017) 012033 doi:10.1088/1757-899X/230/1/012033

Thermodynamic Study of Sn-Bi-Nd, Tb Ternary Systems

F Xu[1], Y T Chen[1], R Ye[1], Y Y Chen[1], X H Su[1], S L Wang[1], C Y Fu[1]

[1]School of Materials Science and Engineering, Southwest Petroleum University, Chengdu 610000,China.
E-mail:wsliang1465@126.com

Abstract. The aim of this study was to investigate the effect of the addition of rare earth elements on Sn-Bi-based alloy, and to study the phase equilibrium of Sn-Bi-Nd,Tb ternary systems by means of establishing the thermodynamic database. Combined with the thermodynamic parameters of relevant binary systems, the thermodynamic database of the Sn-Bi-Nd, Tb ternary systems has been developed to present the significant information for the design of low-temperature lead-free solder alloys.

1. Introduction

With the development of modern technology, low-temperature lead-free solders are used more and more wildly in the field of low-temperature assembly. Phase equilibrium plays an important role in the process of optimizing the composition of low-temperature lead-free solder. Owing to the low melting point and good wettability, the Sn-Bi alloy is treated as one important and representative choice of the low-temperature Pb-free solders [1-3]. The addition of rare earth (RE) elements can control and eliminate the negative effect of impurity elements, which improve the spreading property of Bi-based solder[4, 5]. Therefore, it is essential to understand the phase equilibria in the Sn-Bi-RE system.

In the present work, combined with the assessed relevant binary systems, the thermodynamic database of the Sn-Bi-RE (RE=Nd,Tb) ternary systems have been developed and provides important information about phase equilibria and thermodynamic properties, such as stable and metastable phase equilibrium, phase fraction, mixing enthalpy, activity, Gibbs energy of formation, etc. which is very helpful to design new types of low-temperature lead-free solders.

2. The Sn-Bi-Nd system

In the Sn-Bi-Nd ternary system, the phase equilibria of the Sn-Bi, Bi-Nd and Sn-Nd binary systems have been assessed by Lee et al. [6], Wang et al. [7] and Wang et al. [8] respectively. Combined with the thermodynamic parameters of Ref. [6-8], the phase equilibrium of the Sn-Bi-Nd ternary system has been calculated. The isothermal sections at 150°C and 200°C are predicated and shown in figures 1-2. Figure 3 shows the calculated longitudinal section of the Sn-Bi-Nd ternary system.

The melting temperature of low-temperature lead-free solder is 220°C to 230°C or lower, which is equal or lower than the melting point of the eutectic solder [9]. As is shown in figure 3, with the addition of the rare earth Nd element, the temperature of solidus is 139°C, and the temperature of liquidus up to 1 at.% Nd is under 230°C. In this range, the phase equilibrium can provide theoretical guidance for optimizing the composition of low-temperature lead-free solder. Above 1 at.% Nd, the temperature of liquidus exceed 230°C because of the formation of intermetallic compounds, which is not suitable for low temperature lead-free packaging materials. In addition, the isothermal sections and longitudinal section can provide theoretical guidance for the experimental study of the Sn-Bi-Nd ternary equilibrium.

IOP Conf. Series: Materials Science and Engineering **230** (2017) 012033 doi:10.1088/1757-899X/230/1/012033

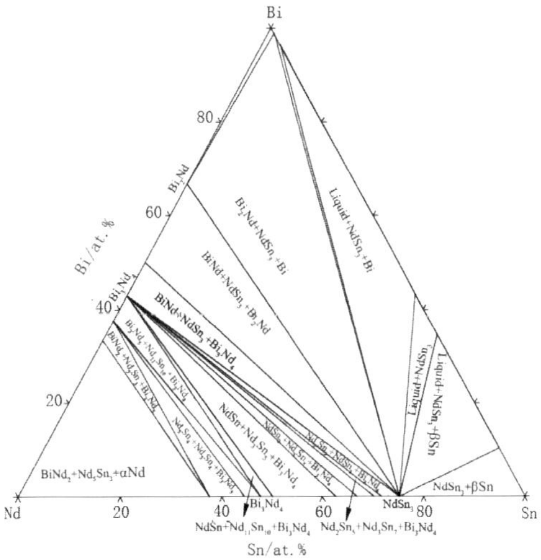

Figure 1. The calculated isothermal-section of the Sn-Bi-Nd ternary system at 150°C.

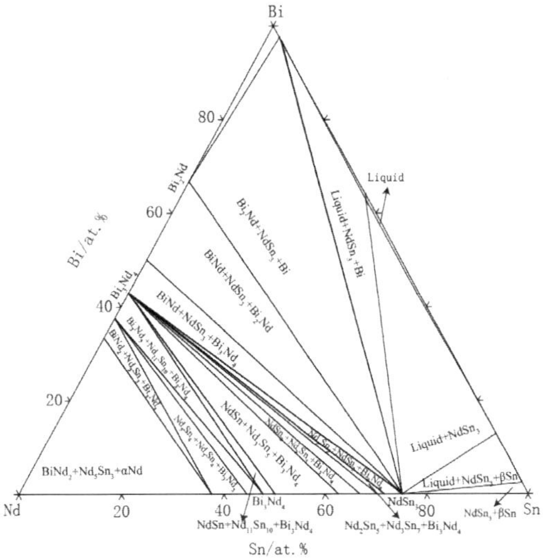

Figure 2. The calculated isothermal-section of the Sn-Bi-Nd ternary system at 200°C.

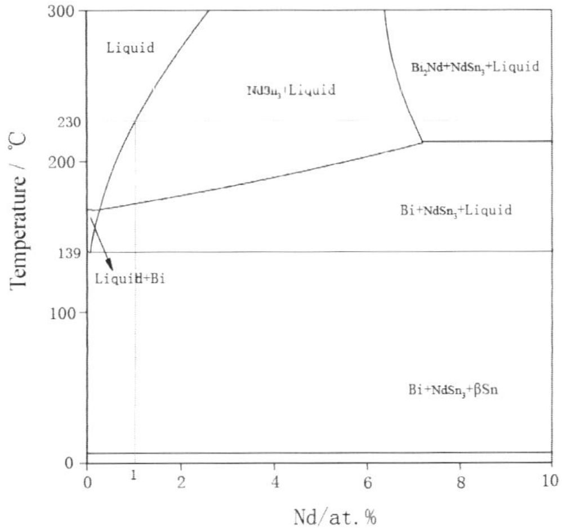

Figure 3. The calculated longitudinal-section of the Sn-Bi-Nd ternary system.

3. The Sn-Bi-Tb system

In the Sn-Bi-Tb ternary system, the thermodynamic assessments of the Sn-Bi, Bi-Tb and Sn-Tb binary systems were carried out by Lee et al. [6], Wang et al. [8] and Wang et al. [10] respectively. On the basis of the assessed thermodynamic parameters of relevant binary systems [6,8,10], the phase equilibrium of the Sn-Bi-Tb ternary system has been calculated. Figures 4-6 show the calculated isothermal section at 150 °C and 200 °C and longitudinal section.

As is shown in figure 6, with the addition of the rare earth Tb element, the temperature of solidus is 127°C, and the temperature of liquidus up to 8.3 at.% Tb is under 230°C. In this range, the phase equilibrium can provide theoretical guidance for optimize the composition of low-temperature lead-free solder. Above 8.3 at.% Tb, the temperature of liquidus exceed 230°C also because of the formation of intermetallic compounds, which is not suitable for low temperature lead-free packaging materials. What's more, the isothermal sections and longitudinal section can provide theoretical guidance for the experimental study of the Sn-Bi-Tb ternary equilibrium.

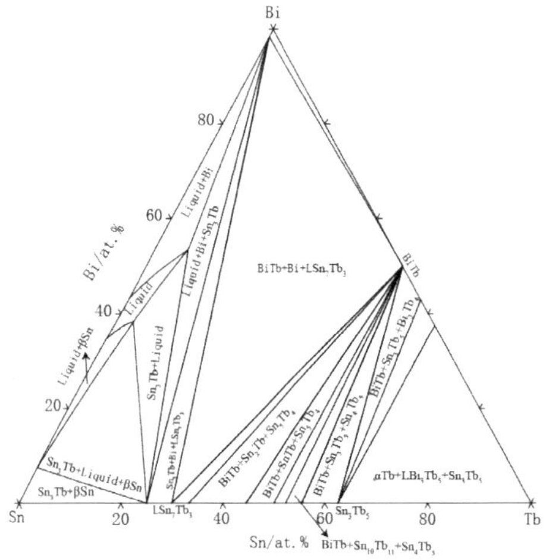

Figure 4. The calculated isothermal-section of the Sn-Bi-Tb ternary system at 150°C.

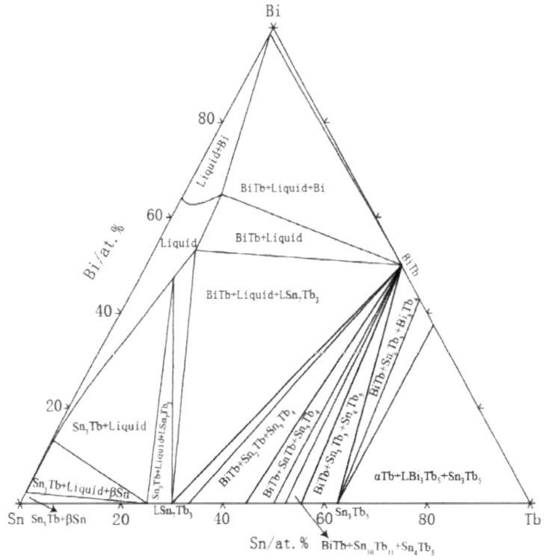

Figure 5. The calculated isothermal-section of the Sn-Bi-Tb ternary system at 200°C.

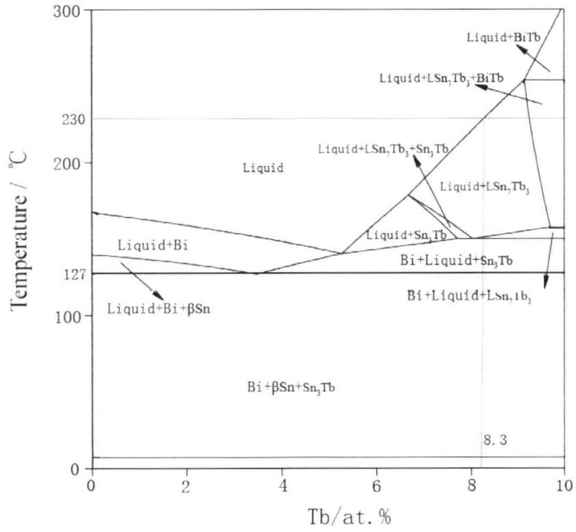

Figure 6. The calculated longitudinal-section of the Sn-Bi-Tb ternary system.

4. Summary

The phase equilibrium of the Sn-Bi-Nd and the Sn-Bi-Tb ternary systems has been calculated on the basis of the thermodynamic parameters of relevant assessed binary systems. And the effect of addition of RE (Nd,Tb) element is theoretical analyzed on the Sn-Bi-based lead-free solder.

5. References

[1] Wang Z J, Lin L U and Wang M 2005 J. Shenyang University of Technology 27 377-380
[2] El-Daly A A, Swilem Y, Makled M H, El-Shaarawy M G and Abdraboh A M 2009 J. Alloys Compd. 484 134-142
[3] Osório W R, Peixoto L C, Garcia L R, Nathalie M N and Amauri G 2013 J. Alloys Compd. 572 97-106
[4] Chen Q H, Chen R G and Xiao L R 2009 Polym. Bull. 62 209-223
[5] Zhang L, Xue S B, Gao L L, Chen Y, Yu S L, Sheng Z and Zeng G 2009 J. Mater. Sci. 20 1193-1199
[6] Lee B J, Oh C S and Shim J H 1996 J. Electron. Mater. 20 471-480
[7] Wang C P, Zhang H L, Tang A T, Pan F S and Liu X J 2010 J.Alloys Compd. 502 43-48
[8] Wang S L, Su X H, Xu F, Wang C P and Liu X J 2017 Calphad. (in press)
[9] Liu P, Long Z Y, Gu X L, Feng J C and Song X G 2014 Electron. Process Technol. 35 198-200
[10] Wang S L, Hu Z B, Gao F, Wang C P and Liu X J 2011 J. P. Equilib. Diffus. 32 441-446

Acknowledgements

This work was supported by the sixteenth stage (2016-2017) college students open experiment in Southwest Petroleum University and Innovative training program (201610615034).

TiO_2 Nanotube Arrays Modified With Zinc and Loaded with Hydrophilic and Hydrophobic Drugs

Fengfen Zhang, Xiaowei Zhan, Xiaojuan Yang and Xiufeng Xiao*

Fujian Provincial Key Laboratory of Advanced Materials Oriented Chemical Engineering, College of Chemistry and Chemical Engineering, Fujian Normal University, Fuzhou, 350007, China
E-mail: xfxiao@fjnu.edu.cn

Abstract. TiO_2 nanotube arrays were prepared by anodization of Ti plat, then modified by hydrothermal method in $Zn(NO_3)_2$ solution and finally loaded alendronate or ibuprofen drugs by vacuum drying. The experimental results show that the amount of two drugs loaded on the modified TiO_2 nanotube array is increased. The release behavior of the modified nanotubes including burst release and sustained relase. high hours before release. The modified TiO_2 nanotube arrays were more slowly released during the sustained release than unmodified nanotubes, indicating that the Zn^{2+} modified nanotubes have certain sustained-release effect.

1. Introduction

Bone disease, such as fracture, osteoporosis and bone cancer, can suffer people from some serious and chronic disease[1], in the field of oral surgery and orthopedic surgery, experts have made a lot of efforts and attempts, in order to make bone healing and bone growth rapidly at the period of osteogenesis component and bone model reconstruction[2-4].During the bone metabolism, trace elements play an important role in the in vitro evaluation, which can accelerate bone regeneration. Zinc is one of essential trace elements, which has been proven to have direct effects on bone mineralization, especially in nucleation and the growth of minerals, in addition, by stimulating the assessment of the proliferation and mineralization of bone cells, the zinc in bone metabolism had demonstrated good assimilation effect[5-8]. so it plays an important role in skeletal development. Based on the important role of zinc ions in bone formation, a kind of zinc modified implant has been developed, for the zinc ions will released from the surface of the implant, which made it have special functions.

TiO_2 nanotube arrays have the regular hollow structure, so it has been used in the field of drug loading. In this paper, TiO_2 nanotube arrays were hydrothermal modified in zinc nitrate solution,and then loaded with hydrophilic drugs and hydrophobic drugs, the release behavior of TiO_2 nanotube array drug loading system in phosphate buffer solution (PBS) was investigated.

2. Materials and Method

2.1. Preparetion of TiO_2 nanotube arrays

Titanium foils were cut into 1cm*10cm and successively polished by 280#, 320#, 400#, 600# abrasive paper until Ti surface was smooth and flat. TiO_2 nanotube arrays can be synthesized by anodicoxidation in the electrolyte contains 0.50 wt% NH_4F, glycerol and 10vol% H_2O at 60V and

30℃. After this, titania nanotube surfaces were rinsed with deionized water and then dried for subsequent use.

2.2. Hydrothermal treatment

Titanium sheet after anodic oxidation should be place in 70 ml zinc nitrate solution, and be treated at 200℃ for a certain time. After this process, the hydrothermal reactor should be cooled down at room temperature for 2h, then the titanium foils should be taken out, cleaned by deionized water and dried in 50℃. The next step was to change the hydrothermal treatment time (0.5h, 1.5h, 2h), in order to research and discuss the influence of different processing time on the drug loading amount. The surface morphology of the samples were studied by scanning electron microscopy (JSM-7500F).

2.3. Drug Loading and release

Add 10μL the alendronate and ibuprofen-ethanol solution respectively onto the surface of modified TiO_2 nanotube arrays, then dry these samples in a vacuum drying oven for 1 hour at room temperature (25°C). To make sure that an amount of drug was loaded into the nanotube, the loading and drying steps were repeated over 5 times.
Modified TiO_2 nanotube arrays loaded with drugs were put in 10ml PBS solution in a sample bottle. Then put the sample bottle in a constant-temperature water bath oscillator at 37.0℃ at the speed of 100r/min. After a certain time interval, 0.5ml solution was removed to test the drug contents by UT1900 type double beam UV spectrophotometer.

3. Results and discussions

3.1. Morphology of TiO_2 nanotube arrays containing zinc titanate

Figure 1a shows the surface topography of the pure TiO_2 nanotube arrays prepared by anodic oxidation method. Figure 1b is the surface morphology of the above samples hydrothermal treated by zinc nitrate solution at 200 °C. Compared with the figure 1, it can be found that the tube structure of TiO_2 nanotube array after the hydrothermal treatment can be seen clearly and almost unchanged, but the nanotube wall became obviously thicker. Figure 1d is the bottom of the TiO_2 nanotube arrays prepared in the above system, by contrast, figure 1c is the TiO_2 nanotube array after hydrothermal treatment. The results show that the whole interior of the nanotube had been successfully modified. The demonstration process is as follows: titanium and its oxide will form $HTiO_3^-{\cdot}nH_2O$ in alkaline solution($TiO_2+OH^-+nH_2O \rightarrow HTiO_3^-{\cdot}nH_2O$), under the condition of high temperature and high pressure, the reaction between the formation of titanic acid molecules and the precipitation molecules of Zn^{2+} around the feet can generate $ZnTiO_3$. Therefore, the collision reaction has no relation to the form of Zn^{2+} precipitation, the various forms of precipitation of Zn^{2+} can react with the titanic acid to produce zinc titanate, due to the large occupied space, the tube wall became thicker and the tube diameter became smaller.
In order to study the influence of hydrothermal time on the surface of TiO_2 nanotube arrays, TiO_2 nanotube arrays were hydrothermal treated in 0.01M zinc nitrate solution at 200℃ respectively 0.5 h, 1.5 h, 2h. The surface morphology of the prepared samples is shown in figure 2. With the increase of hydrothermal time, the diameter of TiO_2 nanotube arrays were chang smaller and smaller, and it is almost impossible to see the TiO_2 nanotube arrays (figure 2c) when the hydrothermal time reaches 2h. The reason is probably because along with the prolonging of hydrothermal time, ZnTiO3 are formed with the reaction of Zn^{2+} and TiO_2 nanotube arrays. Finally, a large number of $ZnTiO_3$ accumulated at outside tubew ith the continuous formation of $ZnTiO_3$, out of the pipe piles ZnTiO3. Based on the above analysis, it is found that the hydrothermal time has a significant influence on the formation of $ZnTiO_3$ in the TiO_2 nanotube arrays.
In order to study the influence of hydrothermal temperature on the morphology of TiO_2 nanotube arrays. TiO_2 nanotube array were hydrothermal treated at 100 ℃, 150℃ and 200℃, as shown in figure. 3. With the increase of hydrothermal temperature, the diameter of TiO_2 nanotube arrays is more and

IOP Conf. Series: Materials Science and Engineering **230** (2017) 012034 doi:10.1088/1757-899X/230/1/012034

more small, and even some pipe is blocked, especially when the water temperature reaches 150℃. The reason may be that with the increase of hydrothermal temperature, more $ZnTiO_3$ are formed with the reaction of Zn^{2+} and TiO_2 nanotube. The higher the temperature, the faster the reaction, resulting in a large amount of ZnTiO3 outside the tube.

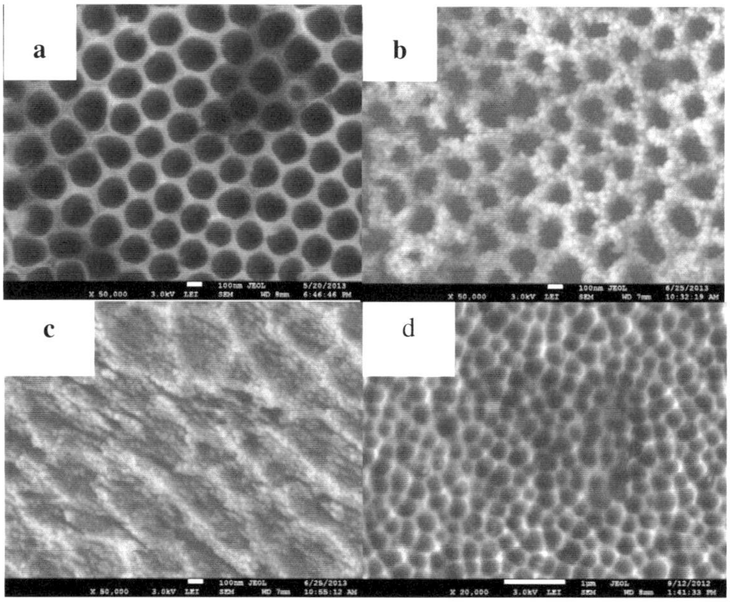

Figure 1. SEM images of modified and unmodified TNTs ,(a)(c)0.01M $Zn(NO_3)_2$ modified TNTs at 200°C with 0.5h; (b) (d) unmodified TNTs.

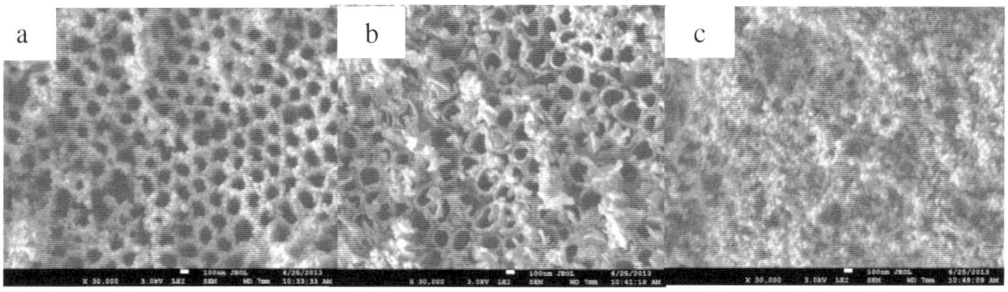

Figure 2. SEM images of modified TNTs hydrothermal at 200°C for 0.5h(a), 1.5h(b) and 2h(c)

Figure 3. SEM images of modified TNTs hydrothermal at 100℃(a), 150℃(b) and 200℃ (c) for 2h

3.2. Elemental analysis of TiO_2 nanotube arrays.

Figure 4 is EDS of TiO_2 nanotube arrays after 200℃ hydrothermal treatment of 0.01M zinc nitrate solution at 0.5h. Analysis showed that the zinc(3.05%) had appeared in the amorphous TiO_2 nanotubes after hydrothermal treatment of zinc nitrate solution, which meaned that the particles formed in the nanotube walls had already contained zinc element after the hydrothermal process. So it can be speculated that zinc had already been doped sucessfully into nanotubes.

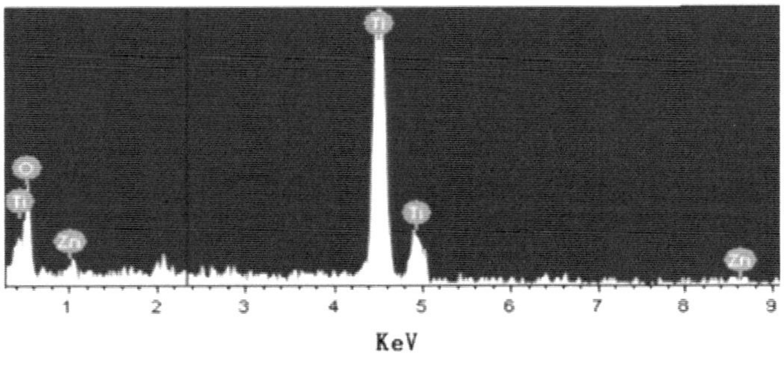

Figure 4. Energy spectrum analysis of TiO_2 nanotube array

3.3. Drug release process.

The alendronate and ibuprofen drug release curve of the zinc modified TiO_2 nanotube arrays and unmodified TiO_2 nanotube arrays were shown in Figure 5. Also, According to the two graphs, the release curves of the two loading systems are similar, and the drug release of the TiO_2 nanotube arrays modified with zinc titanate have the two stages: the early rapid release stage and the later slow release stage. The main reason of rapid release stage is that some of the drugs in the final vacuum drying failed to enter the nanotube interior, which had stacked on the nanotube nozzle and nanotube array surface, resulting in a sudden release. While the reason of the slow release is mainly from nanotube inside and drugs adsorbed in the tube wall. However, the slow release of the unmodified TiO_2 nanotubes was very poor, the drug was released completely at almost 24h. The reason is that the zinc titanate formed on the inner surface of the TiO_2 nanotube makes the inner part of the nanotube rough, and the drug was loaded inside the nanotube.

As for the ibuprofen drug release, there was a slow release after 150h, and the total amount of the drug had increased constantly. The reason may be that the diameter of the tube becomes smaller, the inside of the nanotube becomes rough, and the rough inner wall will make a positive obstruction in the drug release. All of these factors make it difficult to release ibuprofen particles inside the nanotube.

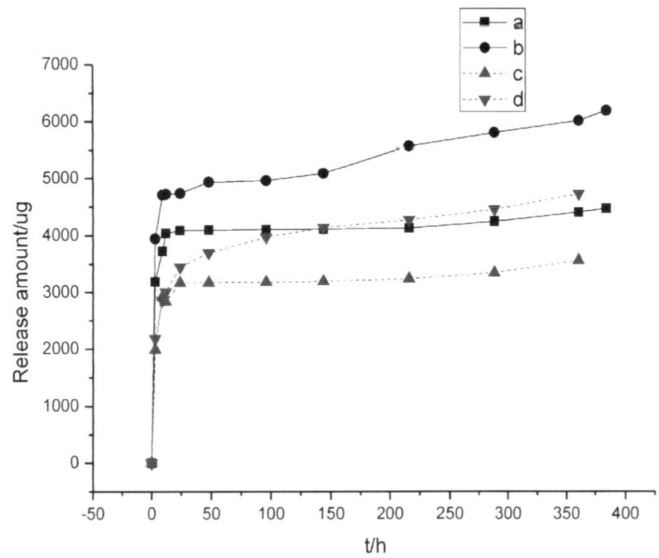

Figure 5. Comparative release of ibuprofen from modified TNTs (a) and unmodified TNTs(b), ibuprofen released from modified TNTs (c) and unmodified TNTs(d)

4. Summary

Modified TiO_2 nanotube arrays with zinc titanate were prepared by hydrothermal treatment,which can be loaded with hydrophilic and hydrophobic drugs via vacuum drying method, then the adsorption and release situation of hydrophilic alendronate and hydrophobic ibuprofen in modified TiO_2 nanotube arrays and unmodified TiO_2 nanotube arrayswere compared. The sustained release effect of hydrophilic alendronate and hydrophobic ibuprofen in PBS solution were more obvious in TiO_2 nanotube arrays modified with zinc titanate, and had increased about 30%.

5. References

[1] Sakamoto K, Nakamura T and Hagino H 2006 *Journal of Orthopaedic Science*. 11 467.
[2] Yusa K, Yamamoto O, Fukuda M 2011 *Biochemical and biophysical research communications*. 412 273.
[3] Iwaniec U, Magee K, Mitova-Caneva N 2003 Bone 33 380.
[4] Kawaguchi H, Nakamura K, Tabata Y 2001 *Journal of Clinical Endocrinology & Metabolism*. 86 875.
[5] Wang X, Ito A, Sogo Y 2010 *Acta biomaterialia* 6 962.
[6] Miao S, Cheng K, Weng W 2008 *Acta biomaterialia*. 4 441.
[7] Miao S, Lin N, Cheng K 2011 *Journal of the American Ceramic Societ* 94 255.
[8] Saino E, Grandi S, Quartarone E 2011 *European Cells and Materials*. 21 59.

Acknowledgments

This work was financially supported by key clinical specialty discipline construction program of fuzhou, and Fujian Province Nature Science Foundation (Grant no.2016Y0025, 2016J01480).

Effect of Controlled Rolling and Cooling On Microstructure and Mechanical Properties of 30crmnti Wire Rod

Shipeng Ruan[1,2,3, a], Qing Dong[2,3,b], Lei Zhang[2,3,c] and Lijun Wang[2,3,d]

[1]Metallurgy Engineering Research Institute; University of Science and Technology Beijing, Beijing 100083, China
[2]Xingtai Iron and Steel Corp., Ltd. Xingtai 054027, China
[3]Hebei Engineering Research Center for Wire Rod, Xingtai 054027, China
E-mail: [a]shipengruan@163.com, [b]jszx2014@xtsteel.com, [c]xcyyyj@xtsteel.com, [d]wljun0604@126.com

Abstract. The effect of controlled rolling and cooling on microstructure and mechanical properties of alloy structure steel 30CrMnTi wire rod with diameter 6.5mm was studied. The results show that the lower finish rolling temperature resulted in a decrease in tensile strength but an increase in elongation and reduction of area. When the finish rolling temperature decreases from 950℃ to 850℃, the tensile strength value decreases from 750MPa to 660MPa, and the elongation increases from 21% to 30%, the reduction of area increases from 64% to 71%. The grain size also refines from 20 μm to 9.9 μm when the finish rolling temperature decreases from 950℃ to 850℃. The decrease of tensile strength is due to the change of microstructure which evolved from more bainite to ferrite and pearlite.

1. Introduction

30CrMnTi is widely used in the manufacture of high quality pin shaft parts[1-2]. As a kind of medium carbon alloy structural steel, 30CrMnTi has high tensile strength which results in the difficulty of subsequent cold drawing. This steel wire rod is easy to obtain quenched microstructure such as bainite or martensite due to its high alloying elements of chromium and manganese. Based on the study of the continuous cooling transformation behavior of austenite in 30CrMnTi steel, bainite occurs when the cooling rate reaches 1℃/s[3]. Controlled rolling and controlled cooling, which is often called as Thermo-Mechanical Controlled Processes (TMCP), can improve the quality of wire rod by controlling the recrystallization or grain growth of austenite and subsequent transformations.
In this paper, the influence of controlled rolling and controlled cooling process on the microstructure and properties of 30CrMnTi hot rolled wire rod with diameter 6.5mm was studied.

2. Experimental material and procedure

Table 1 shows the main chemical composition of experimental 30CrMnTi steel, which were supplied by Xingtai Iron & Steel Corp., LTD. The experimental steel was first cast into billet with the section size of 160*160mm2, and then rolled into wire rod with diameter of 6.5mm by heating, rough rolling, medium rolling and final rolling process.
Table 2 shows the tested controlled rolling and cooling process of 30CrMnTi steel wire rod. The finish rolling temperature of process 1# is 950℃ which is in recrystallization region, and 900℃ of process 2#, and 850℃ of process 3#. The laying head temperature is about 50℃ lower than the finish rolling temperature. All the cooling modes are slow cooling.

Table 1. Chemical composition of experimental 30CrMnTi steel（wt，%）

Steel grade	C	Si	Mn	P	S	Cr	Ti
30CrMnTi	0.24-0.32	0.17-0.37	0.80-1.10	≤0.03	≤0.03	1.00-1.30	0.04-0.10

Table 2. The controlled rolling and cooling process of 30CrMnTi wire rod

Process number	Finish rolling temperature/℃	Laying head temperature/℃	Cooling process
1#	950	900	Slow cooling
2#	900	850	Slow cooling
3#	850	800	Slow cooling

The mechanical properties of the wire rod were tested by WDW50 universal testing machine, and the microstructure was observed by Leica DM2700M optical microscope (OM).

3. Results and analysis

3.1 Mechanical properties of different processes

Table 3 shows the mechanical properties of experimental 30CrMnTi wire rod with three different processes. The tensile strength decreases with the decrease of finish rolling temperature and the laying head temperature, but the plasticity (such as elongation and reduction of area) increase gradually. When the finish rolling temperature is 950℃, the average tensile strength is 750MPa, the average reduction of area is 64%, the average elongation is 21%; when the finish rolling temperature is 900℃ (Process 2#), the average tensile strength decreased to 710MPa, the average reduction of area is 68%, the average elongation is 26%; when perform process 3#, the average tensile strength is further reduced to 660MPa, the average reduction of area is 71%, the average elongation is 30%. Compared with the process 1#, the tensile strength of process 3# was reduced by 90MPa, the reduction of area increased by 7%, the elongation increased by 9%.

Table 3. The mechanical properties of the 30CrMnTi wire rod with diameter 6.5mm

Process number	Tensile strength / MPa	Reduction of area / %	Elongation / %
1#	715-825（750）	60-69（64）	16-28（21）
2#	677-750（710）	67-70（68）	25-28（26）
3#	644-705（660）	68-72（71）	27-32（30）

Note: the values in parentheses are average value

3.2 Microstructure of different processes

Figure 1-3 shows the microstructure of experimental 30CrMnTi wire rod with three different processes. The microstructure is quite different with different. With the process 1#, the microstructure of the steel is mainly bainite and a small amount of ferrite and pearlite shown in figure 1.

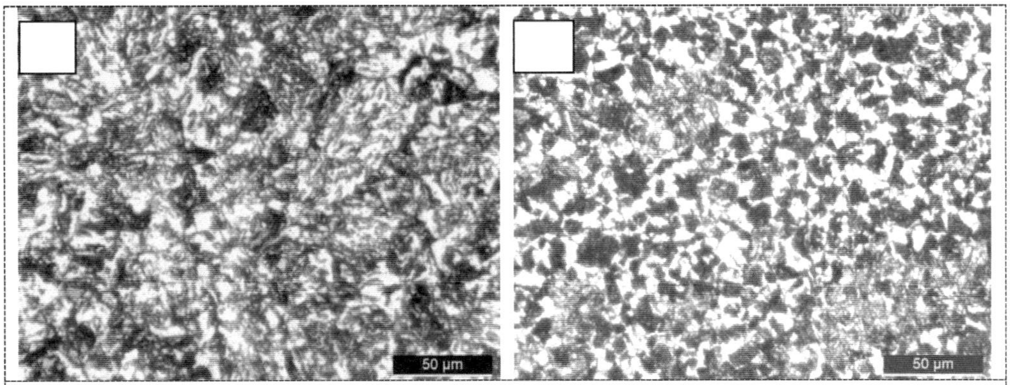

Figure 1. The microstructure of 30CrMnTi with process 1#.

With the process 2#, when the finish rolling temperature decreases, the bainite was significantly reduced and found only at individual non overlapping locations as shown in figure 2c. The microstructure of tested steel with process 3# is mainly ferrite and pearlite, no bainite can be found.

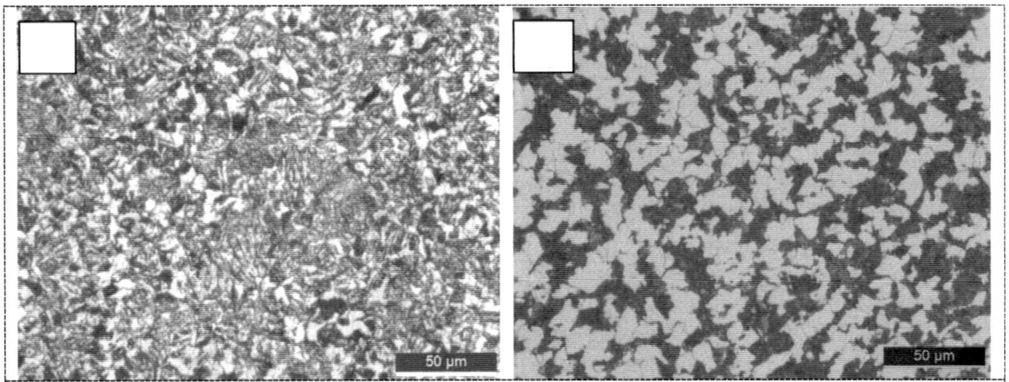

Figure 2. The microstructure of 30CrMnTi with process 2#.

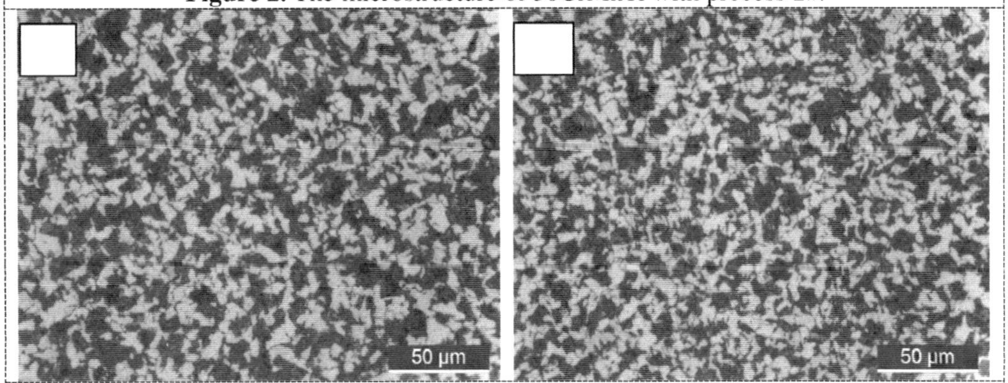

Figure 3. The microstructure of 30CrMnTi with process 3#.

The grain size and ferrite content in different processes were calculated, as shown in table 4. Obviously, the average grain size decreases with the decrease of finish rolling temperature. The average grain size is 20 μ m with process 1#, when the finish rolling temperature become 900℃ （Process 2#）, the average grain size become 14 μ m; when the finish rolling temperature become 850℃, the average grain size become 9.9 μ m. The ratio of ferrite content increases from 35-40% with process 1# to 46-50% with process 3#.

Table 4. The average grain size and ferrite content of the 30CrMnTi wire rod

Process number	Grain size number	Average grain size /μm	Ratio of average ferrite content /%
1#	8	20	35-40
2#	9	14	44-48
3#	10	9.9	46-50

3.3 Discussion

In the case of the same chemical composition, the microstructure of the steel is affected by the controlled rolling and controlled cooling process [4-5]. During the traditional process rolling, the recrystallization and grain growth of deformed austenite at high temperature, large grain size is obtained due to the high driving force. During TMCP, with the decrease of deformation temperature, the driving force of grain growth is reduced, and it's easy to obtain the fine recrystallization grain. When deformation temperature is further reduced, the deformation is in the non-recrystallization zone rolling, recrystallization does not occur during the deformation process, deformed austenite is elongated, and a large amount of energy is stored in the deformed austenite grain boundaries or sub grain boundaries, which will be preferentially nucleated in the subsequent phase transformation process, and the uniform and fine grains can be obtained after the phase transformation.

In this study, the process 1# is traditional process rolling with high deformation temperature, and the grain size is relatively larger after the transformation; The large grain size can improve the hardenability of steel, and because of the high Cr, Mn and other alloying elements, the hardenability of 30CrMnTi can also be improved. For this reason, it is easy to produce bainite structure in the slow cooling process of 30CrMnTi steel. The existence of bainite leads to the increase of tensile strength and the decrease of plasticity for process 1#.

The deformation temperature of process 2# is lower than that of process 1#, the driving force of grain growth is reduced, the grain size decreases after the transformation, and the microstructure is mainly composed of ferrite and pearlite. This kind of microstructure can reduce the tensile strength and increase the plasticity obviously. But in the non-overlapping position of the wire rod, there are also some bainite exists because of the cooling rate is faster than the overlapping position.

The deformation temperature is further reduced by process 3#, and the grain size is smaller after transformation than process 2#. The low deformation temperature makes the CCT curve move to the left[4], which reducing the hardenability of the steel. In the same slow cooling process, the microstructure of process 3# is totally ferrite and pearlite with no bainite. And the tensile strength is further reduced to 660MPa, the elongation and reduction of area were further improved.

The experimental wire rod of Process 3# was drawn directly from 6.5mm to 4.48mm with annealing-free. The properties after drawing can meet customer requirements.

4. Conclusions

Based on the study of the effect of different controlled rolling and controlled cooling processes on the microstructure and properties of 30CrMnTi, the following conclusions can be obtained:

(1) The finish rolling temperature decreased from 950℃ to 900℃, the tensile strength of 30CrMnTi wire rod with diameter 6.5mm is decreased from 750MPa to 710MPa, the elongation increased from 21% to 26%, the reduction of area increased from 64% to 68 %.

(2) The finish rolling temperature decreased from 900℃ to 850℃, the tensile strength of 30CrMnTi wire rod with diameter 6.5mm is decreased from 710MPa to 660MPa, reduction of area and elongation have been further improved.

(3) The CCT curve is shifted to the left by the lower finish rolling temperature, which makes the transformation process of 30CrMnTi steel avoid bainite transformation region, and the uniform ferrite and pearlite structure can be obtained.

5. Reference

[1] Xiao J X 2006 Improvement of loading and clamping for heat treatment of 30CrMnTi Steel Rim Support Shaft *Metal heat treatment*.**12**(2006) p89-91

[2] Yu X S, Shen K, Liao S L and Ma H 2016 Studying and practicing of 30CrMnTi high speed wire rolling process *Shanghai metals*. **2**(2016)p12-17

[3] Wang L J, LI B X, Chen J L, Zhai J P, Feng Z X and Wang N T 2016 Continuous cooling transformation behavior of undercooled austenite of 30CrMnTi steel *Heat treatment of metals*. **8**(2016)p40-44

[4] Hata H, Yaguchi H, Shimotusa M, Toyama M 2000 Development of high quality wire rod through thermomechanical control process, *Kobe steel engineering reports*. **1**(2000)p29-32

[5] Ruan S P, Wang L J, Chen J L, Guo M Y, Zhai J P 2016 Effect of control-rolling-cooling process on structure and properties of steel SCM435 rod coil *Special Steel*. **5**(2016)p45-47

Acknowledgments
The authors wish to thank Mr H L Ma and Z Y Wu (wire rod factory of Xingtai Steel) for their help in the arrangement of the experiment.

IOP Conf. Series: Materials Science and Engineering **230** (2017) 012036 doi:10.1088/1757-899X/230/1/012036

Fabrication and Properties of Al-Cu-Mg-Zn Series Alloys with Low Hot Cracking Tendency for Liquid Forging

Y G Sun[1], Y N Su[1], Z M Du[1], L H Chen[2], C S Wang[2]

[1]School of Materials Science and Engineering, Harbin Institute of Technology, Harbin, 150001, China
[2]Beijing North Vehicle Group Co. Ltd., Beijing, 100072, China

E-mail: duzm@263.net (Z M Du)

Abstract. Twelve new kinds of Al-Cu-Mg-Zn alloys with different components were designed on the basis of optimizing main alloying elements, including Zn, Mg and Cu to resolve hot crack problems of Al-Cu-Mg-Zn series alloys during liquid forging. Hot cracking susceptibility (HCS) was adopted to characterize hot cracking tendency, evaluated using constrained rod casting in a steel mould. The influence of those three main alloying elements on the hot cracking tendency and the mechanical properties of new alloys were researched. The results show that most of those new alloys have better hot crack tendency performances than 2024 alloy and 7075 alloy. The 8th alloy, Al-5Cu-4.5Mg-2.5Zn, shows best mechanical properties after liquid forging, which hot crack tendency index is the lowest, only reaching 40. The tensile strength is 327 MPa and the elongation to fracture is 2.7%, which Brinell hardness reaches 107 N/mm^2.

1. Introduction

As the lightweight requirements in the field of aerospace, weaponry and transportation enhanced increasingly, it is expected more and more intensively to use aluminum instead of steel, especially in those important mechanical parts and structural components with high demand of lightweight that equipped in spacecraft, tanks, artillery, aircraft, high-speed rails and cars [1-2]. In recent years, the trend to use aluminum and aluminum alloys to replace original steel structures is growing dramatically. For example, aircraft components, automobile hubs, bumpers, tank road wheels, helicopter rotating rings and stationary rings, train cylinders and pistons are manufactured by forging used aluminum alloy. However, traditional die forging technology cannot completely meet the requirements to manufacture and use those aluminum alloy parts. And the fact gives more and more attentions on the research of liquid forging and semisolid forging forming technology [3].

Liquid forging forming technology is a process that a certain amount of molten metal liquid is poured directly into metal mold, fills the mold, solidifies and crystallizes with some plastic deformation under special pressure, to obtain net-shape parts which structure are closely knitted. Its advantages are saving time and energy, efficient utilization. What's more, the parts with the final size show good performances close to hot forging parts. But as a result of liquid forging forming mainly companied with solidification, it is hard to avoid the production of hot cracks caused by the uneven stress field during the end of solidification, which seriously restrict the development of liquid forging forming technology[1,4].

Combined the advantages and disadvantages of Al-Cu series alloys and Al-Zn series alloys, it was researched and discussed how to resolve the production of hot cracks during the aluminum alloy liquid forging forming process, from the perspective of alloy element ratios, to fabricate aluminum alloys with less hot cracks and better comprehensive performances for liquid forging forming technology. Hot cracking tendency and mechanical properties of new designed aluminum alloy were studied.

2. Experimental procedure

2.1. Component design of aluminum alloy and alloy melting

Commercial extruded 2024 aluminum alloy rods of 30 mm diameter were used as matrix materials. According to the single variable method, some new aluminum alloys with low hot cracking tendency were designed and fabricated by adding industrial pure zinc and magnesium, and Al-Cu (50%) intermediate alloy. Hot cracking susceptibility of new aluminum alloys, 2024 aluminum alloy and 7075 aluminum alloy were compared. The chemical components of 2024 and 7075 aluminum alloys were shown in table 1.

Table 1. Chemical components of 2024 and 7075 aluminum alloys

Elements (*wt. %*)	Cu	Mg	Zn	Cr	Mn	Ti	Fe	Si	Al
2024	4.90	1.13	0.10	0.01	0.82	0.04	0.20	0.20	others
7075	1.54	1.29	6.18	0.17	0.28	-	0.35	0.25	others

Combined with Al-Zn-Mg-Cu series alloys phase diagram [5], it must ensure that content of zinc element is lower than 6%, that content of Mg element is higher than 2%, and that content of Cu element is higher than 1.5%, to reduce the alloy hot cracking tendency and satisfy the aluminum alloy usage requirement of the strength, hardness and toughness. Main content of alloy elements were optimized by the theoretical limit. And $1^{\#}$ ~ $12^{\#}$ new aluminum alloys were designed by only changing one of three main alloying elements component proportions, as shown in table 2.

Table 2. Component proportions of twelve kinds of aluminum alloys

Number	Mass fractions /Wt. %		
	Zn	Mg	Cu
$1^{\#}$	1.1-2.0	3.0-3.9	4.5-5.4
$2^{\#}$	2.1-3.0		
$3^{\#}$	3.1-4.0		
$4^{\#}$	4.1-5.0		
$5^{\#}$	5.1-6.0		
$6^{\#}$	2.1-3.0	1.0-1.9	4.5-5.4
$7^{\#}$		2.0-2.9	
$8^{\#}$		4.0-4.9	
$9^{\#}$	2.1-3.0	3.0-3.9	1.5-2.4
$10^{\#}$			2.5-3.4
$11^{\#}$			3.5-4.4
$12^{\#}$			5.5-6.4

The amount of alloys were calculated and weighed according to the design as shown in table 2. Melting furnace was heated to 300 °C ~ 400 °C, and certain amounts of 2024 aluminum alloy and Al-

Cu (50%) intermediate alloy were added and heated up to 750 °C, waiting to melt. And a small amount of NaCl & KCl covering agent was sprinkled. Then furnace temperature was cooled to 700 °C, pure magnesium and pure zinc was added respectively and melted. When the furnace temperature was back to 710 °C ~ 730 °C, melted alloy was refined twice adding refining and degassing agent, C_2Cl_6.

2.2. Hot cracking tendency test

Considering the causes of hot cracks, hot cracking susceptibility was adopted to characterize hot cracking tendency, evaluated using constrained rod casting in a steel mold [6-7], which was compared with 2024 and 7075 aluminum alloy performances, to design a new type of aluminum alloys with the lowest hot cracking tendency. Figure 1 and figure 2 show the schematic and physical pictures of hot cracking rod mold respectively. The mold was preheated to 300 °C before the melted alloy was poured in when the temperature of melt reached 730 °C. The melt filled the cavity and solidified. During this process, melted alloy solidification and contraction were limited by the mold cavity, to make the alloy constriction impeded and produce hot cracks. The hot cracking tendency of different aluminum alloys were compared from the position and the crack depth. Cracking rods were constrained from two parts, the runners and the balls, in the process of solidification. By observing hot cracking rod samples and recording, hot cracking susceptibility indexes (HCS) were calculated by the Formula 1 as shown.

$$\text{HCS} = \sum(\omega_{\text{crack}} \cdot f_{\text{length}} \cdot f_{\text{location}}) \quad (1)$$

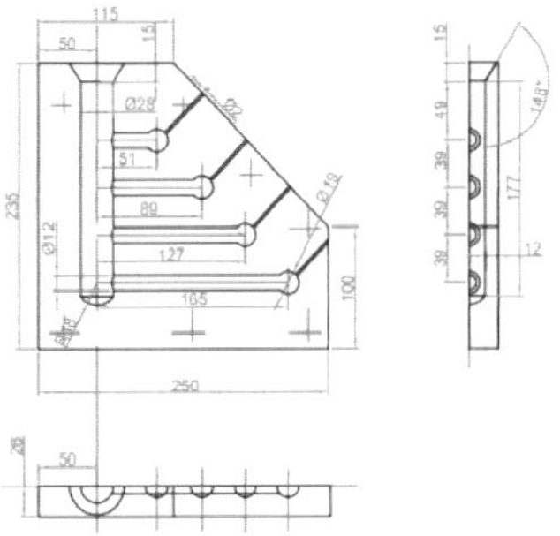

Figure 1. The schematic picture of hot cracking rod mold (mm)

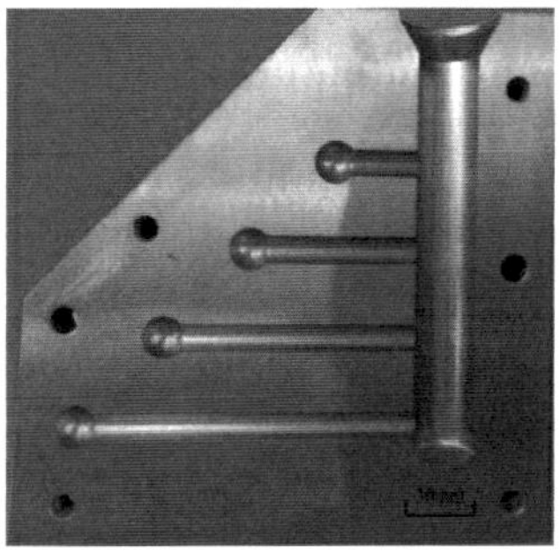

Figure 2. The physical picture hot cracking rod mold

2.3. Liquid forging forming process

Liquid forging forming mold was assembled and preheated to 300 °C, which surface was coating with graphite as mold release agent. The melted alloy heated to 730 °C was poured into cylindrical mold, loaded quickly by forging hydraulic machine and maintained a while. The loading pressure was 30 MPa, and the holding time was 30 seconds. Cylinder forming parts which were 65 mm in diameter and 60 mm in height were prepared after solidification. The microstructures and properties were researched after those forming parts sampled.

3. Results and discussion

3.1. Analysis of hot cracking tendency test

In hot cracking tendency test, all designed aluminum alloys formed completely, and the situation of insufficient pouring or no-fully filling did not appear.

figure 3 shows hot cracking constrained rod hairline and fracture positions on parts of the new aluminum alloys, including 2024 and 7075 aluminum alloys. It can be seen that most of aluminum rod fractures and hairlines exist on the root of the main runner. The fractures concentrate on the longest bars, and the hairlines and partial fractures concentrate on the second and third longest bars. However $10^{\#}$ and $12^{\#}$ aluminum alloys are different, that fractures exist on the end of rod near to the ball. 2024 and 7075 aluminum alloy show the same result as most of new aluminum alloy.

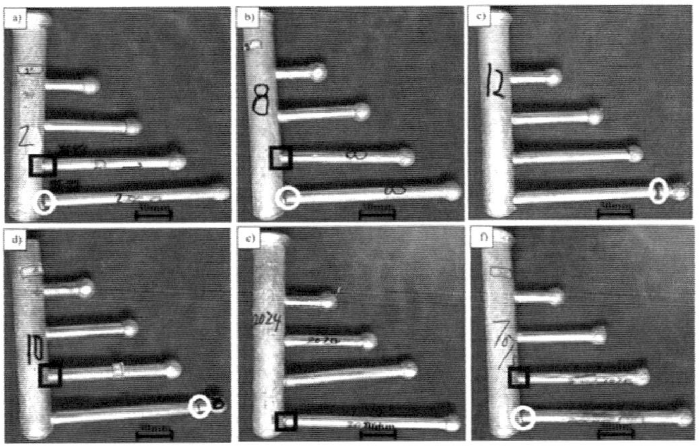

Figure 3. Crack position schematic pictures of parts of hot cracking tendency testing samples
a)2#; b)8#; c)12#; d)10#; e) 2024 alloy; f) 7075 alloy
2# is Al-5Cu-3.5Mg-2.5Zn alloy; 8#is Al-5Cu-4.5Mg-2.5Zn alloy;
12# is Al-6Cu-3.5Mg-2.5Zn alloy; 10# is Al-3Cu-3.5Mg-2.5Zn alloy

According to the principle of hot cracking tendency test and the situation of alloy hot cracking rod cracks, hot cracking susceptibility indexes are calculated, using hot cracking susceptibility formula. The results of hot cracking tendency test are obtained, as shown in table 3.

Table 3. Results of hot cracking tendency test

Number	HCS	Number	HCS
7075	72	6#	56
2024	68	7#	48
1#	40	8#	40
2#	40	9#	96
3#	56	10#	88
4#	56	11#	44
5#	64	12#	56

Figure 4 shows the influence of zinc, magnesium, and copper alloying element percentages on hot cracking tendency of aluminum alloy. It can be found that hot cracking tendency of the designed aluminum alloys are lower than 7075 aluminum alloy and 2024 aluminum alloy which are used more in liquid forging forming technology. Also, it can be seen that the hot cracking tendency of some alloys which contain less copper element percentages performance higher than 2024 and 7075 aluminum alloys.

As it is introduced above, there are limit percentages of the three alloying elements. From figure 4, with the decrease of zinc element percentages during the limit, hot cracking tendency are also gradually reduced. When zinc element percentage is less than 3%, hot cracking tendency keeps constant basically. And when magnesium element percentage is 3% ~ 5%, the hot cracking tendency is also consistent. But when the percentage of magnesium element is less than 3%, the hot cracking tendency becomes increase instead of decreasing when reduce magnesium content. Copper element is the main influence factor, and the influence rule is different. When the percentage of copper element is more than 4%, the cracking tendency reduced following the decrease of copper element. But when copper content is lower than 4%, the hot cracking tendency of aluminum alloy increases sharply as the same change, even being higher than the HCS of 7075 aluminum alloy. The HCS of aluminum alloys are up to 88 and 96.

The hot cracking tendency test shows that the content of Cu element has a big effect on hot cracking tendency of aluminum alloy, and that hot cracking tendency increases sharply when the percentage of copper element keeps in a low level. The reason is that a large number of *S* (Al_2CuMg) phase eutectic organizations are generated when the concentration of copper element is lower. *S* (Al_2CuMg) phases are bulky and have a strong fracturing effect on aluminum alloy, and numbers of crack sources are produced, to make the hot cracking tendency high.

The lowest hot cracking susceptibility index is 40 among all of the new designed aluminum alloys and three kinds of aluminum alloys reach this index. They are 1# (Al-5Cu-3.5Mg-1.5Zn) aluminum alloy, $2^{\#}$ (Al-5Cu-3.5Mg-2.5Zn) aluminum alloy and $8^{\#}$ (Al-5Cu-4.5Mg-2.5Zn) aluminum alloy, respectively. And the hot cracking susceptibility index of $11^{\#}$ (Al-4Cu-3.5Mg-2.5Zn) aluminum alloy is also relatively lower, only 44.

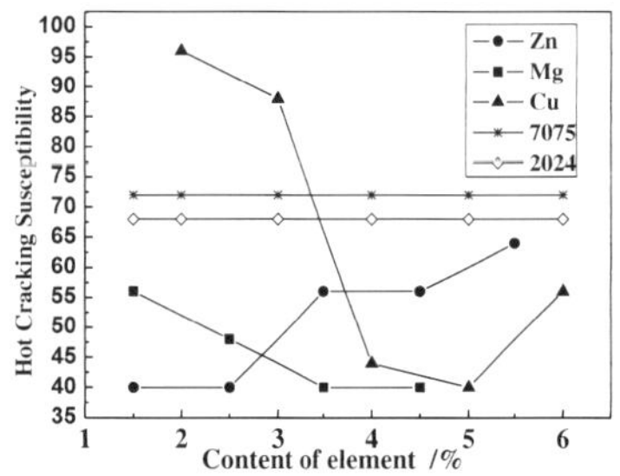

Figure 4. Influence of zinc, magnesium, and copper alloying element on hot cracking tendency of aluminum alloy

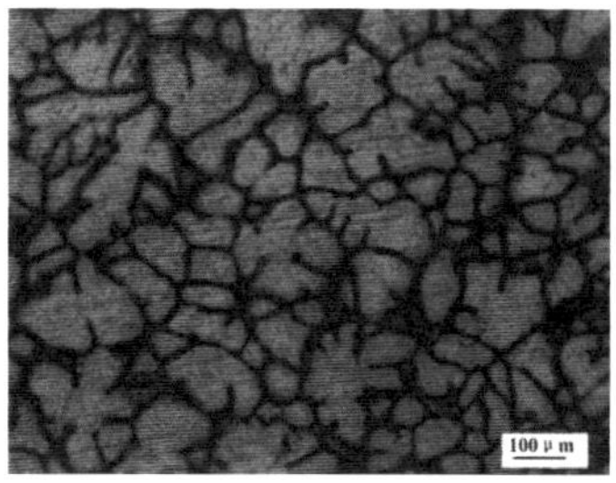

Figure 5. Microstructure of aluminum alloy formed by liquid forging

3.2. Analysis of organizations and mechanical properties

Figure 5 is the microstructure of new aluminum alloy parts formed by liquid forging. The organization of new aluminum formed by liquid forging is a mixed organization consisting of free dendrites and equiaxed grains produced from dendrites deformation and breakage, just like typical casting organization. What's more, there are numerous net-like and thick non-equilibrium eutectic structures distributed between grain boundaries. And most of dendrites present on trend to become equiaxed grains. There was a large different temperature distribution on alloy melt during the liquid forging process. Alloy melt on lower temperature rapidly nucleated and grew, forming fine equiaxed grains because of numerous nucleation sites. Because internal heat of the melt was not easy to loss, temperature was higher and nucleation particles were less. The release of the crystallize latent heat caused crystal nucleus to grow up rapidly and form dendrites. On the effect of outside loads, dendrites were deformed and broken into rose-shape organizations. At the same time, the solidification speed under outside load of alloy melt was too fast, and high percentages of alloy elements distributed unevenly. So the organization deviated from the equilibrium state in different levels, and numbers of non-equilibrium eutectic organizations formed and segregated in the grain boundary.

It is known that there is a big difference on solubility of three elements in aluminum, Zn, Mg and Cu [8-9]. So with the difference of Zn, Mg and Cu percentages, all the designed aluminum alloys contain different intermediate phases and eutectic organizations, showing different mechanical properties.

The four new designed aluminum alloys were sampled, including $1^{\#}$ (Al-5Cu-3.5Mg-1.5Zn) alloy, $2^{\#}$ (Al-5Cu-3.5Mg-2.5Zn) alloy, $8^{\#}$ (Al-5Cu-4.5Mg-2.5Zn) alloy and $11^{\#}$ (Al-4Cu-3.5Mg-2.5Zn) alloy. Tensile strength, hardness and elongation tests have completed, and the results are shown as figure 6.

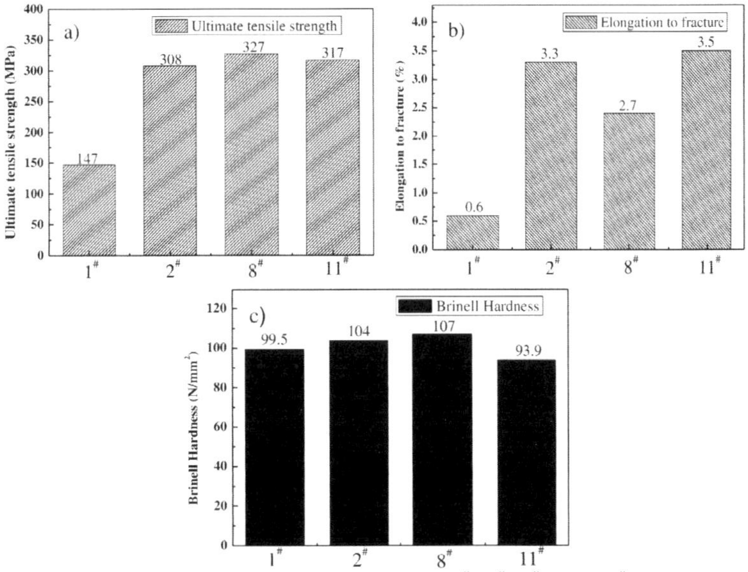

Figure 6. Mechanical properties test results of 1#, 2#, 8# and 11# aluminum alloys
a) Tensile strength; b) Elongation; c) Hardness
1# is Al-5Cu-3.5Mg-1.5Zn alloy; 2#is Al-5Cu-3.5Mg-2.5Zn alloy;
8# is Al-5Cu-4.5Mg-2.5Zn alloy; 11# is Al-4Cu-3.5Mg-2.5Zn alloy

It can be found that mechanical properties of 1# (Al-5Cu-3.5Mg-1.5Zn) aluminum alloy are obviously lower. After analyzing the microstructure and tensile fracture surface morphology (as figure 7), it is concluded that a large amount of shrinkage defects and inclusions existing lead to worse mechanical performances of 1# aluminum alloy.

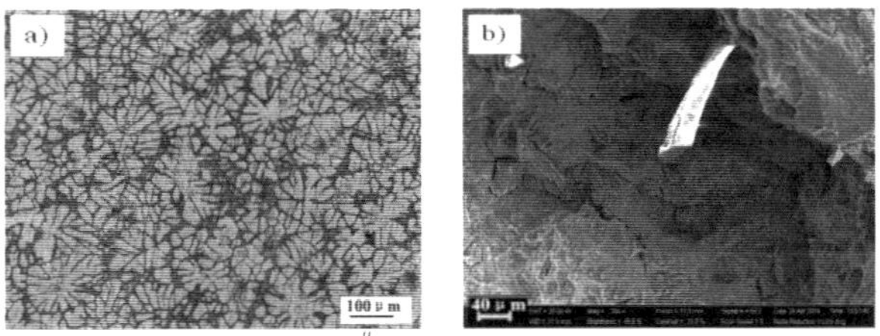

Figure 7. Microstructures of 1# (Al-5Cu-3.5Mg-1.5Zn) aluminum alloy
a) Metallographic structure; b) shrinkage defects and inclusions

From figure 6, 2#, 8# and 11# aluminum alloys except 1# aluminum alloy show excellent mechanical properties: tensile strength is about 310 MPa, Brinell hardness is about 100 N/mm^2 and elongation to fracture reach to 3% at the same time.

Compared with 2# alloy, it is found that the strength and hardness of 8# alloy is higher, but the elongation is lower. The reasons are analyzing and discussing, combined with microstructures (as figure 8a, 8b, 8c, 8d) and XRD results (as figure 9). Compared with 2# alloy, the percentage of Mg element in 8# alloy is on the high side. The increasing content of η ($MgZn_2$) phases in 8# alloy enhances dispersion-strengthening effect and leads to ascension of strength and hardness of 8# alloy. At the same time, the number of net-like eutectic organization among grain boundaries is bigger, causing the lower alloy elongation [10-11].

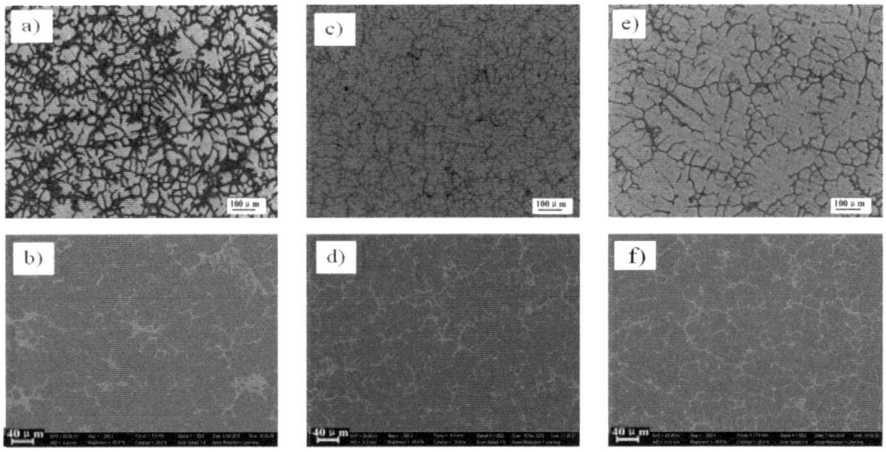

Figure 8. Metallographic image and Scanning Electronic Microscope image of $2^{\#}$, $8^{\#}$ and $11^{\#}$ aluminum alloys
a), b) $2^{\#}$ alloy; c), d) $8^{\#}$ alloy; e), f) $11^{\#}$ alloy
$2^{\#}$is Al-5Cu-3.5Mg-2.5Zn alloy; $8^{\#}$ is Al-5Cu-4.5Mg-2.5Zn alloy;
$11^{\#}$ is Al-4Cu-3.5Mg-2.5Zn alloy

Compared $11^{\#}$ and $2^{\#}$ alloy, $11^{\#}$ alloy shows higher strength and elongation, but its hardness is relatively lower. In terms of components, the content of Cu element in the $11^{\#}$ alloy is lower. The content of θ (Al_2Cu) phases reduces or even disappears (as figure 9), but the S (Al_2CuMg) phases formed by Al, Mg and Cu elements disperses in $11^{\#}$ alloy (as figure 8e, 8f), resulting the higher strength. And in $2^{\#}$ alloy, solution strengthening effect caused by S (Al_2CuMg) phases and θ (Al_2Cu) phases, leads higher hardness in $2^{\#}$ alloy (as figure 9). And also, the presence of net-like eutectic organization must make the toughness and plasticity be weakened, and elongation is lower.

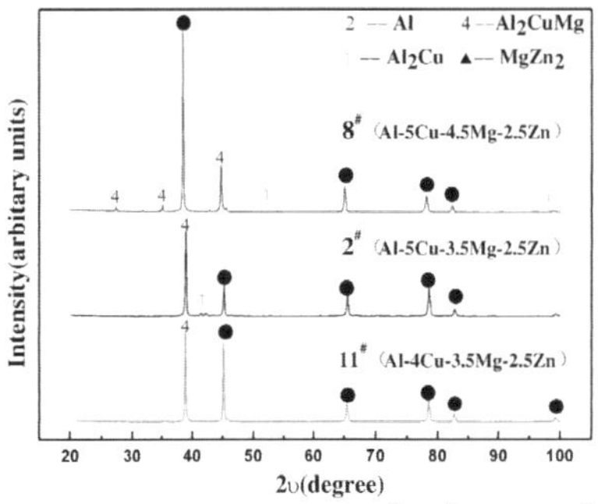

Figure 9. X-Ray Diffraction results of $2^{\#}$, $8^{\#}$ and $11^{\#}$ aluminum alloys
$2^{\#}$is Al-5Cu-3.5Mg-2.5Zn alloy; $8^{\#}$ is Al-5Cu-4.5Mg-2.5Zn alloy;
$11^{\#}$ is Al-4Cu-3.5Mg-2.5Zn alloy

4. Conclusions

The different contents of Zn, Mg and Cu elements determine the differences of microstructures and mechanical properties in aluminum alloys. New high-strength aluminum alloy using in liquid forging with low hot cracking tendency and great comprehensive performances was fabricated by designing different percentages of alloying elements.

(1) The hot tearing susceptibility indexes of 7075 aluminum alloy and 2024 aluminum alloy are 72 and 68 respectively, evaluated using constrained rod casting in a steel mold. The hot cracking tendency of most of new designed high-strength aluminum alloy is lower than them.

(2) Contents of Zn, Mg and Cu elements affect the hot cracking tendency. During a certain range, the hot cracking tendency indexes decrease as the content of zinc element reducing and the content of Magnesium element increasing. When the percentage of copper element is more than 4%, the cracking

tendency reduced following the decrease of copper element. But when copper content is lower than 4%, the hot cracking tendency of aluminum alloy increases sharply following the decrease of copper.
(3) Among all of new designed aluminum alloy, the smallest hot tearing susceptibility index is 40, and three designed alloys reach this level. They are $1^{\#}$ (Al-5Cu-3.5Mg-1.5Zn) alloy, $2^{\#}$ (Al-5Cu-3.5Mg-2.5Zn) alloy, $8^{\#}$ (Al-5Cu-4.5Mg-2.5Zn) alloy, respectively.
(4) The microstructure of new types of Al-Cu-Mg-Zn alloys after liquid forging forming are similar, which consist of dendrites, equiaxed grains and net-like eutectic organizations and intermediate phases dispersed-distributing among grain boundaries.
Considering the influence of main element percentages on hot cracking tendency, microstructures and mechanical properties, $8^{\#}$ (Al-5Cu-4.5Mg-2.5Zn) aluminum alloy shows the optimal performances among the twelve kinds of designed aluminum alloys. Its hot tearing susceptibility (HCS) index is 40, and tensile strength is 327 MPa. And the elongation to fracture is 2.7%, which Brinell hardness reaches 107 N/mm^2.

Acknowledgements
The authors express their appreciation for the financial support of International Cooperation Project of the Ministry of Science and Technology of China under Grant No. 2014***50320.

References

[1] Li S, Li H W, Zhang X, et al. Advances of Liquid Forging Technology for Forming High-performance Aluminum Parts [J]. *New Technology & New Process*, 2015: 129-134.
[2] Liu J A, Sheng C L. Development and Application of Forging Technology and Aluminum Alloy Forgings [J]. *Nonferrous Metals Processing*, 2012(1): 12-17.
[3] Hong S Z. Application and Development of Liquid Die Forging for the Aluminum Alloy [J]. *Journal of Net Shape Forming Engineering*, 2012, 4(3): 64-66.
[4] Qi P X. Development of Squeeze Casting Processing Technology [J]. *Die Casting*, 2007, 28(S1): 13-17.
[5] Li N K, Cui J Z. Effect of Al-Zn-Mg-Cu Series Alloy Structures on the Properties [J]. *Light Alloy Processing Technology*, 2008(1): 5-10.
[6] Kamguo K H, Larouche D, Bournane M, et al. Hot Tearing of Aluminum-copper B206 Alloys with Iron and Silicon Additions [J]. *Materials Science and Engineering: A*, 2010, 527 (27-28): 7413-7423.
[7] Cao G, Kou S. Hot Cracking of Binary Mg-Alloy Castings [J]. *Materials Science and Engineering: A*, 2006, 417 (1-2): 230-238.
[8] Mukhopadhyay A K. Microstructure and Properties of High Strength Aluminum Alloys for Structural Applications [J]. *Transactions of the Indian Institute of Metals*, 2009, 62(2): 113-122.
[9] Li X L. The Effect of Returns on Microstructures and Mechanical Properties of ZL205A Alloy [D]. *Harbin Institute of Technology*, 2008: 65.
[10] Chen X, Huang S, Wu R C. Influence of Zn, Mg and Cu Contents on Strength and Hot Crack of Cast Al-Zn-Mg-Cu Alloys [J]. *Special Casting & Nonferrous Alloys*, 2012,32(4): 306-310.
[11] Li T S, Xu H, Zhang H. Effect of Content of Mg, Cu and Zn on Mechanical Properties of Al-alloy Under Liquid State Forging [J]. *Hot Working Technology*, 2009(23): 145-147.

Study on Microstructure and Properties of 105mm Thick 5083 Aluminum Alloy Hot-Rolled Plate

Xinwei She[1,2], Xianquan Jiang[1,2,*], Bao Qi[2], Quan Li[1], Jiangyu Li[3], Jun Li[4], Zhende Ran[4]

[1] Chongqing Academy of Science and Technology, Chongqing 401123, China
[2] Faculty of Materials and Energy, Southwest University, Chongqing 400715, China
[3] Guangxi Liuzhou Yinhai Aluminium Co., Ltd, Guangxi 545006, China
[4] Chongqing Gearbox Co., Ltd, Chongqing 402263, China
*Corresponding author e-mail: 574754298@qq.com

Abstract. The microstructure and tensile property of 105mm thick 5083 aluminum alloy hot-rolled plate in different thickness layers were investigated by optical microscopy, scanning electron microscopy, transmission electron microscopy and tensile tests. The results show that along the thickness direction, the microstructure and tensile property are inhomogeneous, and a "double low" phenomenon, which has low strength and plasticity, occurs in the central area. There are many sub-micron sized second phases in the center and near-center area, which cause two completely different dislocation distribution states. they can effectively prevent the dislocation movement during plastic deformation, and improve the strength of the alloy.

1. Introduction

With the development of global industrialization, the requirement for mechanical properties of 5083 aluminum alloy thick plate become higher and higher. In order to meet the need of large deformation, a wider and thicker plates are produced. However, in actual manufacture, the larger the size is, the more difficult it will be. This is principally because the solidified structure is inhomogeneous for the existence of a large temperature gradient along the direction of thickness and width. At present, researches on 5083 aluminum alloy plate primarily focus on the microstructure, heat treatment process, and the relationship between deformation and properties[1-4], while there are few reports about the inhomogeneity of microstructure and properties, and no test standard for the performance of plates beyond 20mm thick at home. So, it is very necessary to carry out the research in this field, and improve the detection mechanism as soon as possible.

In this paper, the microstructure and tensile properties of 105mm thick 5083 aluminum alloy hot-rolled plate were studied to discuss the factors affecting the inhomogeneity, so that it can provide a theoretical basis for further improving the quality of 5083 aluminum alloy plate.

2. Experiments

The material used in this experiment is 2500mm×105mm(width×thickness) 5083 aluminum alloy hot-rolled plate with a free machining state, and its chemical composition is shown in Table 1.

Table 1. Chemical composition of 5083 aluminum alloy hot-rolled plate(mass fraction/%)

Elements	Si	Mg	Mn	Cr	Fe	All other element	Al
content	0.163	4.434	0.678	0.104	0.238	<0.05	Balance

First, take samples from one end of the plate to the center along the transverse direction, and sign as sample A, B, C and D, respectively. Then, equally divide each sample into ten pieces of blank-parts along the thickness direction, and number them from the top to the bottom, as shown in figure 1. Finally, make tensile specimens, according to the GB/T228-2002 standard.

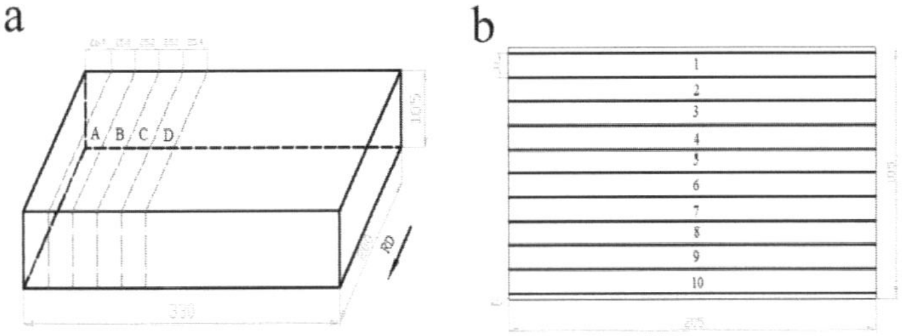

Figure 1. Sketch map of sampling:
a-hot-rolled plate; b-pre-stretching blanks along the sample thickness direction

The tensile tests are carried out at room temperature with a constant displacement rate of 1 mm/min to determine the average value of tensile strength, yield strength and elongation. Cut off metallographic samples from the clamped end of 1-5# tensile specimens, and have a observation on the metallurgical microscope and scanning electron microscope which equips with EDS. Take some wafers near the deformation zone of 4# and 5# tensile specimen, and have a microscope observation by Tecnai G2F20 transmission electron microscope after mechanical thinning.

3. Results and analysis

Figure 2 shows the microstructure of 5083 aluminum alloy. From the figure 2a~2e, the second phases and impurities are mainly distributed in the grain boundaries and elongated along the rolling direction. The phases precipitated in the surface layer of hot-rolled plate is small, dense and continuous, and those in the center is relatively large, dispersive and intermittent. The reason for the result is that in the beginning of hot rolling, the large deformation in the surface layer of plate and the cooling effect of rolling oil emulsion can make the second phase broken and prevent them from growing up suddenly by heat, and in the later period of rolling, the second phase in the center will grow up rapidly because of the high temperature. A further observation can be found the alloy is mainly composed of lamellar grains, which are also elongated along the rolling direction. From the figure 2f~2j, the grains in the center(position 5 in figure 1b) and near-center area(position 4) are nearly equiaxed, and the grain boundary edges are serrated, which means the partial recrystallization occurs in the hot rolling process, and the closer to the center, the greater the degree of recrystallization is.

figure 3 is the SEM image of 5083 aluminum alloy. it can be found that the alloy is composed of α-Al, (Fe, Mn)Al_6, $Al_{12}Fe_3Si$, and a little Mg_2Si via energy spectrum analysis[5]. The results of EDS analysis are shown in Table 2.

Figure 4 shows the dislocation distribution along the crystallographic direction $\langle 111 \rangle_\alpha$ in the center and near-center area of 5083 aluminum alloy. It can be observed there are a number of sub-micron sized second phases, and a serious phenomenon of dislocation congestion in these areas. Through the energy spectrum analysis, it is found that these sub-micron sized second phases are mainly AlMg(Mn, Cr). The distribution of dislocation in the central area is very uneven, and approximately persents a characteristic of cell structure. While, the distribution of dislocations in the near-center area is uniform, and it is characterized as Taylor lattice distribution[6]. Comparing figure 4a and 4b, it is observed the number of the sub-micron sized second phase in the center of plate is less than that in the near-center, and the distribution of dislocation clusters is scattered, which may be the reason for the different distribution of dislocations in two adjacent areas.

According to the theory of plate rolling[7], the thicker the aluminum alloy ingot is, the larger the difference of deformation in different stages of hot rolling is. Generally, in the initial stage of rolling, the deformation mainly occurs in the surface layer of the plate. In this case, the dislocation motion

IOP Conf. Series: Materials Science and Engineering **230** (2017) 012037 doi:10.1088/1757-899X/230/1/012037

throughout the central area of the plate is mainly planar slip, and the dislocation state presents Taylor crystal lattice distribution, as shown in figure 4b. In the later period of rolling, deformation is deep into the central area. At this point, dislocation motion is dominated by dislocations climb and cross-slip, and cell structure starts to appear. However, taking into account the partial recrystallization, the opposite sign dislocations cancel out in the cell wall, which results in the weakening of the characteristic of cell structure, as shown in figure 4a.

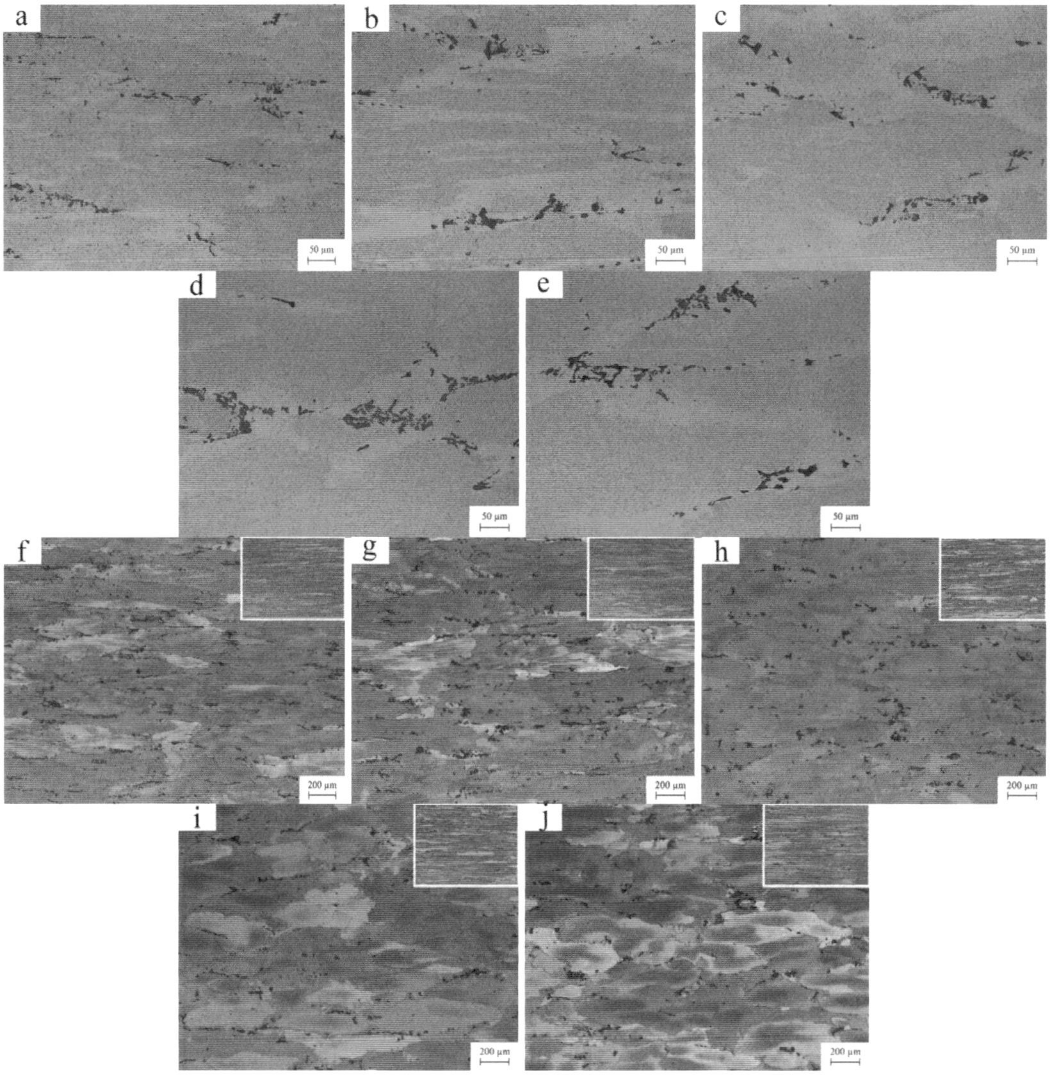

Figure 2. Microstructure of 5083 aluminum alloy: a~e-the second phase precipitation of 1~5# postion; f~j-grain structure of the corresponding postion and grain distribution of the longitudinal section

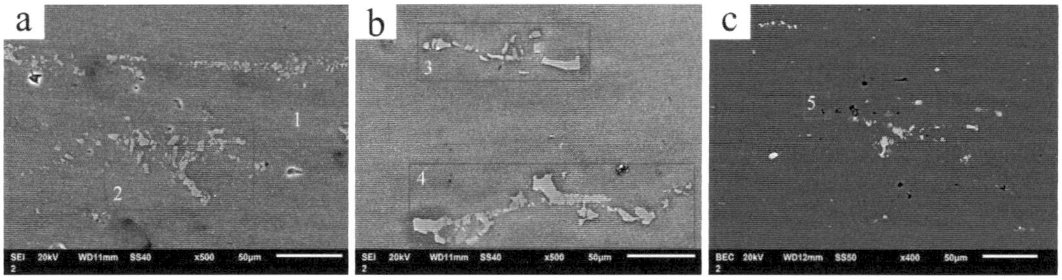

Figure 3. Primary second phases of 5083 aluminum alloy:
a-matrix and skeleton phase; b-irregular-shape phase; c-lump phase

Table 2. EDS analysis of the second phase of 5083 aluminum alloy(atomic/%)

Elements	Al	Mg	Mn	Cr	Fe	Si	Cu	S
1	94.88	5.12	/	/	/	/	/	/
2	87.70	0.98	4.48	0.43	6.41	/	/	/
3	79.12	/	6.23	1.04	8.82	4.79	/	/
4	78.67	/	6.23	0.55	9.25	5.30	/	/
5	38.34	41.48	0.24	0.79	/	19.15	/	/

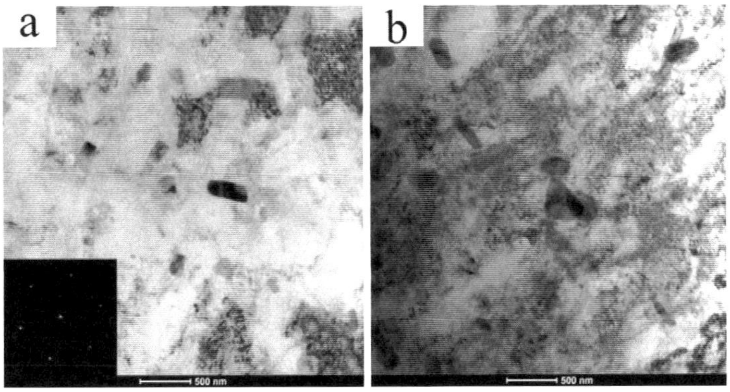

Figure 4. TEM images of 5083 aluminum alloy: a-the central area; b-the near-center area

Figure 5 is the conventional tensile properties of 5083 aluminum alloy hot-rolled plate with different thickness. Along the rolling direction, the tensile properties in different thickness layers are uneven. From the surface to the center, the strength of the plate shows a trend of declining-rising-declining, and the elongation increases at first, then decreases. In particular, a "double low" phenomenon, which has low strength and plasticity, occurs in the central area.

Compared with the high strength of the plate in the near-center area, there are two important reasons for the decrease in the center. On the one hand, the second phase in the central area is more coarser and easy to cause stress concentration. Meanwhile, a large number of inclusions and voids can reduce the effective area of the external load[8, 9]. On the other hand, recrystallization occurs in the hot rolling, which makes the microstructure softening.

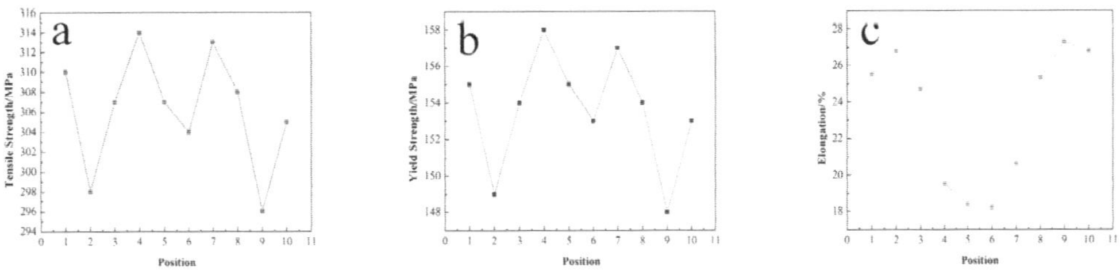

Figure 5. Conventional tensile properties of 5083 aluminum:
a-tensile strength; b-yield strength; c-elongation

From the interaction between dislocations and particles[10], when dislocations meet a couple of particles that are close to each other, they will not be able to cut through these paticles at the same time, and can not be back, so turn to be local bending, and eventually trapped by these particles, forming a relatively independent “inner packing” dislocation group, as shown in figure 4a. The experimental results show that the effect of recrystallization on the elimination of dislocation groups is weak. The size of the dislocaiton group is much bigger than that of a single particle, therefore, it has a weak blocking effcet on the dislocation motion. This dislocation group is easier to cause stress concentration, resulting in a decrease in the strength of the plate.

4. Conclusions

(1) The microstructure and and tensile property of 5083 aluminum alloy hot-rolled plate along the thickness direction is inhomogeneous. The “double low” phenomenon appears in the center of the plate with low strength and low plasticity.
(2) there are two important reasons for the appearance of the “double low” phenomenon: on the one hand, a larege number of coarse phases and impurities gathered in the central area, and on the other hand, recrystallization in this area is severer.
(3) There are a number of sub-micron sized second phases throughout the central area, they have a precipitation strengthening effect on the 5083 aluminum alloy hot-rolled plate. However, in the center of the plate, dislocations are trapped by these particles and formed a relatively independent “inner packing” dislocation group, resulting in a decrease in the strength.

5. References

[1] Y. Chen, H. Ding, J.Z. Li, Z.H. Cai, J.W. Zhao, W.J. Yang, Influence of multi-pass friction stir processing on the microstructure and mechanical properties of Al-5083 alloy, Mater. Sci. Eng., A 650 (2016) 281-289.
[2] S.L. Xia, M. Ma, J.X. Zhang, W.X. Wang, W.C. Liu, Effect of heating rate on the microstructure, texture and tensile properties of continuous cast AA5083 aluminum alloy, Mater. Sci. Eng., A 609 (2014) 168-176.
[3] K. Masaki, H. Matsuoka, Y. Murakami, N. Oiwa, Effect of post-weld heating on microstructure and mechanical properties of friction stir welded thick 5083 Al alloy, Proc. 1st Inter. Joint Symp. Joining Weld. (2013) 165-168.
[4] D. Singh, P.N. Rao, R. Jayaganthan, Effect of deformation temperature on mechanical properties of ultrafine grained Al-Mg alloys processed by rolling, Mater. Des. 50 (2013) 646-655.
[5] Z.T. Wang, R.Z. Tian, Handbook of aluminum alloys and their processing, Central South University Press, Changsha, 2000, pp. 292-310.
[6] Z.G. Wu, M. Song, Y.H. He, Effects of rare-earth element Er on microstructure and mechanical properties of an Al-Mg alloy, Mater. Sci. Eng. Powder Metall. 181 (2008) 2262-2266.
[7] L. Zhong, Y.Y. Ma, Y.C. Xie, Production technology of aluminum alloy plate, Metallurgical Industry Press, Beijing, 2009, pp. 52-53.
[8] X.M. Zhang, N.M Han, S.D. Liu, F.X. Song, R.L. Zeng, L.Y. Huang, Inhomogeneity of texture, tensile property and fracture toughness of 7050 aluminum alloy thick plate, Chin. J. Nonferrous Met. 20 (2010) 202-208.
[9] G. Ran, J.E. Zhou, Y.F. Wang, Study on tensile properties and fractography of cast A356 aluminum alloy, Rare Metal Mat. Eng. 35 (2006) 1620-1624.
[10] D.K. Shi, Fundamentals of materials, China Machine Press, Beijing, 2003, pp. 338-341.

Acknowledgements

The authors would like to thank the International Science & Technology Cooperation Program of China under grant No. 2011DFR50950 and Liuzhou Science and Technology Plan Project under No. 2015C020101.

IOP Conf. Series: Materials Science and Engineering **230** (2017) 012038 doi:10.1088/1757-899X/230/1/012038

Finite Element Analysis on the Crack Growth and Stress Intensity Factor for the Contact Fatigue

P X Du[1,3,4], Y L Liang[1,3,4,*], C H Yin[2,3,4], L Q Zhong[2,3,4], and H J Fan[1,3,4]

[1]College of Materials and Metallurgy, Guizhou University, Guiyang, Guizhou, People's Republic of China
[2]School of Mechanical Engineering, Guizhou University, Guiyang, Guizhou, People's Republic of China
[3]Guizhou Key Laboratory for Mechanical Behavior and Micro Structure of Mateials, Guizhou University, People's Republic of China
4 National & Local Joint Engineering Laboratory for High-performance Metal Structure Material and Advanced Manufacturing Technology, People's Republic of China.
E-mail: liangyilong@126.com

Abstract. In order to simulate contact fatigue crack propagation, the ANSYS APDL and Franc3D were used. The rolling contact model was established by the ANSYS APDL, then the distribution of the contact stress field under various loads was analyzed. The contact stress value demonstrated an error within ± 6%, compared to the obtained results from the Hertz theory. The contact stress field was imported into the Franc3D with a preset radical crack, the crack tip stress intensity factor distribution was calculated and the growth path of the three-dimensional contact fatigue cracks was analyzed. The comparisons with the experimental results demonstrated that the fatigue crack profile path from the peeling position of the subcutaneous layer was consistent with the simulated path, which was in a circular arc form. The main reason for this result was the fatigue crack change from the opening type I to the sliding type II. The a-N curves of the crack growth life under various load conditions were predicted based on the Franc3D, whereas the results displayed that the crack growth life decreased as the compressive stress increased.

1. Introduction

The contact fatigue failure phenomena occur frequently during the service of gears, rails, rolling bearings and other components and parts, mainly due to the surface fatigue damage of the workpieces which sustain a long-term cycle contact stress. The contact fatigue can cause serious damage. In 1998, the German high speed train derailment with serious casualties was caused by the wheel contact fatigue[1-2]. For a long time, the contact fatigue life extension of the workpieces has effectively been studied by a high number of scholars both locally and abroad. With the development of computer technology in recent years, the finite element method has been applied for the contact problems study, which can predict the contact fatigue life effectively and also minimize a high amount of manpower and financial resources.

Wu[3] simulated the surface cracks under stretching and pure bending fatigue loads according to the Paris fatigue crack growth law, through the crack tip stress intensity factor(SIF) distribution fitted a cubic spline function curve, solving that the crack tip SIF distribution is difficulty measured. Shi[4]

established a thermal fatigue crack model of the brake disc by the Franc3D. The crack growth speed and the stress intensity factor change were studied with various crack ratios and multiple cracks shape, then the distribution regularity was obtained under each conditions. Wang et al.[5] simulated the crack initiation life of face gear under various working conditions and with various surface properties based on the MSC.Fatigue software. Then the crack growth life on the tooth root was simulated based on the Franc3D, finally discovering the bending life prediction of the face gear. When bending moment and stress ratio were constant, Janusz Lewandowski et al.[6] studied the crack growth behavior of a plate specimen with the two sides passivation notch and stripe weld respectively using the Franc3D. The fatigue crack growth life was calculated according to the fatigue mechanism and boundary element method. The results proved that the fatigue crack growth life was improved significantly subsequently to welding. In condition of increased temperature and sustained loading, Erik Storgärds et al.[7] expanded the crack growth results of the nickel-based Inconel 718 from the one dimensional to the three dimensional in the Franc3D. The crack growth life was calculated when the stress intensity factor on crack tip changed, which was consistent with the experimental results. A series of engineering problems have been solved by a high number of scholars worldwide in the Franc3D, however, only a low number of studies were focused on the contact fatigue crack growth behavior. The study of crack tip stress intensity factor change trend n means of the finite element analysis method is proven of guiding significance for the contact fatigue crack growth behavior research.
In this paper, the rolling contact model was established in both ANSYS APDL and Franc3D. After the initial crack was preset in the contact area, the change of stress intensity factor of crack under various contact stress fields were calculated. the crack growth path was simulated and the predicted results were compared to the actual contact fatigue test of the failure sample. By these settings and calculations, the Franc3D simulation accuracy of the contact fatigue crack growth was verified. In addition, the contact fatigue crack growth life was predicted in the Franc3D, which provided the basis for the numerical simulation of the contact fatigue crack growth and the crack growth life assessment.

2. Experimental method

2.1. Finite element analysis

The test material was 42CrMo with the elasticity modulus of $E=2.12\times10^5$MPa, the density of $\rho=7850$kg/m^3 and the Poisson ratio of $\lambda=0.28$. The solid 185 was utilized as the finite element and the control unit size was 1 mm. The Sweep method of volume mesh was utilized for the model mesh generation. The fix constraint was set for the bottom of the lower cylindrical, whereas the bottom of the upper cylindrical (See figure 1)was set as 1 kN, 1.5 kN, 2 kN, 2.5 kN and 3 kN(loads selected according to the contact fatigue experiment) of the normal compressive stress, respectively. The other boundary conditions were given, the point-surface contact pair was set, the upper cylindrical surface was the contact surface and the lower cylindrical surface was the target surface.

2.2. Crack preset method

The contact stress field obtained from ANSYS was imported into the Franc3D, two different types of cracks were placed in the radial and the axial direction of the sub model at the contact point. The crack diameter was 1 mm, as presented in figure 2.

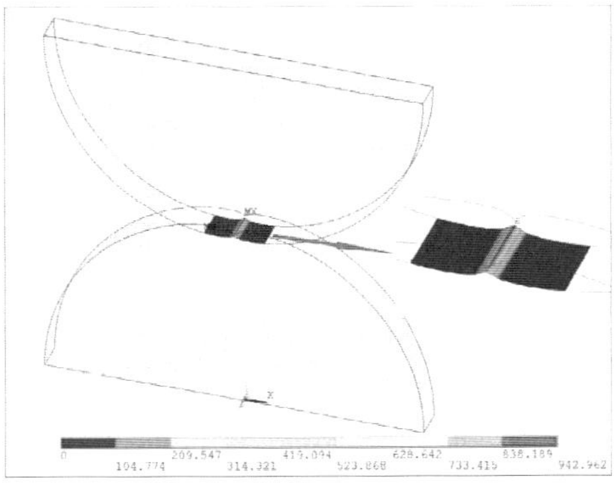

Figure 1. Contact status

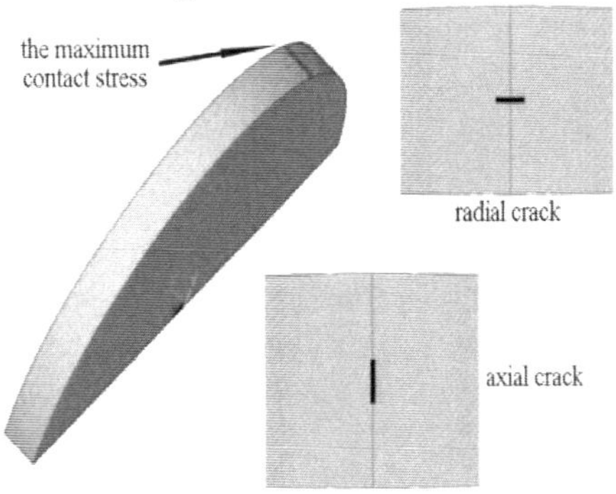

Figure 2. Crack preset direction

2.3. Crack growth analysis
Firstly, select the appropriate crack preset direction. According to the Paris-equation, the radial crack growth with tensile stress was simulated by the Franc3D. The extension step length was set at 0.1 mm. The Paris-equation was written as:

$$\frac{da}{dN} = C(\Delta K)^n \tag{1}$$

where C and n are the basic coefficients for the crack growth properties description. The values of C and n in the 42CrMo crack growth were reported in reference[8], whereas the average values (C=1.7636×10^{-8}, n=2.0703) were selected in this study.

2.4 Contact fatigue test
The 42CrMo material with the same coefficients as in the simulation model was utilized for thermal refining, oil quenching at 850°C and tempering at 580°C. The chemical composition are listed in table 1.

Table 1. Chemical composition of the 42CrMo(wt%)

Element	C	Si	Mn	Cr	Mo	S	P	Fe
Content	0.41	0.24	0.68	1.03	0.18	0.011	0.006	Bal.

According to the YB/T 5345-2006 "Metal rolling contact fatigue test method" standard, five groups of contact fatigue specimens were processed in the same size with the simulation model. The contact fatigue test machine MJP-30 was utilized, a 10% slip was set, the given load was 200kg (corresponding simulation load was 2kN), the contact fatigue tests were executed and the failure sample was analyzed by the field emission scanning electron microscope.

3. Results

3.1. Contact stress field calculation

The calculated results of the ANSYS APDL are listed in Table 2, and the distribution of the contact stress field is presented in figure 2.

Table 2. Contact stress under various loads

Load[kN]	1	1.5	2	2.5	3
Contact stress[MPa]	733.42	838.19	942.96	1047.74	1152.51

With the assumption that in the simulation state, the instantaneous contact of the sample at any time was a pure theoretical contact, either clearance or wear did not exist at the contact center. The contact stress can be calculated according to the Hertz theory[9]:

$$\sigma_{\max} = 0.418\sqrt{\frac{FE}{l}\left(\frac{R_1+R_2}{R_1R_2}\right)} \tag{2}$$

$$E = \frac{2E_1E_2}{E_1+E_2} \tag{3}$$

where E_1,E_2 are the elastic modulus of the upper and lower cylinder model material, respectively, $E_1=E_2=2.12\times10^5$MPa; R_1,R_2 are the radii of the upper and lower cylinder model, respectively, $R_1=R_2=30$ mm; l is the contact width, $l=5$ mm. The theoretical contact stress values are listed in Table 3, the errors between the simulation value and the theoretical value were within ±6%. The error generated in the actual simulation occurred because the deformation existed between two cylindrical contacts, however, the contact stress field under the deformation behavior was not considered in the theoretical calculation.

Table 3. Theoretical value of contact stress and error

Load[kN]	1	1.5	2	2.5	3
Theoretical value[MPa]	702.84	860.72	993.87	1111.18	1217.24
Error[%]	-4.41	2.62	5.12	5.71	5.32

3.2. Crack preset direction

The calculation results of the KI of the crack tip in the radial and axial directions are presented in figure 3 (a) and (b), respectively. It could be observed that the type I stress intensity factor of the radial crack tip was positive, whereas the value was negative for the axial crack tip. The horizontal axis was represented by the crack front normalized coordinate X, whereas the corresponding crack surface points of X=0 and X=1 are presented in figure 3(c). According to the linear elastic theory, the KI stress intensity factor could be normally written as

$$\mathrm{K_I} = \mathrm{Y}\sigma\sqrt{\pi a} \tag{4}$$

where σ is the nominal stress, Y is the shape coefficient related to the geometry boundary with the crack body, where in general case Y≥1[10]. It was proved that the radial crack front nominal stress was tensile stress, which accelerated the crack growth; the axial crack front nominal stress was compressive stress and the crack closure phenomenon was generated. It proved that the radial direction is the appropriate direction.

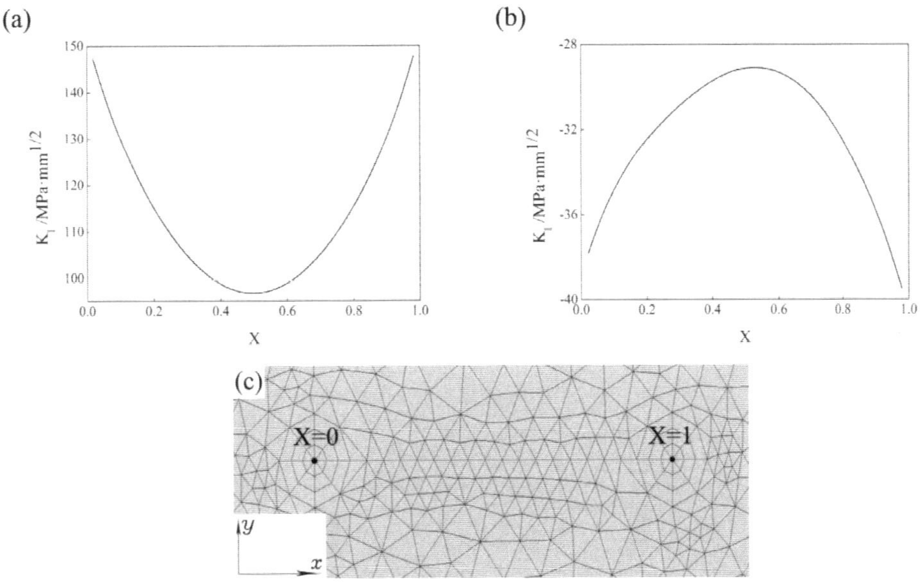

(a) Radial crack; (b) axial crack; (c) normalized crack length coordinate.

Figure 3. KI values in various crack directions

3.3. Contact fatigue test results analysis

The failure mode of five groups of contact fatigue specimens peeled from the subcutaneous layer, is presented in figure 4(a). In general case, the origin of the peeling crack is on the surface. When the crack originated from the surface, the crack could be formed rapidly. Moreover, the oil wedge effect caused by the invasion of the lubricating oil formation also accelerated the crack growth, which would shatter fast finally[11].

The cross-section of the peeling is presented in figure 4(b), whereas it was measured that the deepest failure depth was approximately 195μm to the surface. No crack was observed in the subcutaneous layer, which proved that the failure crack initiated from the surface, because the surface was not strengthened, leading the crack to an easier growth at the surface. On the other hand, the failure crack occurred in the sub-surface easily by the surface strengthening [12].

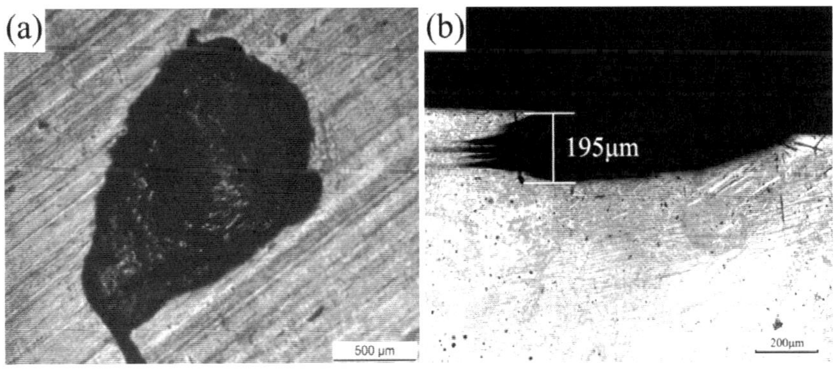

Figure 4. Peeling failure macro morphology

Figure 5(a) demonstrates the crack morphology where the peeling initiated and the crack type was an intergranular fracture. The bottom morphology of the peeling is presented in figure 5(b), which was overall flat and a layered step pattern partially existed. This indicated that the crack at the peeling bottom still grew in a fatigue form, which was mainly developed by both the crack and the crack group.

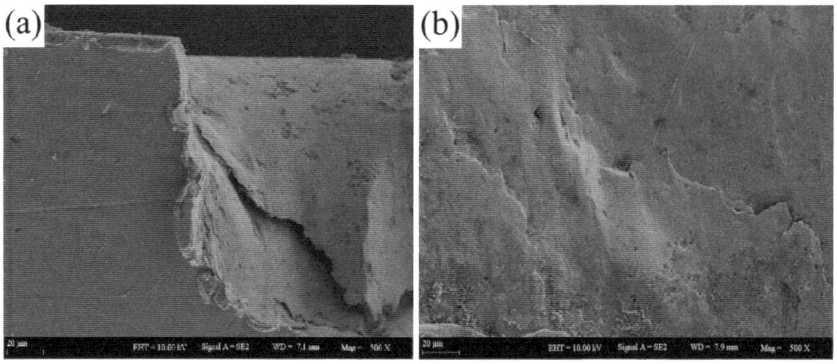

Figure 5. Peeling failure morphology

3.4. Crack path analysis

Figure 6(a) demonstrates the crack growth by the Franc3D, where it could be observed that the crack expanded along the depth direction and the crack path was of a circular arc shape. Figure 6(b) presents the received SEM images from the cross-section of the peeling, where the section crack shape was in accordance with the Franc3D simulation result.

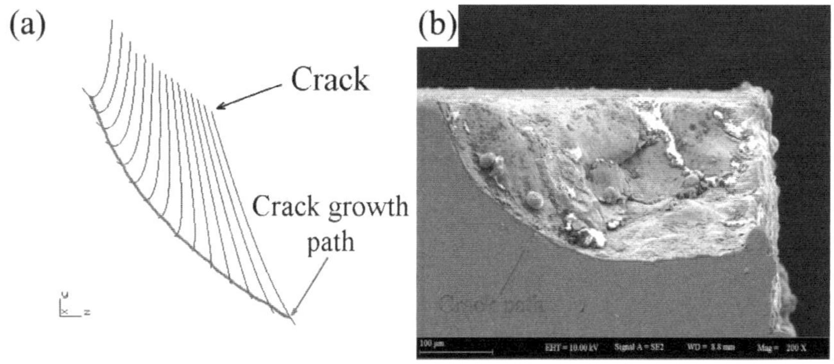

Figure 6. Crack path(a,Franc3D and b,contact fatigue test)

Figure 7(a) demonstrates the distribution of the KI stress intensity factor during crack growth. The change trend of KI at the position of the 1-15 steps crack growth path is presented in figure 7(b). It could be observed that in the 1-6 steps of the crack growth, the KI increased gradually, whereas in the 6-15 steps, the KI decreased entirely. figure 7(c) presents the distribution of the KII stress intensity factor during crack growth, the change trend of KII at the position of the 1-15 steps crack growth path is presented in figure 7(d). In the 1-6 steps of the crack growth, no apparent change could be observed for the KII, while the KII increased gradually with the expansion in the 6-15 steps. This was interpreted that the contact fatigue crack grew in the open type I from the beginning and continues growing subsequently to becoming the open type II crack, leading the crack growth direction changed and the crack growth section was finally of a circular arc shape.

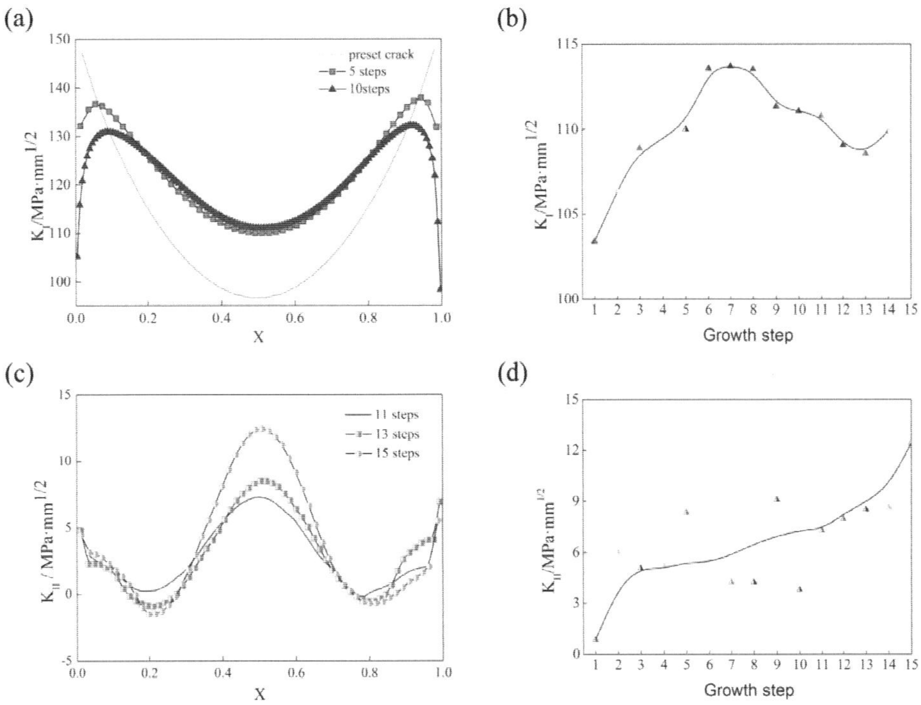

Figure 7. Distribution of stress intensity factor KI(a and b) and KII(c and d)

3.5. Crack growth life numerical calculation

In order to predict the contact fatigue crack growth life under various load conditions, the crack growth life at the cyclic load （$R = -\infty$） was simulated in the Franc3D, and the results are presented in figure 8. It could be observed that following the crack growth of 1 mm, the latter entered the rapid expanding region, when the contact fatigue specimen was considered as invalid. The final crack length was within 1.5 mm. The crack growth cycle number with the same steps under various load conditions is presented in Table 4. The crack cycle number decreased as the stress pressure increased, which indicated that the stress pressure increase speeded up the contact fatigue crack growth [13].

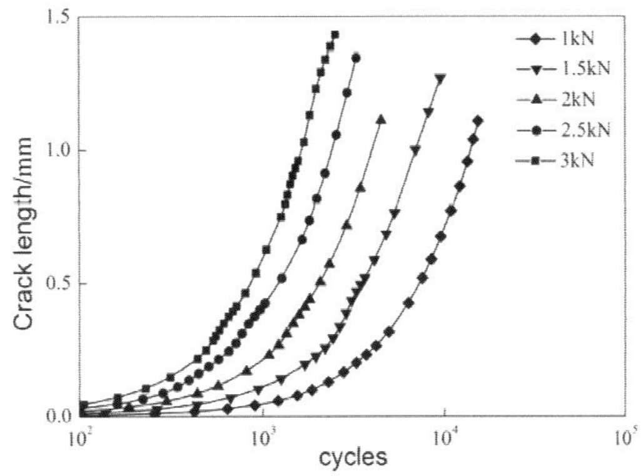

Figure 8. Relationship between crack growth length and cycle number under various load conditions

Table 4. Crack growth cycle number with the same steps under various load conditions

Load[kN]	1	1.5	2	2.5	3

Growth life	15280	9565	4573	3326	2529

4. Conclusions

The main conclusions of the present work are as follows:

(1) The rolling contact stress value under various stress pressure states was calculated by the finite element calculation; the error between the calculated results and the Hertz theory results was within $\pm 6\%$.

(2) According to the calculation results of the stress intensity factor for various directions based on the Franc3D, the results showed that the stress intensity factor of the axial crack tip was negative. This proved the nominal stress was compressive stress which generated a closure phenomenon into the crack. While the stress intensity factor of the radial crack tip was positive. Namely the nominal stress was tensile which accelerated the crack growth.

(3) The results on the crack growth under various loads and the analysis of the contact fatigue failure demonstrated that the fatigue failure form was peeled off from the subcutaneous layer. The crack initiated from the surface and the fracture morphology was the intergranular fracture. Both numerical and experimental results of the crack growth paths were in a circular arc shape. The a-N curves of the crack growth life demonstrated that the entire crack length was within 1.5 mm. Also, the crack cycle life decreased as the stress pressure increased.

5. References

[1] Zhu J C 1998 *J. Chinese Railways* 8 48
[2] Dong X M 2000 *J. Chinese Railways* 11 46-49
[3] Wu Z X 2007 *J. Chinese Journal of Applied Mechanics* 1 42-46
[4] Shi X L, Li Q and Song Z X 2016 *J. Journal of Mechanical Engineering* 10 126-132
[5] Wang Y Z, Tian Z M and Hou L W 2014 *J. Journal of Beijing University of Aeronautics and Astronautics* 2 148-153
[6] Lewandowski J and Rozumek D 2016 *J. Theoretical & Applied Fracture Mechanics*
[7] Storgärds E, Simonsson K and Sjöström S 2015 *J. Theoretical & Applied Fracture Mechanics* 81 2-10
[8] The editorial board of mechanical engineering material performance data manual 1995 *M. Mechanical engineering material performance data manual* 232-236.
[9] Gao C Q and Liu J J 1989 *M. Material of adhesive wear and fatigue wear*
[10] Zhong QP, Zhou Y and Zhang Z 2014 *M. Study of crack*
[11] Hong H, Sun J Z and Ma C Y 2001 *J. Journal of Shenyang University of Technology* 23 463-466
[12] Ma L, Wang M Q and Shi J 2009 *J. Chinese Journal of Materials Research* 23(3) 251-256
[13] Song X 2009 *D. Harbin University of Science and Technology.*

Acknowledgment

The authors are especially grateful to the Scientific Cooperation Major Project of Guizhou Province, China (Grant No. [2014]6012).

Study on Warm Forging Prosess of 45 Steel Asymmetric Gear

Yushi Qi[1, a], Zhiming Du[1, b], Hongsheng Sun[1], Lihua Chen[2], Changshun Wang[2]

[1]School of Material Science and Engineering, Harbin Institute of Technology, Harbin 150001, China
[2]Technology Center, Beijing North Vehicle Group Corporation, Beijing 100072,China
E-mail: [a] qys_gd@sina.cn [b] duzm@263.net

Abstract. Asymmetric gear has complex structure, so using plastic forming technology to process the gear has problems of large forming load, short die life, bad tooth filling, and so on. To solve these problems, this paper presents a radial warm extrusion process of asymmetric gear to reduce the forming load and improve the filling in the toothed corner portion. Using the new mold and No. 45 steel to conducting forming experiments under the optimal forming parameters: billet temperature is 800°C, mold temperature is 250°C, the forming speed is 30mm/s, and the friction coefficient is 0.15, we can obtain the complete asymmetric gear with better surface and tooth filling. Asymmetric gears' microstructure analysis and mechanical testing showed that the small grain evenly distributed in the region near the addendum circle with high strength; the area near the central portion of the gear had a coarse grain size, uneven distribution and low strength. Significant metal flow lines at the corner part of the gear indicated that a large number of late-forming metal flowed `into the tooth cavity filling the corner portion.

1. Introduction

Gear is a mechanical transmission component commonly used in the mechanical structure. It has been widely used in many areas of industrial production such as automotive, aerospace, marine, machine tools, engineering machinery, light industry machinery, metallurgical machinery and instruments[1-3]. The bearing capacity of the gear depends on its' pressure angle, the greater the pressure angle is, the higher the bending fatigue strength is. The tooth top of the gear comes to a point due to the increase of the pressure angle, which will lead to decrease of capacity of resistance to impact load. However asymmetric gear is a new type of gear which is designed for improving the bending strength and its characteristic is having two different pressure angle, corresponding to the two asymmetric involute. This kind of gear has many advantages such as small volume, light quality, large carrying capacity, etc. [4-7]

The gear processing methods are divided into machining and plastic forming. Gear machining has some defects, such as damage of metal flow line, resulting in a decline in gear strength, low material utilization rate, a considerable part of the billet is removed or becomes into an iron scrap in machining process. After machining the blank also need heat treatment to improve its mechanical properties. This procedure will increase the complexity of the process, and maybe affect the precision of gears. Using plastic forming process to produce gear can improve the utilization of raw materials and production efficiency, increase the mechanical properties of gears. As a consequence, exploring the plastic forming process of asymmetric gear has wide application prospects[8-12].

This artical studies on precision forging process of 45 steel asymmetric gear, presents a radial warm extrusion process of asymmetric gear to reduce the forming load and improve the filling in the toothed corner portion

2. Experimental procedure

45 steel without quenching used as experimental blank with specifications of Φ32×20. Using four-column hydropress with pressure of 200 tons as forming equipment. Firstly, mould was preheated to 250°C (about 2.5 h) and the mould surface was sprayed by graphite coating. Billet will be heated to 800°C used by a induction heating equipment with water-cooled copper crucible, then quickly put the heated billet in a tooth profile extrusion die to forming. Subsequently, the microstructure and mechanical properties were characterized of forming specimen were investigated.

3. Results and discussion

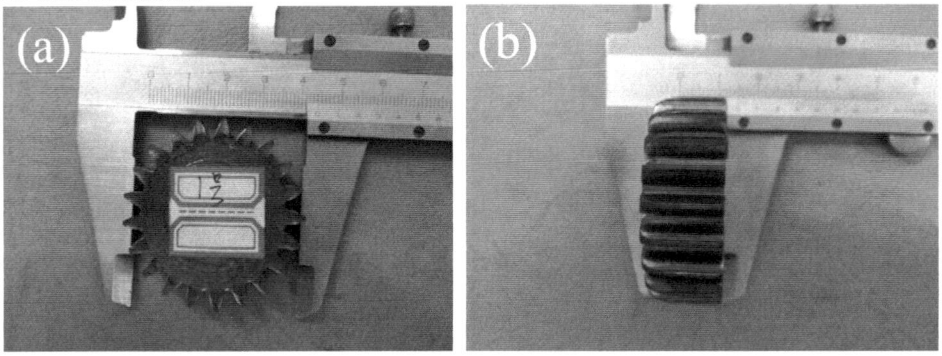

Figure 1. Digital image of the formed asymmetric gear

Digital image of formed asymmetric gear was shown in figure 1. From the image we can see that the top and bottom of gear's corner were partly filled completely, asymmetric tooth profile was obvious. The effect of overall forming was excellent in addition to the slightly uneven surface because of the mould structure. Excess metal in the rectangular shunt cavity could be seen from the side image, redundant rectangular columns were formed above the addendum circle. Excess metal flowed into the shunt cavity which is on the top is more, and flowed into the shunt cavity at the bottom of gear was smaller, relatively.

3.1 Microstructure of the formed asymmetric gear

Figure 2 shows the microstruture of metal flow line in different part. The direction of metal flow line could be found in the picture. From the figure 2(a) which is the part of asymmetric gear addendum circle, it can be seen that grains are pressed flat along the radial direction, which flow to the angle, and the flow lines on the top are also along the radial direction. That reveals that in the process of metal filling cavity, excess metal flow to shunt cavity, and the shunt cavity on the top is not completely filled by metal, it will make that there are still many free flow of metal at the end of forming, process which lead to reduce the forming force greatly. Figure 2(b) is the middle part of the involute, the metal flow line in this part is obvious than which in (a), direction is also radial. Image (c) shows the rounded part between dedendum circle and involute, the metal flow can be seen here is very clear, this is due to the violent impact between die and metal blank of which flow direction will be changed under the effect of die. Deformation of the metal in this part is more obvious, its mechanical properties will be different from the other part.Image (d) is near to the transition fillet, within the grain was flattened, but the direction is different due to that the metal in this part had upsetting deformation and the gear filling process at the same time. In the upsetting deformation, this part of the metal had a trend of circumferential flow due to the increase in billet diameter. In the filling process, the metal flow to the cavity part, which caused this part of the metal had a trend of the radial flow. All above makes the metal flow direction in this area was not unified. (e) and (f) belong to the inner part of the billet. The images reveal that the grain did not occur to extrusion deformation, and the metal flow was not obvious, this is because the two parts of metal's deformation is very small in the forming process of asymmetric gear, for another the metal flow is also lesser.

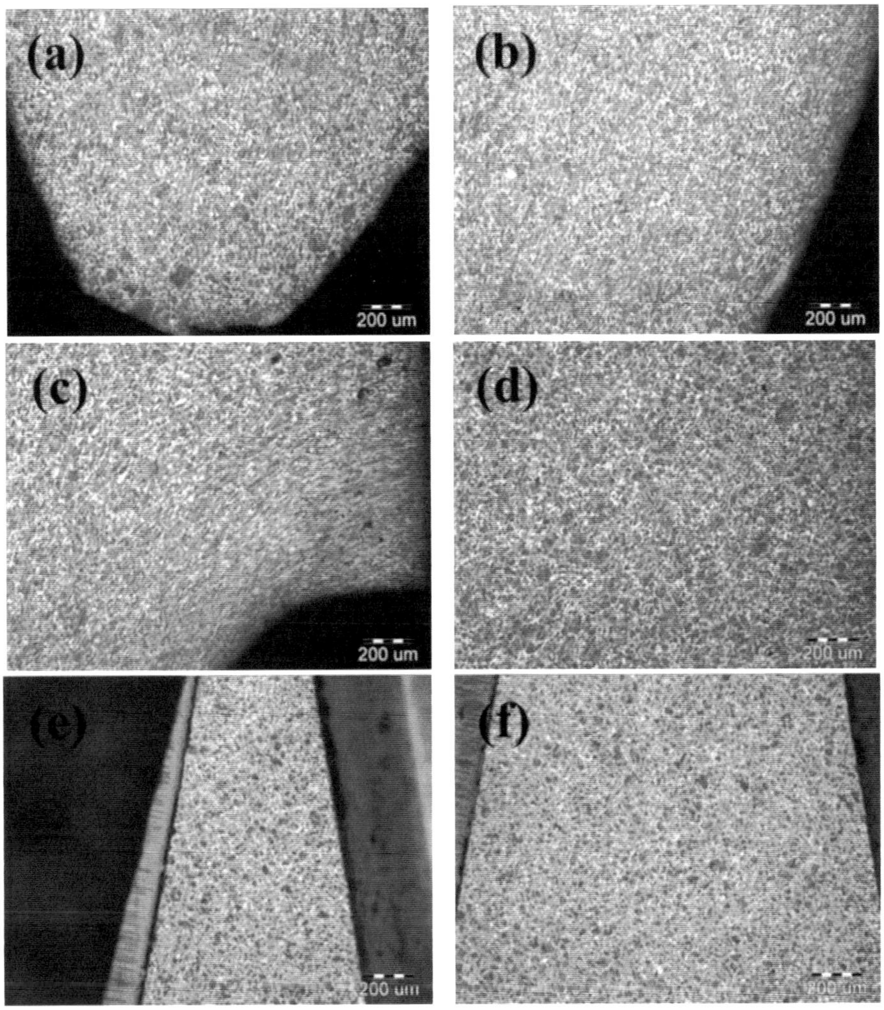

Figure 2. Microstruture of metel flow line in different part

Using higher-expansion microscopic analysis to observe grain size and distribution. Figure 3 shows the metallographic image of gear which was magnified 500 times. In terms of grain size, we can find that the grain size of (b) and (c) is relatively smaller than that of (a) and (d), and the grain size of (e) and (f) is the most massive. From the uniformity of the grain size distribution, (b) and (c) is the optimal, (a) and (d) is a bit poor, (e) and (f) is uneven. This reveals that in the process of radial extrusion of asymmetric gear, deformation in the part of rounded and involute tooth profile is larger, and its grain was isometrically extend along the radial under the action of external force. At the same time the structure became refinement and forming a fine organization, this deformation process of this part of metal will be accompanied by a large number of dislocation multiplication, which would make the deformation resistance of the metal increases, this part will be discussed in the next section.The metal in the area (a) mainly flowed belong radial under the action of terrace die because of the shunt cavity, which different from the process of (b) and (c). So the grains size in (a) is smaller than, but bigger than the (b) and (c). is There was no significant of metal flow deformation in region (e) and (f), therefore its grain was bulky, and distribution of grain was uneven.

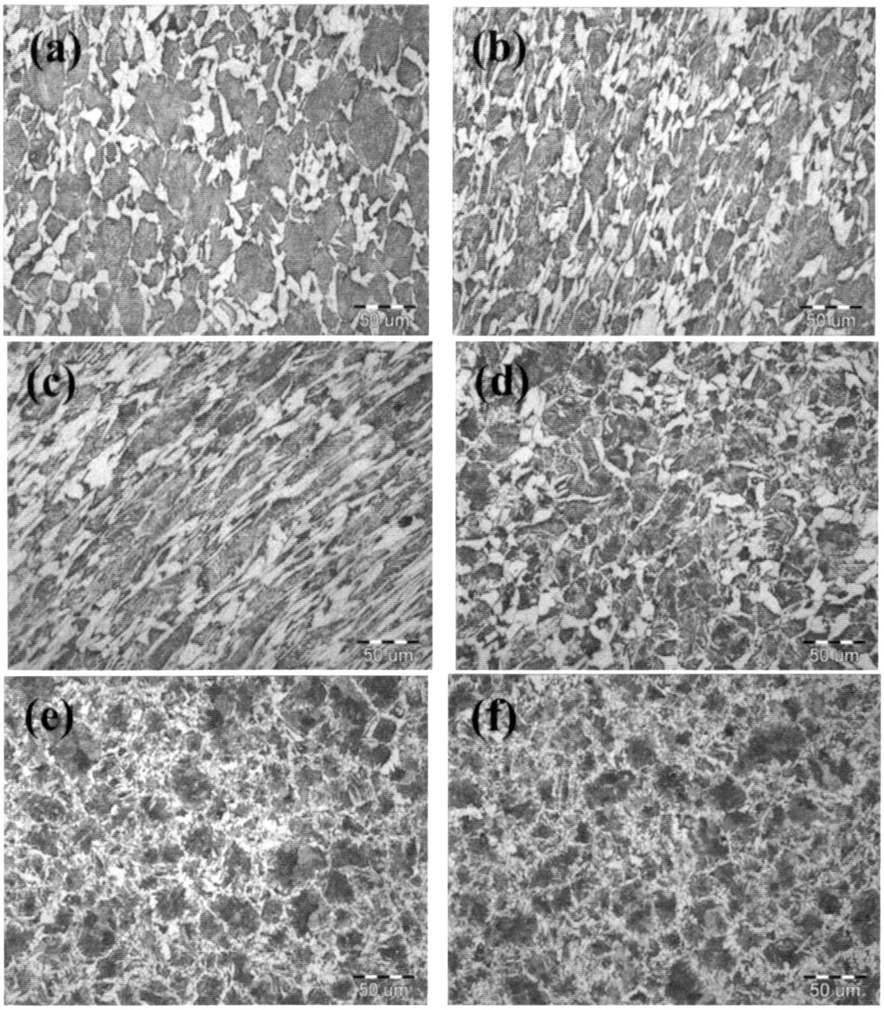

Figure 3. Microstruture of grain in different part

Observing the grain structure, we find that the organization of each area is pearlite grains, which is a mixture of ferrite and cementite organization in the warm forging experiments of asymmetric gear, there is no presence of the martensite structure. The warm forging temperature is 800°C, the austenitizing have occurred at this temperature, and the adoption of the cooling system is air cooling after forging, the cooling speed is less than the critical cooling rate, so the organization is pearlite.

3.2 Mechanical properties of the formed asymmetric gear

Table 1. The hardness in different region

	1st/HV	2st/HV	3st/HV	average /HV
region1	180.22	175.57	184.36	180.05
region2	172.13	164.04	168.84	168.34
region3	199.82	203.45	208.65	203.97
region4	148.79	151.84	145.93	148.85
region5	131.20	136.75	124.58	130.84
region6	125.54	120.21	131.53	125.76

Figure 4 shows the hardness in different region on the basis of Table 1. It can be obviously found that the hardness is higher in where the deformation of asymmetric gear forming is large in the process. Because the warm forging temperature is 800°C, above the recrystallization temperature of 45 steel, so a lot of original rough grains become fine uniform after recrystallization, combined with the structure refinement when the metal occurs to plastic deformation, and the proliferation of dislocation result in the increase in strength. The greater the deformation, the more obvious the strain hardening process , the higher the hardness.

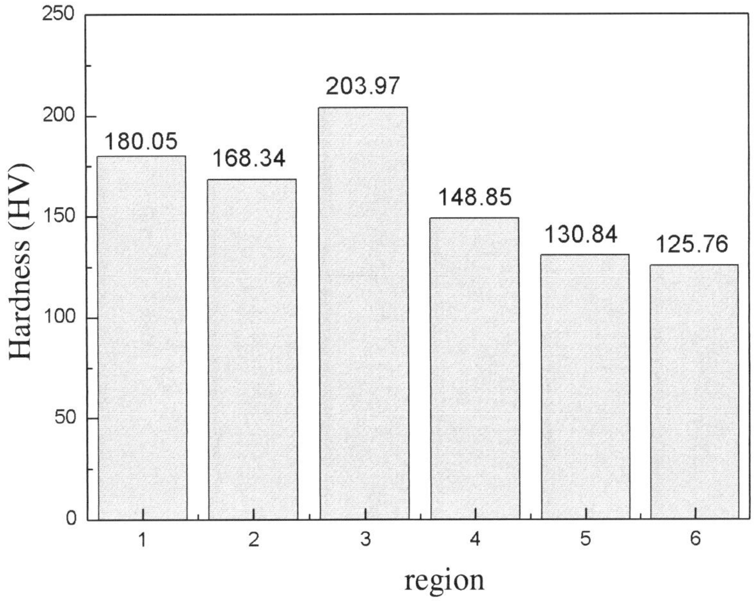

Figure 4. The hardness in different region

Table 2. The compression results in different region

	Yield Strength/MPa	Ultimate Strength /MPa	Compressive	Compressive Strain/%
Sample1	739.87	2757.49		71.30
Sample2	690.16	1857.36		60.03
Sample3	688.91	1307.61		39.80

There are some parts are not suitable for tensile experiment due to the smaller and irregular shape, so experiment test of the compression strength is adopted. We get sample 1, 2, 3 from the tooth part, middle part and tooth part of the gear, respectively. tooth part is the sample 1, the middle section is sample 2, tooth heart part for sample 3. Compression performance of three samples tested by electronic universal testing machine, the result is shown in figure 5. Table 2 is the compression results of the three samples.

Both figure 5 and table 2 illustrate the gear mechanical properties (compressive strength, ultimate compressive load and the ultimate compression strain) become poor from the top circle part to tooth heart gradually. The tooth shape parts compared with the tooth of the heart, of which the compressive strength increased by 7.40%, the ultimate compressive strength increased by 110.8% and the biggest compressive strain increased by 79.15%. Three samples' flow stress is not quite difference before happen to yield, but the flow stress of sample 2and 3 increases with the increase of deformation due to work hardening after yield, and sample 1 has long yield platform until its deformation reaches 43%.

The three samples' intensity has no significant difference, but the difference of plastic is larger caused by the microstructure. Grain of the sample 1 is fine and uniform distribution, deformation in more grain, strain in grain boundaries and internal is similar, so the deformation is uniform. Sample 2 and 3 have an uneven distribution of grain which is relatively bulky and its deformation is prone to stress concentration and fracture, so the plastic is poorer. Since all three samples are pearlitic structure without martensite structure, its yield strength is smaller.

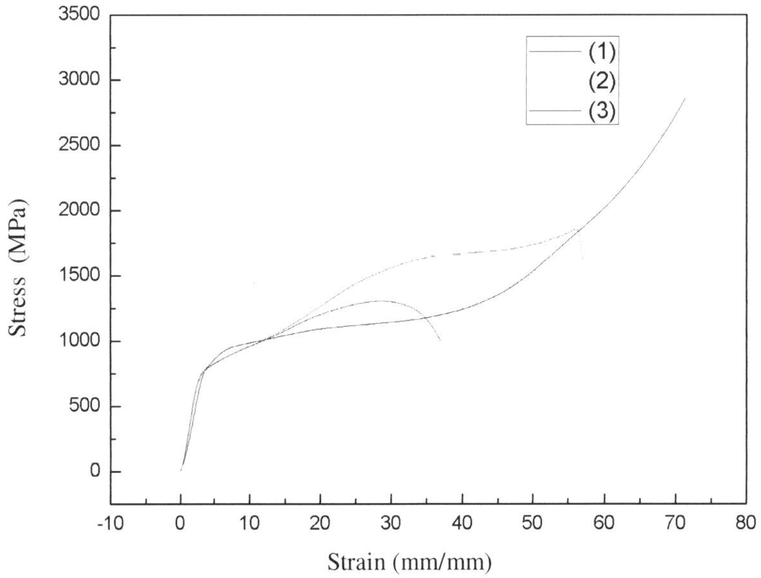

Figure 5. The hardness in different region

4. Conclusion

This paper presents a radial warm extrusion process of asymmetric gear to reduce the forming load and improve the filling in the toothed corner portion Asymmetric gears' microstructure analysis and mechanical testing showed that the small grain evenly distributed in the region near the addendum circle with high strength; the area near the central portion of the gear had a coarse grain size, uneven distribution and low strength. Significant metal flow lines at the corner part of the gear indicated that a large number of late-forming metal flowed `into the tooth cavity filling the corner portion.

5. References

[1] Yoerkie C A, Chory A G. Acoustic vibration characteristics of high contact ratio planetary gears[J]. *The Journal of American Helicopter Society*, 1984, 40: 19-32.

[2] Lange K. Some aspects of the development of cold forging to a high-tech precision technology[J]. *Journal of materials processing technology*, 1992, 35(3): 245-257.

[3] Hu C, Wang K, Liu Q. Study on a new technological scheme for cold forging of spur gears[J]. *Journal of Materials Processing Technology*, 2007, 187-188: 600-603.

[4] Sheljaskow S. Tool lubricating systems in warm forging[J]. *Journal of Materials Processing Technology*, 2001.

[5] Narayanasamy R, Srinivasan P, Venkatesan R. Computer aided design and manufacture of streamlined extrusion dies[J]. *Journal of materials processing technology*, 2003, 138(1): 262-264.

[6] Ponalagusamy R, Narayanasamy R, Srinivasan P. Design and development of streamlined extrusion dies a Bezier curve approach[J]. *Journal of Materials Processing Technology*, 2005, 161(3): 375-380.

[7] Cai J, Dean T A, Hu Z M. Alternative die designs in net-shape forging of gears[J]. *Journal of Materials Processing Technology*, 2004, 150(1-2): 48-55.
[8] Ohga K, Kondo K. Research on precision die forging utilizing divided flow(First report, Theoretical analysis of process utilizing flow relief-axis and relief-hole)[J]. *Bulletin of JSME*, 1982, 209(25): 1828-1835.
[9] Ohga K, Kondo K, Jitsunari T. Research on precision die forging utilizing divided flow(Second report, Experimental study of process utilizing flow relief-axis and relief-hole)[J]. *Bulletin of JSME*, 1982, 209(25): 1836-1842.
[10] Ohga K, Kondo K, Jitsunari T. Research on precision die forging utilizing divided flow(Third report, Study on an optimum combination by two step method proposed)[J]. *Bulletin of JSME*, 1982, 209(25): 1843-1850.
[11] Ohga K, Kondo K, Jitsunari T. Research on precision die forging utilizing divided flow(Forth report, Influence of restricting a centripetal flow)[J]. *Bulletin of JSME*, 1983, 218(26): 1434-1441.
[12] Ohga K, Kondo K, Jitsunari T. Research on precision die forging utilizing divided flow(Fifth report, Application to actual machine parts)[J]. *Bulletin of JSME*, 1985, 244(28): 2451-2459.

Acknowledgements
The authors express their appreciation for financial support of Ministry of science and technology project of international cooperation under Grant(No.2014***50320).

IOP Conf. Series: Materials Science and Engineering **230** (2017) 012040 doi:10.1088/1757-899X/230/1/012040

Study of Cold Coiling Spring Steel on Microstructure and Cold Forming Performance

Y Jiang[1,3,4] , Y L Liang[2,3,4,*], Y Ming[2,3,4] and F Zhao[2,3,4]

[1]School of Mechanical Engineering, Guizhou University, Guiyang, Guizhou, People's Republic of China
[2]School of Material Science and Metallurgical Engineering, Guizhou University, Guiyang, Guizhou, People's Republic of China
[3]Guizhou Key Laboratory for Mechanical Behavior and Micro Structure of Mateials, Guizhou University, People's Republic of China
[4]National & Local Joint Engineering Laboratory for High-performance Metal Structure Material and Advanced Manufacturing Technology, People's Republic of China.
E-mail address: liangyilong@126.com (Y L Liang).

Abstract. Medium-carbon cold-coiling locomotive spring steels were treated by a novel Q-P-T (quenching-partitioning-tempering) process. Scanning electron microscopy (SEM), transmission electron microscope (TEM) and X-ray diffraction (XRD) were used to characterize the relevant parameters of the steel. Results show that the microstructure of tested steel treated by Q-P-T process is a complex microstructures composed of martensite, bainite and retained austenite. The volume fraction of retained austenite (wt.%) is up to 31% . After pre-deforming and tempering again at 310℃, the plasticity of samples treated by Q-P-T process is still well. Fracture images show that the Q-P-T samples are ductile fracture. It is attributed to the higher volume fraction of the retained austenite and the interactions between the multi-phases in Q-P-T processed sample.

1. Introduction

Numerous researches have been performed to develop new advanced high strength steels for the application in lightweight automotive structures and heavy load, high speed railways. From dual phase (DP) steel, transformation induced plasticity (TRIP) steel in 1990s[1-3] to quenching-partitioning (Q-P) steel[4-7] and quenching-partitioning-tempering (Q-P-T)[8-10] steel in recent years, the comprehensive properties of steels have promoted greatly in contrast with traditional steels. For example, steels treated by Q-P process show good mechanical properties which are obtained through controlling the percentage of stable retained austenite at room temperature. But one of its strengthening preconditions is to prevent the carbide precipitation which is not corresponding with the actual experimental status in some cases[11, 12]. Based on this, Xu[9] proposed a new process called Q-P-T process which not only ensures the stability of retained austenite but also makes fine carbide precipitation in the matrix by adding carbide forming elements like silicon. Following this, the product of strength and elongation of steels with 0.4 wt.% carbon content reaches to 31627 MPa•% which meet the requirements of new generation high strength steel's strength and elongation product≥3×104 MPa•% predicted by Matlock and Speer[9, 13]. The high value of the product of strength and ductility suggests that steels treated by Q-P-T process also have good comprehensive properties. The processes aforementioned are usually used in steels with carbon content lower than 0.5 wt.%. While spring steels used in high speed locomotives usually contain more than 0.5 wt.% carbon content. It has been confirmed that spring steels treated by Q-P-T process possess better comprehensive properties compared with that of the traditional Q-T process[14]. But studies mentioned above are mainly for hot coiling spring steel with the diameter larger than 20 mm which needs heating during coiling process. Study on cold coiled spring steels treated by Q-P-T process is rarely reported. The diameter of the cold coiled spring steel is smaller than 20 mm which does not need any heating during coiling process. So it has increasingly wide utilization for its higher precision and less energy conservation than hot ones. But the heat treatment parameters may have differences for its size effects and application differences comparing with hot ones. In this paper, the novel process used in medium-carbon cold-coiling spring steel is discussed to find suitable parameters for actual production in which the hardness is between 42-46 HRC. On the basis of this, the strengthening mechanism of the spring steel treated by Q-P-T process compared with the traditional Q-T process are discussed by setting up the pre-strain and tempering treatment after deformation, which simulate the actual production process.

2. Experimental procedures

Commercial hot-rolled spring steel with Φ12mm diameter bar was used in the investigation (chemical compositions, wt. %: 0.59C, 1.61Si, 1.15Cr, 0.66Mn, 0.16V, Fe). And the phase transformation temperatures were measured by DIL 805A/D dilatometer, as shown in Table 1.

Table 1. Phase transformation temperatures of the tested steel

NO.	AC_1	AC_3	M_s	M_f
Temperature (℃)	780	865	262	-27

The Q-P-T process was performed as shown in figure 1. Firstly, the sample of the test steel was austenitized at 880℃for 18min; then it was quickly cooled with -80℃/s to 210℃ for 12s, followed by held at different isothermal temperatures (300℃, 350℃, 380℃, 400℃) for 90min. after that, the sample was tempered at 310℃ for 90min. Secondly, samples treated by the optimal parameters of Q-P-T with enough strength and good plasticity and suitable hardness were pre-strained and then tempered at 310 ℃ for 90min to simulate the productive process of the cold coiling spring.

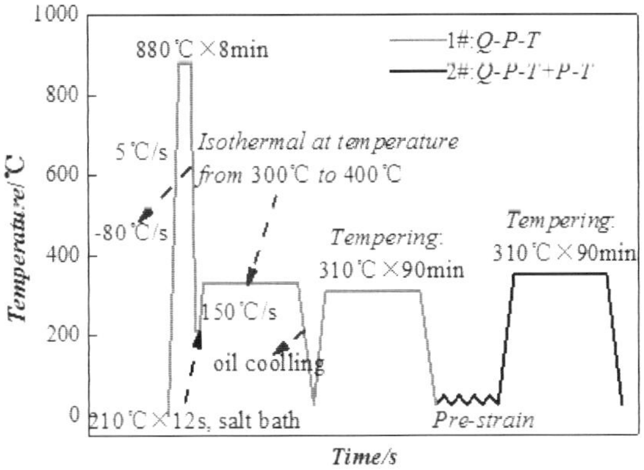

Figure 1. schematic diagram of Q-P-T process(a) and Q-T process(b)

Uniaxial tensile tests were conducted as standard dog-bone tensile test samples on MTS testing machine at room temperature at 2mm/min, and the hardness was tested on the micro hardness tester (HVS-1000). Meanwhile, the microstructure of the steel was characterized by OM (PMG3-OLYMPUS), SEM (SUPPA40), TEM (Tecnai G2F20S-TWIN, 200KV) and the fraction of the retained austenite was detected by XRD (350 X- ray, Cu Ka radiation), respectively.

IOP Conf. Series: Materials Science and Engineering **230** (2017) 012040 doi:10.1088/1757-899X/230/1/012040

3. Results

3.1 Microstructure

As shown in figure2, the typical microstructure of Q-P-T process is composed of martensite (M), bainite(B) and retained austenite (R) in which the size of the initial martensite(IM) is about 0.5μm while the size of the fresh martensite(FM) is 0.2~0.3μm which is consistent with researches in lieraure[15] and there are small steps in bainite which is a typical characteristic for lower bainite.

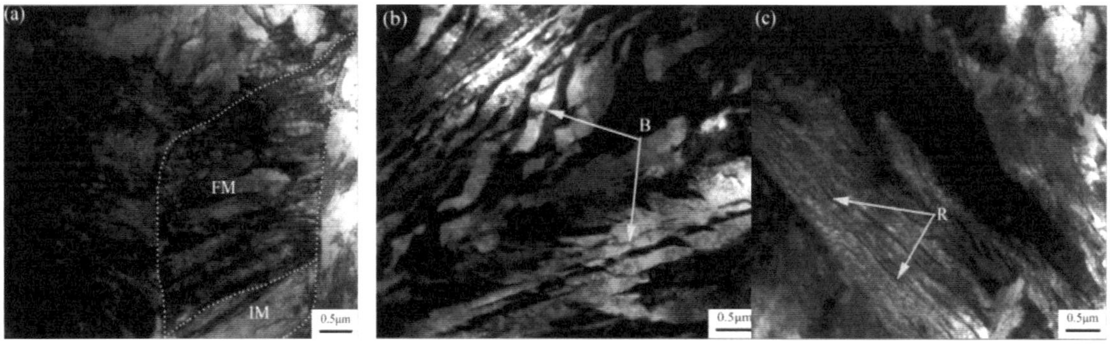

Figure 2. Microstructure composed of M(a), B(b) and R(c) with isothermal temperature at 380℃ in spring steel treated by Q-P-T

Corresponding to figure 2, the SEM microstructure (figure 3) of the tested steel is composed of dark gray martensite (IM), blocky martensite/retained austenite island(M/A) and bainite (B) which is present in the form of long, thin sheaves with a dominant crystallographic orientation within a single plate. At the same time, the white film or sheet structural retained austenite dispersed on the matrix.

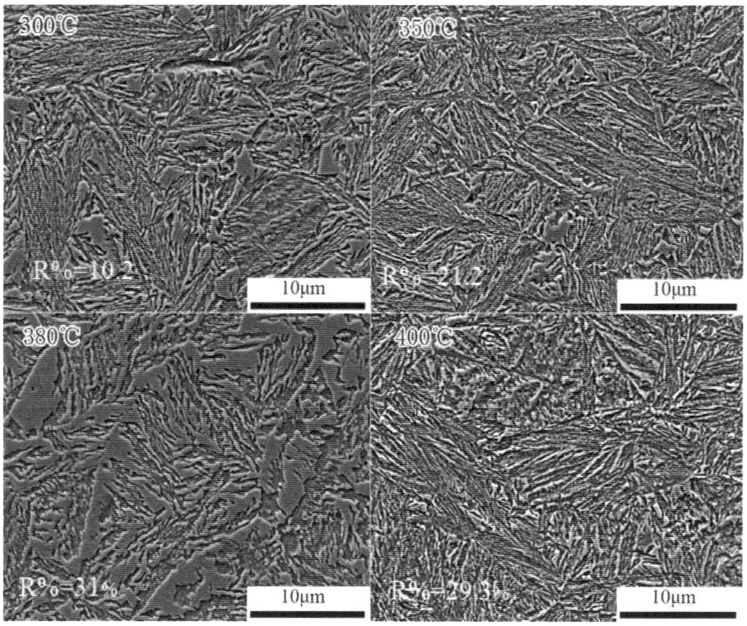

Figure 3. SEM image of samples isothermal at different temperature

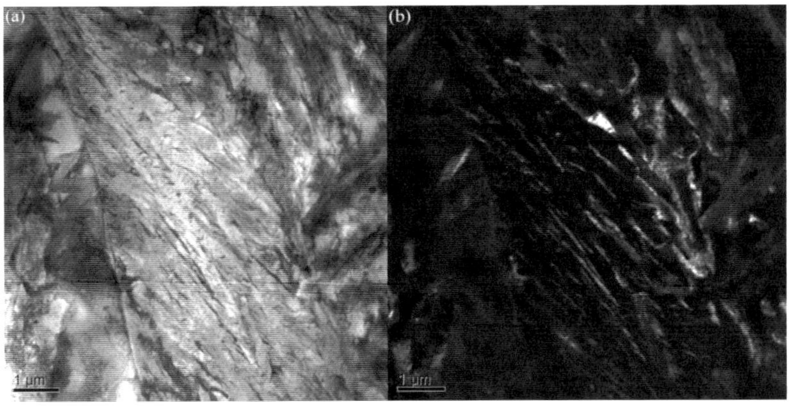

Figure 4. Typical microstructure of tested steel treated by Q-P-T process: (a)BF, (b) DF

3.2 Volume Fraction of Retained Austenite before Pre-strain Deformation

Typical residual autenite microstructure was show in figure 4 lots of film like and some bulk retained austenite can be seen in the DF image. And as shown in figure 3, the volume fraction of the retained austenite (wt.%) increased at first and then decreased with the increasing of the isothermal temperature. The maximum value appears at 380℃ which is correspond with the variation tendency of the volume fraction of M/A island. It is a result of competition between carbon partition and the decomposition of retained austenite. On the one hand, with the increasing of the isothermal temperature, more and more carbon distributed from initial martensite to retained austenite which makes the austenite more and more stable; on the other hand, with the increasing of the isothermal temperature, the thermal stability of the retained austenite become more and more worse. Before 380℃, the partition mechanism plays main role of the process, while after 380℃, the thermal stability is a leading role.

3.3 Mechanical properties

3.3.1 Mechanical Properties before Pre-strain Deformation. The variation tendency of mechanical properties (Rm, Rp0.2, hardness, reduction of area and elongation) for the samples treated by Q-P-T process in table 2 . For Q-P-T process, it can be seen that the strength (Rm and Rp0.2), the hardness and the reduction of area are decreased with the increasing of the isothermal temperature while the elongation of the tested steel increased at first and then decreased a little when the temperature exceeds 380 ℃. It is inconsistent with the change rule of the retained austenite. In addition, the product of strength and ductility changed from 1.86×10^{4}MPa•% at 300℃ to 3.49×10^{4} MPa•% at 380℃ and decreased to 1.56×10^{4}MPa•% at 400 ℃. The higher the value, the more energy of the tested steel absorbed during the deformation process or more work done by tensile force. Over all consideration, 380 ℃ is the isothermal temperature that is best suited for cold coiling process for Q-P-T process.

Table 2. Mechanical properties and the volume fraction of retained austenite in the tested steels treated by Q-P-T.

Temperature (℃)	**Z (%)**	**A (%)**	**Rp0.2 (MPa)**	**Rm (MPa)**	**Hardness (HRC)**
300	47.7	10.2	1527.66	1822.75	54

350	45.88	10.7	1331.08	1589.42	48
380	45.58	23.32	1207.93	1493.51	43
400	43.31	13.67	1174.38	1443.25	42

3.3.2 Mechanical Properties after Pre-strain Deformation. After pre-strain and tempering treatment mentioned above for Q-P-T process, the samples isothermal at 380℃ was clamped on the MTS again for concurrent tensile test. The change of the ultimate mechanical properties are shown in Figure 5 there is substantial improvement in yield strength and the ultimate tensile strength exceed 1700 MPa.

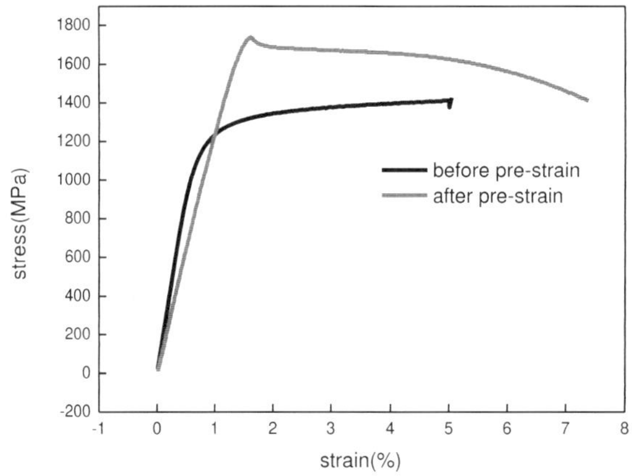

Figure 5. Stress-strain curves treated by Q-T and Q-P-T process

3.4 Fracture image

Figure 6 shows images of the tensile fracture of the processes after pre-strain treatment. The depth of dimples is deeper and the sizes are uniform (~0.6μm) in fibrous area. And there are few micro-cracks in fibrous area. There are holes connection characteristics in radiated area.

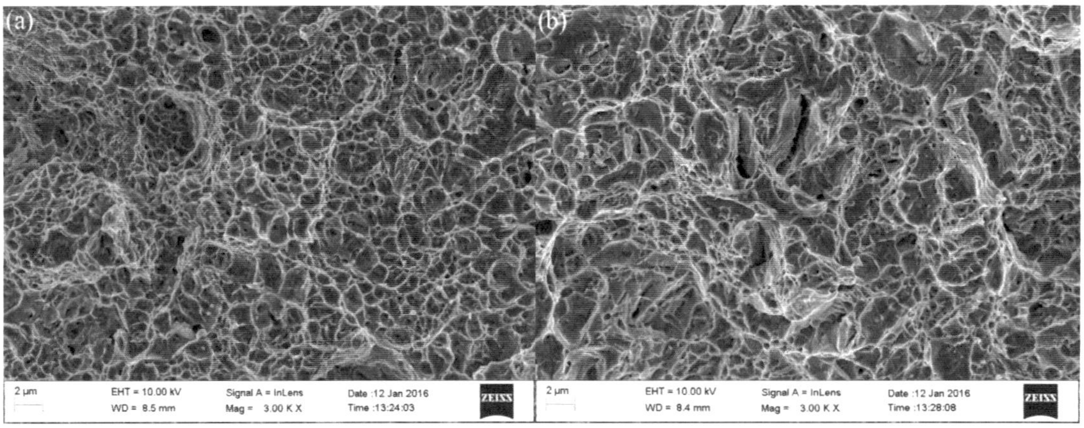

Figure 6. SEM images of tensile fracture surface: (a) fibrous and (b) radiated area of Q-P-T process.

4. Discussion

After pre-strain treatment, tensile strength and the yield strength is significantly improved concurrently. This is the result of Bauschinger effect[15]. The plasticity after pre-strain is still good which is due to the exsistence of the large amount of the retained austenite. It has been proved that the residual austenite can significantly improve the plasticity of steels. Firstly, residual austenite can split martensite area space, reduce the effective grain size, stop the crack expansion[16]. Secondly, residual austenite can transform to martenite during tensile deformation and produces TRIP effect which effectively alleviate the local stress concentration and delay the formation of cracks, which delays the onset of necking [17, 18]. Lastly, it is pointed out that the average dislocation density in martensite decreases during uniform deformation which is lower than the value before deformation while the dislocation density in the retained austenite increases rapidly. It has been observed in TEM that dislocations in martensite moved to the adjacent residual austenite which makes martensite in an 'undeformed' state and greatly enhances the deformation capacity of hard phase, which was called dislocation absorption by retained austenite (DARA) effect [19].

So there is interaction between each phase and under the triple action, i.e. Bauschinger effect, TRIP effect and DARA effect, the plastic deformation capacity of the novel process shows significant enhancement compared with traditional ones which means materials treated by Q-P-T process is more suitable for cold forming.

5. Conclusions

The main conclusions of the present work are as follows:

(1) Muti-phase microstructure composed of martensite, bainite, retained austenite is obtained for the tested steel treated by Q-P-T process.

(2) Comprehensively considering the factors suitable for cold coiling as mentioned above, samples treated by Q-P-T process have lower yield strength, approximate tensile strength, good plasticity, higher contents of residual austenite and suitable hardness. So the optimal partitioning temperature is 380℃.

6. References

[1] Kawagoishi N, Nisitani H, Toyohiro T and Kamifukumoto H Nihon Kikai Gakkai 1993 J. *A Hen/transactions of the Japan Society of Mechanical Engineers Part* A, 752-56
[2] Tomita Y and Morioka K 1997 *J. Materials Characterization* 38 243-50
[3] Bai D Q, Chiro A D and Yue S 1998 *C. Materials Science Forum* 253-62
[4] Speer J, Matlock D K and Cooman B C D 2003 *J. Acta Materialia.* 51 2611-22
[5] Speer J G and Rizzomatlock F C 2005 *J. Materials Research.* 8 253-65
[6] Speer J G, Moor E D and Findley K O 2011 *J. Metallurgical and Materials Transactions* A, 42 3591-601
[7] Knijf D D, Puype A and Föjer C. 2015 *J. Materials Science & Engineering* A 627 182-90
[8] Zhang K, Xu W Z and Guo Zh H 2011 *J. Acta metal sin,* 47 489-96
[9] Hsu T Y, Jin X J and Rong Y H 2013 *J. Journal of Alloys & Compounds*, 577 S568-71
[10] Rong Y. 2011 *J. Acta Metallurgica Sinica,* 47 1483-89
[11] Moor E D, Lacroix S and Clarke A J 2008 *J. Metallurgical and Materials Transactions* A 39 2586-95
[12] Nayak S S, Anumolu R and Misra R D K 2008 *J. Materials Science & Engineering A,*498 442-56
[13] Zhong N 2009 *D. Shanghai Jiao Tong University p* 92-7
[14] GAO W X and LIANG Y L 2013 *J. Materials Review*, 27 81-4
[15] Hu J, Chen B and Smith D J 2016 *J. International Journal of Plasticity* 84 203-23
[16] Wang C, Shi J and Cao W 2011 *J. Acta Metallurgica Sinica* 47 720-26
[17] Sarikaya M, Jhingan A K and Thomas G1983 *J. Metallurgical and Materials Transactions* A, 9 439-450
[18] Hao Q, Qin S and Liu Y 2016 *J. Materials Science & Engineering* A, 662 16-25

[19] Zhang K 2011 *D. Shanghai Jiao Tong University* p 87-97

Acknowledgement
The research work is jointly supported by the National Natural Science Foundation of China (NO. 51461006, NO. 51571066), and Guizhou Science and Technology Project (NO.(2015)7646).

IOP Conf. Series: Materials Science and Engineering **230** (2017) 012041 doi:10.1088/1757-899X/230/1/012041

Study on Plastic Deformation Characteristics of Shot Peening of Ni-Based Superalloy GH4079

L Q Zhong[1,2], Y L Liang[3,*] and H Hu[1]

[1]School of Mechanical Engineering, Guizhou University, Guiyang, Guizhou, People's Republic of China1 College of Mechanical Engineering, Guizhou University, Guiyang, China
[2]School of Mechanical Engineering, Guiyang University, Guiyang, Guizhou, People's Republic of China
[3]School of Material and Metallurgy, Guizhou University, Guiyang, Guizhou, People's Republic of China
E-mail address: liang_yilong@126.com (Y L Liang).

Abstract. In this paper, the X-ray stress diffractometer, surface roughness tester, field emission scanning electron microscope(SEM), dynamic ultra-small microhardness tester were used to measure the surface residual stress and roughness, topography and surface hardness changes of GH4079 superalloy, which was processed by metallographic grinding, turning, metallographic grinding +shot peening and turning + shot peening. Analysized the effects of shot peening parameters on shot peening plastic deformation features; and the effects of the surface state before shot peening on shot peening plastic deformation characteristics. Results show that: the surface residual compressive stress, surface roughness and surface hardness of GH4079 superalloy were increased by shot peening, in addition, the increment of the surface residual compressive stress, surface roughness and surface hardness induced by shot peening increased with increasing shot peening intensity, shot peening time, shot peening pressure and shot hardness, but harden layer depth was not affected considerably. The more plastic deformation degree of before shot peening surface state, the less increment of the surface residual compressive stress, surface roughness and surface hardness induced by shot peening.

1. Introduction

Shot peening has been widely used to improve fatigue property and stress corrosion cracking resistance of materials due to the advantages of simple equipment, convenient operation, energy conservation, time saving, low cost and obvious effect[1]; and the nickel-base superalloy has been widely used in the manufacture of turbine disc, turbine shaft, journal, vane and other high temperature parts in the rocket and the aero-engine, due to well high-temperature strength, heat stability, corrosion resistance, thermal fatigue resistance and creep resistance[2]; so, there is a large amount of literatures[3-8] about the shot peening of nickel-base superalloy, among which, the literatures[3, 6] have studied the residual stresses fields of GH909 alloy and inconel718 alloy shot peening, which show that beneficial residual compressive stress field was obtained by proper shot peening both in GH909 alloy and inconel718 alloy, and partly loosed during high temperature maintenance process. The literatures[4, 5, 8] have studied DZ4 directionally solidification superalloy, GH4169 superalloy and GH907 superalloy shot peening, which show that under high temperature, strengthening effect of shot peening is basically equal to weakening effect, and even the effect on notch specimen is more obvious than on smooth specimen, there is certain function of shot peening on the smooth specimen.

The organization structure in shot peening strengthened layer of GH4169 superalloy has studied in the literature[7], it shows that the thickness of elastic-plastic deformation layer formed by shot peening is about 1.2mm, and the surface roughness Ra is 1.4274μm. Although the scholars have obtained some achievements on studies of the shot peening in nickel-base superalloy, it is pointed out by studies of GAO Yu-Kui[9] that there is one best shot peening specification for one given material. As for the GH4079 which is a nickel-base superalloy just like GH4742, the content of Al, Ti and Nb was further improved in order to form more Ni_3 (AlTiNb) precipitation strengthening. The characteristics of shot peening plastic deformation layer are different when the marks and compositions of materials are different, and the research of shot peening of GH4079 superalloy is few. This paper plans to study the characteristics (surface redidual stress, surface roughness, surface appearance and surface hardness gradient) of GH4079 superalloy shot peenting plastic deformation layer, in order to provide reliable experimental data and certain theoretical foundation for choosing technological parameters of GH4079 superalloy shot peening and confirming surface conditions before shot peening.

2. Experimental

2.1 Material

For the tests the superalloy GH4079 was selected, the chemical composition of it is(wt%): C 0.059, Cr 11.02, Co 14.35, Mo 4.62, Al 3.11, Ti 2.77, Nb 2.68, V 0.62, Fe 0.3, La 0.046, Mn 0.03 and Ni margin. Experimental sample shown as figure 1, 50mmx20mmx10mm, which was provided by a company.

2.2 Methods

Before GH4079 superalloy shot peening, the surface is processed separately by metallographic abrasive, which grinding the surface of sample by abrasive paper for metallograph without any machining, and by turning with cutting parameters shown in Table 1; shot peening parameters shown in Table 2 were adopted to process lapped surface, and No.2 shot peening parameter shown in Table 2 was adopted to turning surface. Surface roughness Ra was tested randomly seven times by CALISUM surface roughometer CR-4032, adopting the mean value as the value of tested surface roughness; using field emission scanning electron microscope(SEM) SUPPA40 to further observe the surface appearance; the state of residual stress in the surface was measured using X-ray stress diffraction (XRD). XRD analysis was conducted using Mn k_α source on a DRON-3M diffractometer with a diffraction angle of 151° in plane {311}, which stress constant is -349 MPa. A $\mathrm{Sin}^2\,\psi$ method with ψ = 0, 45° was used for stress estimation, among which, scanned area 2θ is 158°~143°, scanned step pitch is 0.05°, counting time is 1s, voltage is 18 kV, current is 4 mA, light pipe diameter is Φ 2 mm; and using DUH-211(S) dynamic ultra-small microhardness tester to test the changes of hardness on surface of the samples before and after shot peening, the test load is 196 mN, and retention time is 10s.

Table 1. Cutting parameters of turning

number	Cutting speed v(m/min)	Cutting depth a_p(mm)	Feed f(m/r)
1	560	1.0	0.035
2	280	0.5	0.053
3	280	0.5	0.088

Table 2. Shot peening parameters

parameter	Shot peening 1	Shot peening 2	parameter	Shot peening 1	Shot peening 2
Intensity	0.11	0.15	Coverage	>100%	>100%
Time	90s	120s	Type of pill	A S70	B S70
Pressure	0.22b	0.25b	Diameter of pill	0.125~0.355	0.125~0.355

Hardness of pill	HRC42.5～48.5	HRC48.5～52.5	Nozzle diameter	Φ8 mm	Φ8 mm

3. Result and analysis

3.1 Characteristics of GH4079 superalloy metallographic abrasive + shot peening plastic deformation layer

3.1.1 Surface residual stress and surface roughness

The test results of surface residual stress in X and Y directions and surface roughness of metallographic abrasive and metallographic abrasive + shot peening of GH4079 superalloy are as shown in Table 3. It shows that the surface residual stress both in X and Y directions are compressive stress for the metallographic abrasive surface of GH4079 superalloy, due to the function of grinding force, and the $\sigma_{r(x)}$ is 100MPa slightly larger than $\sigma_{r(y)}$. The surface residual stress of metallographic abrasive + shot peening which parameters shown in table 2 are about -800 MPa in X direction, and -650MPa in Y direction, about 350MPa residual compressive stress was induced by shot peening, which explains that higher surface residual compressive stress of GH4079 alloy can be achieved through shot peening with the parameters adopted in this paper. Comparing residual compressive stresses induced by two kinds of shot peening parameters, it can be find that shot peening parameter 2 is slightly superior to shot peening parameter 1, because shot peening parameter 2 has larger shot peening strength, shot peening time, shot peening pressure and shot hardness, which enlarges deformation degree of surface plasticity, that the induced residual compressive stress is larger than it of shot peening parameter 1. On the other hand, the surface residual compressive stress of shot peening mainly depends on crystal structure, yield strength and work hardening index of material [10], instead of shot peening parameter that is why the superiority is not obvious. After shot peening, the difference of $\sigma_{r(x)}$ and $\sigma_{r(y)}$ was about 100 MPa as well as before shot peening, which illustrates that the residual compressive stress induced by shot peening was isotropy when the surface residual stress status of GH4079 alloy before shot peening has few difference in different direction.

As the surface roughness shown in Table 3, the surface roughness of GH4079 alloy with metallographic abrasive + shot peening is 5 times before shot peening, which is resulted from different sizes shot marks lefted on material surface by high-speed shot flow when shot peening, and before the shot peening, the surface of material with metallographic abrasive look likes very smooth, which is as shown in figure 2(a), we can find from the surface appearance pictures before and after shot peening that the surface with metallographic abrasive + shot peening2(figure 2(c)) is rougher than metallographic abrasive + shot peening1(figure 2(b)), which is consistent with measurement result of roughness.

Table 3. Metallographic abrasive surface residual stress and roughness of before and after shot peening

State of surface	Surface residual stress $\sigma_{r(x)}$ (MPa)	Surface residual stress $\sigma_{r(y)}$ (MPa)	Surface roughness R_a(μm)
Before/shot peening1/shot peening2	-473.7/-793.3/-827.7	-369/-624.3/-718.3	0.155/0.833/0.882

3.1.2 Surface hardness gradient

Seen from the surface hardness gradient curve of metallographic abrasive GH4079 superalloy before and after shot peening shown in figure 3, it can be find that the surface hardness of before shot peening was 560DHV, and beyond about 30μm away from the surface, the value maintains to be about 500DHV. However, the surface hardness of shot peening may reach 700～800DHV, due to the shock of high-speed shot flow generates work hardening to metal on surface layer, and the hardness value gradually decreases along with the increase of layer depth, the value maintains to be 500DHV. However, the surface hardness of shot peening may reach 700～800DHV, due to the shock of high-speed shot flow generates work hardening to metal on surface layer, and the hardness value gradually decreases along with the increase of layer depth, the value maintains to be about 500DHV until the distance from the surface beyond 400μm. The surface hardness and hardness gradient of shot peening 2 are larger than them of shot peening 1 as shown in figure 3, but the depths of hardened layer for both are equivalent. It can be conclude that higher shot peening strength, longer shot peening time and bigger shot peening pressure correspondingly with more surface plastic deformation degree, should cause further the work-hardening capacity.

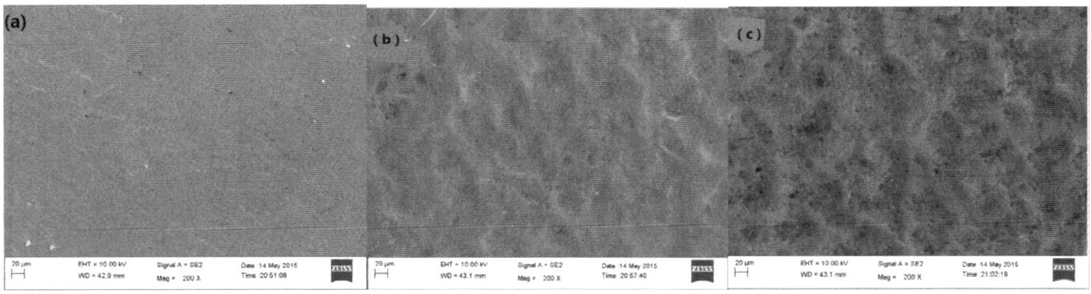

Figure 2. Surface topography(a) Metallographic abrasive,
(b) Metallographic abrasive + Shot peening1,
(c) Metallographic abrasive +Shot peening2

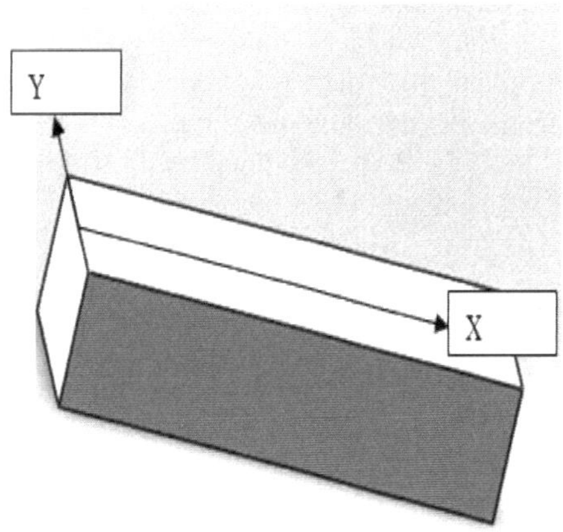

Figure 1. 3-D parts of test sample

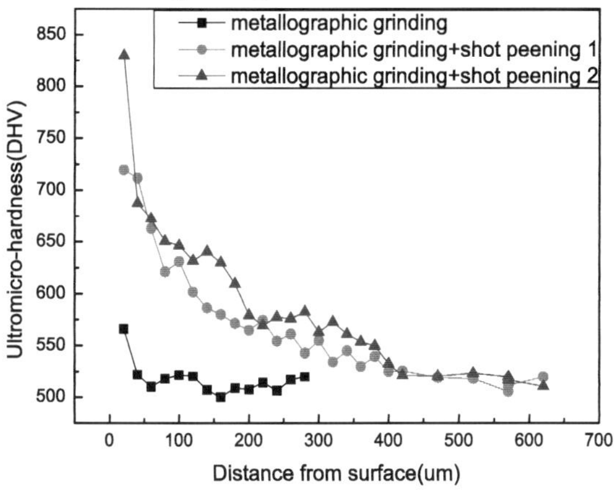

Figure 3. Curves of ultramicro-hardness before and after shot peening of metallographic grinding surface

3.2 Characteristics of GH4079 superalloy turning + shot peening

3.2.1 Surface residual stress and surface roughness

The test results of surface residual stress and surface roughness of GH4079 superalloy before and after shot peening were shown in Table 4, and 0#, 1#, 2# and 3# separately represented the surface state of metallographic abrasive+ shot peening 2, turning 1+ shot peening 2, turning 2+ shot peening 2 and turning 3+ shot peening 2. It easy to see that residual compressive stress was induced by shot peening no matter what the surface status of before shot peening. And the residual stress after shot peening of GH4079 superalloy in X direction $\sigma_{r(x)}$ can reach up to above -800 MPa and in Y direction $\sigma_{r(y)}$ can also be more than -450 MPa, due to the surface residual stress of shot peening mainly depends on crystal structure, yield strength and work hardening of material[10]. And it illustrates that residual compressive stresses induced by shot peening was changing with surface statuses of before shot peening, the more residual compressive stress before shot peening, the less residual compressive stress induced by shot peening, due to there is no obvious change at final surface residual compressive stress.

As shown in Table.4, the surface roughness of before shot peening changes within the 0.155-0.748μm, and within the 0.787-0.957μm after shot peening, which was obviously increased, the reason for this is that when the surface roughness of before shot peening is far smaller than diameter of the shot, passing by the shot with more than 100% coverage rate, the shot mark size left by shock of shot will be larger than the size of machining mark before shot peening and cover the whole surface. The surface roughness of shot peening are almost equivalent under different surface statuses, which illustrates that the surface roughness of after shot peening mainly depends on size of shot diameter. Figure 4 shows us the appearance pictures of turning surface before and after shot peening, the cutting mark and built-up edge on turning surface(figure 4(a),(c),(e)) are obvious, through shot peening, the shot marks on shot peening surface(figure 4(b),(d),(f)) can be seen clearly, and plastic deformation were observed, the cutting marks can just be seen indistinctly due to shock of high-speed shot.

Table 4. Surface residual stress and roughness of before and after shot peening

Number of sample	Residual stress in X direction σ_{rx}(MPa) before/after	Residual stress in Y direction σ_{ry}(MPa) before/after	Surface roughness Ra（μm） before/after
0	-473.7/-827.7	-369/-718.3	0.155/0.882
1	-998.2/-1102.3	504/-448.3	0.748/0.834
2	-546.5/-828	482/-495.5	0.675/0.787
3	-578.5/-829.7	478/-618.5	0.667/0.957

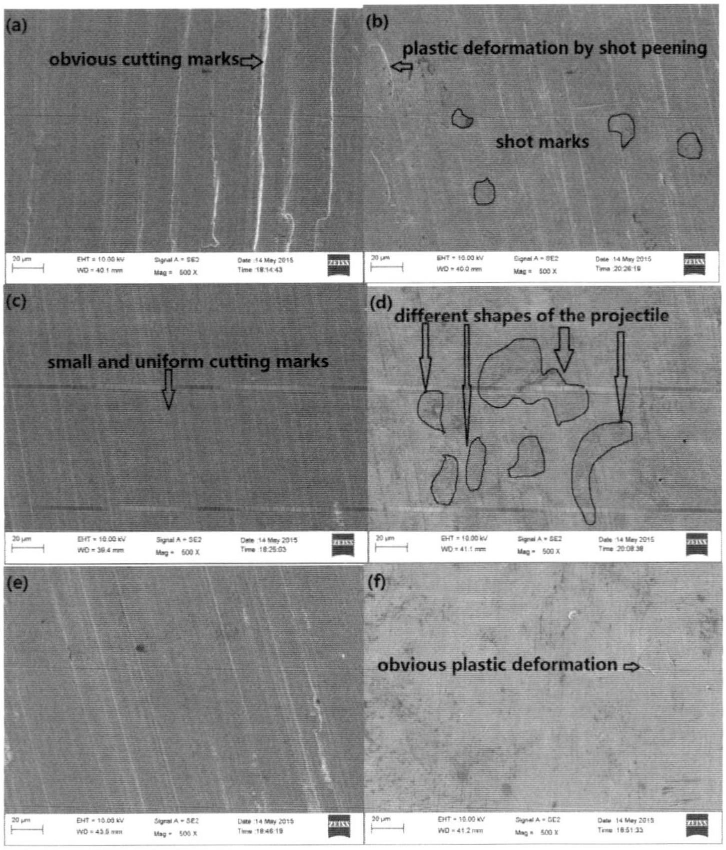

Figure 4. Surface topography (a) turning1, (b) turning1+ shot peening2 (c) turning 2 (d) turning2+shot peening2 (e) turning3 (f) turning3+shot peening2

IOP Conf. Series: Materials Science and Engineering **230** (2017) 012041 doi:10.1088/1757-899X/230/1/012041

3.2.2 Surface hardness gradient

The figure 5 shown the surface hardness gradient curves of GH4079 superalloy before and after shot peening, the variation trend of hardness of GH4079 superalloy decreased along with the increase of distance from the surface, to matrix hardness about 500DHV, which is resulted from work hardening generated from grinding, turning and shot peening, the depth of working harden was different from different surface processing, and in order to compare the effect of surface state on shot peening. Figure 5(a) shown the surface layer hardness gradient curves of turning1 surface before shot peening, we can find that the surface ultra-microhardness of turning1 was about 590DHV, which was harder than metallographic grinding surface, and the hardened depth of turning 1 was about 120μm, deeper than metallographic grinding surface too; the surface ultra-microhardness of turning1 +shot peening2 was about 610DHV, and the hardened depths was about 130μm. Figure 5(b) shown that the surface ultra-microhardness of turning2 was about 575DHV, and the hardened depth was about 50μm, the surface ultra-microhardness of turning2 +shot peening2 was about 725DHV, and the hardened depth was about 250μm. Figure 5(c) shown that the surface ultra-microhardness of turning3 was about 580DHV, and the hardened depth was about 60μm, the surface ultra-microhardness of turning3 +shot peening2 was about 610DHV, and the hardened depth was about 100μm. It is thus clear that the deeper hardened layer before shot peening, the less increment of hardened layer after shot peening, namely less obvious work-hardening phenomenon of shot peening. The reason for this is that harder surface affected by the shot, the larger resistance to be overcome, and the less work-hardening effect.

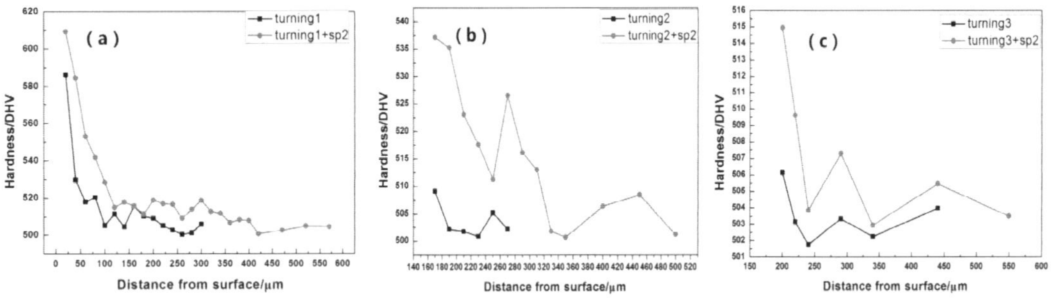

Figure 5. Curves of ultra-microhardness before and after shot peening (a) turning1 (b) turning2 (c) turning3

4 Conclusions

(1) The surface residual compressive stress in X direction of GH4079 alloy shot peening $\sigma_{r(x)}$ can reach up to more than 800MPa, and in Y direction $\sigma_{r(y)}$ can reach up to more than 450MPa, the surface roughness waves around 0.8μm, and the hardened depth of surface layer is more than 100μm.

(2) Increase shot peening intensity, shot peening time, shot peening pressure and shot hardness, the residual compressive stress induced by shot peening increases, surface roughness increases, hardness gradient of surface layer increases, and hardened depth is invariant.

(3) The greater surface compressive stress before shot peening, the less residual compressive stress induced by shot peening; the lower surface roughness before shot peening, the larger increment of surface roughness after shot peening; the deeper hardened layer before shot peening, the less increment of hardened layer after shot peening.

5. References

[1] Gao Y K and Ying Y F 2001 *J. Journal of Materials Engineering* 12 46-8
[2] Xiao H M 2010 *Nanjing University of Aeronautics and Astronautics*
[3] Gao Y K ,Ying Y F and Li X B 2002 *J. Journal of Materials Engineering* 04 40-2
[4] Gao Y K and Jing T 2010 *J. Journal of Aeronautical Materials* 06 35-8
[5] Feng S, Zhao X J and Cheng C H 2008 *J. New technology & New Process* 06 88-90
[6] Tang Z G, Cai D Y and Liu W C 2006 *J. Journal of Yanshan university* 06 503-6
[7] Song Y G, Gao Y K and Lu F 2010 *J. Heat Treatment of Metals* 09 94-7
[8] Ru J L, Wang R Z and Li X B 1997 *J. China Surface Engineering* 04 32-4

[9] Gao Y K, Li X B and Ying Y F 2003 *J. Journal of Aeronautical Materials* S1 132-5
[10] Gao Y K 2004 *J. Rare Metal Materials and Engineering* 11 1209-12

Acknowledgements
The author's acknowledge the surport of the Science and technology project of Guizhou Province, project number (2014) 6012

IOP Conf. Series: Materials Science and Engineering **230** (2017) 012042 doi:10.1088/1757-899X/230/1/012042

Obtaining Thin-Films Based on Chitosan and Carboxymethylcellulose with Antibacterial Properties for Biomedical Devices

Aliya Kapanovna Ospanova[1], Balzhan Esimkhanovna Savdenbekova[1], Mariam Kozybaevna Iskakova[2], Roza Amirzhanovna Omarova[2], Rahmet Nokeevich Zhartybaev[2], Balnur Zhanbolatovna Nussip[1], and Aibek Serikuly Abdikadyr[1]

[1]Al-Farabi Kazakh National University, Republic of Kazakhstan.
[2] Kazakh National Medical University, Republic of Kazakhstan.
E-mail: ospanova_a@mail.ru; iskakova-maryam@mail.ru

Abstract. The physico-chemical studies of produsing nanocoating in the form of multilayers on the basis of chitosan and carboxymethyl cellulose were done. Was justified the use of triclosan, chlorhexidine, silver ions and iodine as potential antibacterial agents in the composition of nano-coating for medical and biological implantable systems. Preliminary studies of antibacterial activity of the resulting multilayers on silicon plate showed good activity against many bacteria.

1. Introduction

A major problem in medicine is the formation of bacterial films observed on the surfaces of implants and other biomedical devices. This leads to the emergence of the infectious complications and the need for premature removal and replacement of such devices. Since the adhesion of bacteria is the initial stage of formation of biofilm, creating conditions that prevent the emergence of bacterial colonies on the surface of implants, can greatly slow down or completely inhibit its formation. A promising direction in traumatology and orthopedics is the creation of such properties of the implant in which does not occur biofilm formation. Currently, a number of researchers try to solve this problem by applying of the implants coating on the surface with different antibiotics or spraying silver. One solution to the problem of implant-associated infections is applying special coatings to implants that provide local time prolonged release of antibacterial substances in immediate proximity from the implant.

Complex nanostructural coating applied to the surface of the implant, should consist of antibacterial chemical agent and inherently biodegradable polymer and have pronounced universal bactericidal effect in the early postoperative period (1-2 weeks) in respect of all clinically relevant microorganisms. Authors [1] have been established optimal conditions of producing the multilayer films based on (Heparin / chitosan) 10 - (polyvinylpyrrolidone / polyacrylic acid)) 10 [(HEP/CHI)10–(PVP/PAA)10]. The interaction between polyelectrolytes has a purely electrostatic character, and the layers (PVP / PAA) 10 of the multilayer film are held by hydrogen bonds. PAA molecules then sewn to form anhydride groups by heat treatment at 110 °C for 16 h.

Obtained nanofilms show enhanced antimicrobial activity and may be promising for application in medical implants, tissue engineering, and etc. Studies [2] allow to develop a method of producing a

multilayer film on a silicon surface catheters by treatment with amino cellulose nanospheres (ACNSs) with hyaluronic acid (HA).
Antibacterial effect of the HA / ACNSs (amino cellulose nanosphere) multilayers was 40% higher in comparison with others. Five bilayer coatings based on HA / ACNSs was enough, to prevent the formation of bacteria biofilm. The paper presents a scheme of action of the antibacterial agent on a signal of emergence of bacteria colonies.
In many papers multilayer coatings formed by synthetic polymers were studied, but natural ionic polyelectrolytes have the greatest potential for application in the field of biomedical coatings. The wide use of polysaccharides in such studies are explained by their biocompatibility.
Today, it offers two alternative approaches to obtain biomedical devices on the surface of antiadhesive nanoscale functional coatings containing of ionic polysaccharides [3] 1) covalent immobilization of chitosan; 2) formation of multilayers of polyelectrolyte complexes of chitosan and carrageenans. Properly studied features of formation of multilayers with the participation of carrageenans in different conformational states optimized conditions for producing coatings with controlled architecture, and substantiated the mechanism of the exponential growth of multilayers.
In adhesion mode antimicrobial and antiadhesive properties of the resulting coatings were studied. It based on the example of enterococcus faecalis bacteria. It was shown that chitosan-containing coatings can significantly reduce bacterial adhesion, thus multilayers providing decreased adhesion 50-100 times, and the less hydrated covalently grafted coating - 10-20 times compared to the initial substrate.
The basic material for producing implants is titanium and therefore studies [4] are related to obtaining antibacterial multilayer coatings on the surface of Ti substrate with chitosan and alginate multilayer by self-assembly method. In this study, the surface of Ti substrate at first hydroxylated, and then treated with 3-aminopropyltriethoxysilane (ATPES) for producing amino functionalized Ti substrates. After this, Ti substrate alternately covered with alginate and chitosan by LBL self-assembly. The clinical significance is that the loading of a minocycline on implants surfaces can impart to implant stable antimicrobial activity. The current state of scientific research in of modified antibacterial coatings by self-assembly (LBL) for medical and biological products are well represented in a number of generic articles [5-6], from which it follows their theoretical and practical importance.
In this regard, this paper presents the results of physico-chemical foundations of obtaining multilayers based on chitosan and sodium carboxymethyl cellulose applied to silicon and titanium plates and containing in polylayers antibacterial agents such as triclosan, chlorhexidine, silver and iodine ions.

2. Experimental section

Materials. Triclosan, chlorhexidine digluconate (0.05 % solution, manufacturer Ltd. "Sultan" Republic of Kazakhstan), silver ions (analytical grade, GOST 1277-75, CRIS Analyte), molecular iodine (chemical grade, GOST TU), sulfuric acid, buffer solutions (4.0 and 9.18, Uralhiminvest, GOST 8,135-2004), glutaraldehyde (50% solution, Mw 100.11, manufacturer Amresco), low molecular chitosan (M_w 50-190 kDa) and sodium carboxymethylcellulose (Mw 700 kDa) (Sigma Aldrich). All chemicals were used without further purification. Filtered water with a resistivity of 18.2 MΩ was used in all experiments. Silicon (110) plates are the highest grade, p-type (the thickness of the dopant - 525 ± 25 μm, the thickness of the oxide layer - 2 mn.) (Cemat Silicon, South Australia). Titanium orthopedic plates.

Applying multilayers. Polyelectrolyte multilayers were applied on silicon and titanium plates.
Silicon plates were immersed in concentrated sulfuric acid solution for 2 h, and then thoroughly washed with Milli-Q water. Further silicon plates were immersed into 0.2 M NaOH solution for 10 min, washed repeatedly with water and dried at room temperature.
Titanium plates were treated with a solution of acetic acid with a concentration of of 5 M, followed by washing with water. Preparation of multilayers on the substrate surface was performed using 0.01 M solutions of polyelectrolytes. Polyanions and polycations were adsorbed sequentially for 10 min. Each deposition step was followed by washing with buffer solution at the same pH. All received multilayers further sewn in glutaraldehyde solution at room temperature for 12 h [7].

The implantation of antibacterial agent into the layers. To load an antibacterial agent, multilayer films were exposed to a 0.01 M solution of chlorhexidine, triclosan, silver ions and molecular iodine for 48 h to achieve complete absorption inside of films, and then plates was dried with nitrogen flow.

Scanning electron microscopy. The surface structure was examined by scanning electron microscopy (SEM). SEM images images were obtained by a scanning electron microscope Auriga. Silicon plates were attached to the SEM rack with conductive tape. Au-Pt alloy was sprayed onto the surface of samples of RF-plasma chamber for 10 s. The applied voltage was varied from 1 to 3 kV.

3. Result and discussion

In this study to obtain multilayers polyelectrolytes such as sodium carboxymethyl cellulose and chitosan were used as a polymer matrix. To btain nanofilms silicon and titanium plates were used as a model of solid substrate. As antibacterial agents were chosen: triclosan, chlorhexidine, silver ions and molecular iodine.
The authors [8] were investigated processes of interacting chitosan hydrochloride with sodium carboxymethyl cellulose in an aqueous solution by potentiometric, conductometric and turbidimetric titration, as well as in the solid state by FTIR spectra, and thermogravimetric analysis.
All methods titration have proved complex formation of in a stoichiometric ratio 1:1 between chitosan and carboxymethyl cellulose, interaction has electrostatic character. The mechanism of electrostatic interaction of the polyelectrolyte in the binary system can be explained by the emergence of a positive charge on chitosan and negative - on the surface of the CMC molecules as follows, figure 1.

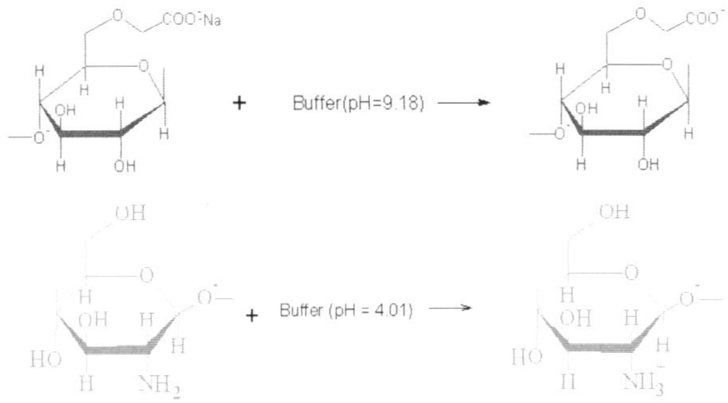

Figure 1. The emergence on the charges of polyelectrolyte macromolecules at handling with buffer solutions.

Obtaining of multilayer nanocoating on the surface of platelets based on chitosan with carboxymethylcellulose with applied antibacterial agents carried out as follows, figure 2.

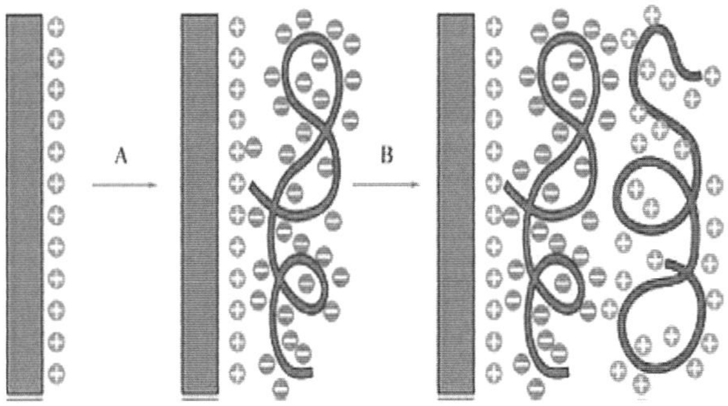

Figure 2. The scheme for obtaining multilayers with antibacterial properties: orange layers- chitosan; blue - sodium CMC, green and red - antibacterial agents.

Study of the dependence of thickness and the roughness of coatings on the number of bilayers revealed that the coating thickness is linearly dependent on the number of multilayers, figure 3, and at the most near to the plate layers bond cleanly, apparently, electrostatic, and with the growth of the layers in formation of multilayers play a major role pure hydrogen bonds. The roughness of the surfaces obtained Nanofilms increases with the number of films, figure 3, which is associated with an increase functional groups of both reagents [9].

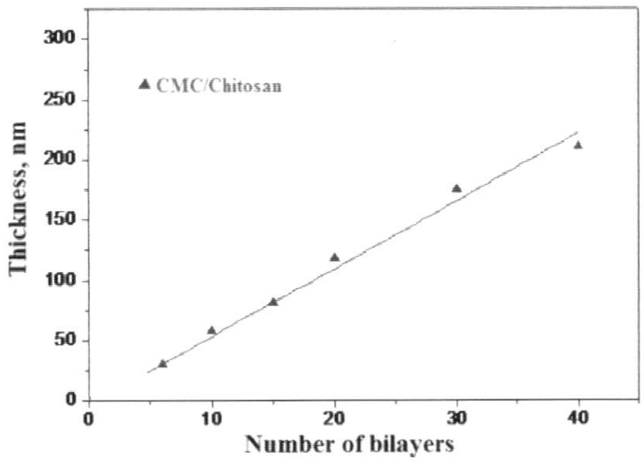

Figure 3. The dependence of the coating thickness on the number of bilayers for the system chitosan – sodium CMC.

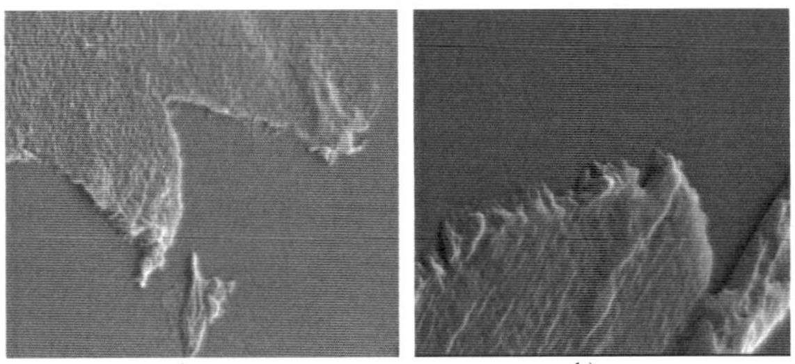

a) b)

Figure 4. The thickness of the films on the silica plate a) 15 bilayers, б) 30 bilayers.

For binding of biologically active compounds multilayer films obtained as described above were immersed in solutions of antibacterial agents concentration (0,05 mol / L) in water at a pH of 2-3 for 12-16 h. In addition for comparing were taken triclosan, chlorhexidine, silver nitrate, iodine-containing drugs, and the silver-containing control agent. After saturation of the film antibacterial agents films are dried at a temperature of 100-110 °C. Further heat-treated films are sewed in the glutaraldehyde solution at room temperature and dried in a nitrogen stream.

Determination of antimicrobial activity of silicon and titanium plates (SP) with antibacterial coatings based on their ability to inhibit the growth of microorganisms. Determination was performed by the method of diffusion in agar-agar on thick nutrient medium by comparing the sizes of zones inhibition of growth test microbes formed during the test of solutions of certain concentrations.

Into a sterile Petri dish was poured 20 mL of nutrient agar. The thickness of the agar layer affects on the results of the determination, therefore strictly observed the specified amount of the nutrient medium. As the nutrient media used Mueller-Hinton medium and Saburo agar.

For obtaining lawns prepared homogeneous suspension of the bacterial cells in a physiological solution, corresponding to standard 0.5 by McFarland turbidity units. Bacterial suspension applied with sterile gauze to the surface of the agar in three different directions. After 5-10 min after inoculation on the dried up surface of the agar applied plate with antimicrobial properties.

Plates for 30 min were left at room temperature, and then, without inverting incubated in a thermostat at a temperature of 28-37 °C for 24-48 h to 5 days. Microbial growth inhibition zones were measured by millimeter ruler. The results of these studies are shown in figure 5.

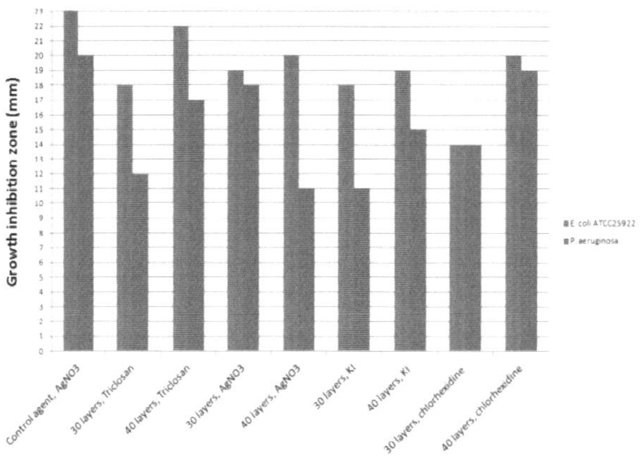

Figure 5. The study of the antimicrobial activity of multilayer films by method of diffusion.

The study of antimicrobial activity of plates by the method of diffusion and comparison with the control experiment showed that all studied dental plates (DP) had antibacterial effect against museum strain E. coli ATSS25922 and wild strain P. aeruginosa; thus the most acting (with a greater zone of inhibition growth of bacteria) on studied museum and wild strains of bacteria were plates with triclosan, showing sensitivity to E. coli ATSS25922 and P. aeruginosa; further antibacterial action had a plate with chlorhexidine, inhibited the growth of E. coli ATSS25922 and P. aeruginosa, while plate DP - 40 layers with chlorhexidine inhibited the growth of S. aureus ATSS29213. Dental plates with silver nitrate showed antibacterial effect against E. coli ATSS25922, plate with 30 layers $AgNO_3$ influenced also on P. aeruginosa. Plate with iodine also affected with big zone of inhibition growth on E. coli ATSS25922 and P. aeruginosa, the impact on microorganisms of plates with potassium and chitosan were about the same - with antibacterial effect action on E. coli ATSS25922 and with an average action (11-15 mm) on P. aeruginosa. The preliminary carried out tests showed good antibacterial activity of multilayers fixed the silicon and titanium plates, indicating the prospect of such research.

4. Conclusion

The study of physical and chemical conditions of obtaining antibacterial coatings on silicon and titanium plates allowed choose the optimum conditions for obtaining nanofilms based on chitosan and carboxymethyl cellulose. The thickness and the roughness of coatings in multilayers increases in direct proportion with the increase of bilayers. Interaction between the polymer carrier on the surface of plates purely electrostatic and antibacterial agents associated with the basics of nanofilms, probably due to the forces of adsorption and hydrogen bonds. Antimicrobial activity such coatings due to the presence of bioactive functional groups on the surface of agents such as triclosan, chlorhexidine, silver ions and iodine.

5. References

[1] Bai-liang Wang, Ke-feng Ren, Hao Chang, Jin-lei Wang and Jian Ji. "Construction of Degradable Multilayer Films for Enhanced Antibacterial Properties" *Applied Materials and Interfaces*, vol. 5, pp. 4136-4143, May 2013.

[2] A. Francesko, M. Fernandes, K. Ivanova, S. Amorim , Rui L. Reis, I. Pashkuleva, E. Mendoza, A. Pfeifer, T. Heinze and T. Tzanov, "Bacteria-responsive multilayer coatings comprising polycationic nanospheres for bacteria biofilm prevention on urinary catheters" *Acta Biometrialia*, vol. 33, pp. 203-212, March 2016.

[3] S. Bratskaya, "Multilayer and covalently grafted functional coatings based on polysaccharides for prevention of bacterial adhesion" *Bulletin FEB RAS,* vol. 2, pp. 84-92, 2009. (In Russian).

[4] Hongbin Lv, Zhen Chen, Xiaoping Yang, Lian Cen, Xu Zhang and Ping Gao. "Layer-by-layer self-assembly of minocyclineloaded chitosan/alginate multilayer on titanium substrates to inhibit biofilm formation" *Journal of Dentistry*, vol. 42, pp. 1464-1472, Jun. 2014.

[5] S. Sukhishvili and E. Kharlampieva. "Hydrogen-Bonded Layer-by-Layer Films" *Macromolecular science*, vol. 46, pp. 377-395, Nov. 2006.

[6] Lydie Séon, Philippe Lavalle, Pierre Schaaf and Fouzia Boulmedais. "Polyelectrolyte Multilayers: A Versatile Tool for Preparing Antimicrobial Coatings" *Langmiur,* vol. 31, pp. 12856-12872. October, 2015.

[7] A. Ospanova, R. Omarova, U. Abdurazakov, R. Jartybaev, M. Iskakova, B. Savdenbekova and M. Amkhadova "Method of multilayer assembly as alternative to antibacterial coverings of medico biological implants" *Stomatology*. vol. 95, pp. 9-13. 2016. (In Russ.).

[8] Carmen Rosca, Marcel Ionel Popa, Gabriela Lisa and Gabrielle Charlotte Chitanu, "Interaction of chitosan with natural or synthetic anionic polyelectrolytes. The chitosan–carboxymethylcellulose complex" *Carbohydrate Polymers*, vol. 62, pp. 35–41. 2005.

[9] A. Mentbayeva, A. Ospanova, Zh. Tashmuhambetova, V. Sokolova and S. Sukhishvili. "Polymer-metal complexes in polyelectrolyte multilayer films as catalysts for oxidation of toluene" *Langmuir, vol. 28, pp.* 11948-11955, July 2012.

IOP Conf. Series: Materials Science and Engineering **230** (2017) 012043 doi:10.1088/1757-899X/230/1/012043

Effect of Experimental Parameters on Nanofiber Diameter from Electrospinning with Wire Electrodes

Guocheng Zhu[1,2,3], L Y Zhao[2], L T Zhu[2], X Y Deng[2] and W L Chen[2,4]
[1] Key Laboratory of Advanced Textile Materials and Manufacturing Technology, Ministry of Education, Zhejiang Sci-Tech University, Hangzhou, 310018, China,
[2] College of Materials and Textiles, Zhejiang Sci-Tech University, Hangzhou, 310018, China,
[3] Hubei Key Laboratory of Advanced Textile Materials & Application, Wuhan Textile University, Wuhan 430200, China
[4] Corresponding author. Email: wlchen193@163.com; zgc100100@hotmail.com

Abstract. Polyvinylidene Fluoride (PVDF) nanofibers were electrospun by Nanospider equipment with wire electrodes. The parameters which would influence the fiber diameter were investigated in terms of solution concentration, cartridge speed (feed rate of solution), voltage, electrode distance, rotating wire speed, winding speed and slit diameter. The morphology and diameter of fibers were observed by scanning electron microscope. The results revealed that the solution concentration had significant influence on both fiber morphology and fiber diameter; the cartridge speed, voltage, electrode distance and slit diameter had slight effect on fiber diameter since the standard deviations were large; the rotating wire speed and the winding speed had insignificant influence on fiber diameter.

1. Introduction

A nanofiber is a fiber with a diameter of 100 nanometers or less[1]. The most attractive feature of nanofiber is its very high specific surface area, which provides it some remarkable properties, i.e., high capacity for the attachment or release of functional groups, absorbed molecules, ions, catalytic moieties and nano-scale particles[2], great effect on the chemical, biological reactivity, electro-activity[3], better rectification properties as well as better transport[4], higher tensile strength and strain energy[5, 6].

The birth of nanofiber can be traced back to 1934 when Formhals produced nanoscale fibers by electrospinning for viscoelastic solutions[7]. But the nanofiber had not stirred up much interest for its limitation in production and applications until 1990s when D.H. Reneker's group demonstrated that many organic polymers could be electrospun into nanofibers [8-10].

Even though several techniques appeared for generating nanofibers in the past several decades, i.e., conjugate spinning, chemical vapor deposition, drawing, template synthesis, self-assembly, meltblown, and electrospinning [2]. Among these techniques, only meltblown and electrospinning can reach the industrial scale production, which make the practical applications possible. But the diameters of fibers from meltblown technology often have a broad range from nano-scale to above 20 μm[11], which limits the usage of meltblown nanofibers when the small diameters of fibers and uniformity of fiber diameters are required. Electrospinning has been proven to be an ideal method of producing continuous nanofibers. The biggest advantages of this technology is the great simplicity of the electrospinning process[2, 11], its versatility in processing a wide range of polymers, and the consistency in producing nanofibers with controllable fiber diameter and morphology[12].

The principle of electrospinning is to produce nanofiber by electrostatic forces[13]. The process of electrospinning is to draw electrically charged jets from polymer solution or melts in an electric field caused by high voltage towards to an oppositely charged collector[14]. In a conventional electrospinning setup system, a needle nozzle/syringe is often used as a jet outlet. But the production rate of nanofiber is very low which limits its commercial applications. In order to improve the productivity of nanofibers, several significant innovations in electrode have been done i.e., multi-needle system[15], cylindrical porous hollow tube system[16], roller electrode system [17], bubble free-surface system[18], and wire electrodes system [19]. The fiber production from needle, cylinder and coil technology was introduced by [20-23], but the fiber from wire technology hasn't been reported. Poly (vinylidene fluoride) (PVDF) has been widely used as a membrane material thanks to its good chemical resistance, great mechanical properties and high thermal stability [24-27]. It is broadly applied in ultrafiltration (UF), microfiltration (MF), membrane distillation (MD) and other membrane processes[27, 28]. Therefore, the diameter of nanofibers from wire electrospinning technology was investigated under different conditions for better controlling the quality of fibers and final products.

2. Materials and methods

Polyvinylidene Fluoride Powder (average molecular weight 600,000~700,000) was provided by Arkema Group, Dimethylacetamide (DMAc) was provided by PENTA Chemicals. Tetraethylammonium bromide (TEAB) was obtained from Sigma Aldrich.

PVDF was dissolved in DMAc with concentration of 8%, 10%, 12% and 14% in w/w, and before electrospinning process, 2% TEAB was added into the PVDF solution. Nanofibers were electrospun from PVDF solution under different operating conditions in terms of cartridge speed, rotation wire speed, winding speed, electrodes distance, slit diameter, and voltage. The temperature and humidity were controlled at 23±0.5 ℃ and 18±1 %. The specific process parameters are given in table 1.

The conductivity of solution was measure by SCHOTT instruments (Model: k00090-L), and the solution viscosity was measured by HAAKE Viscotester E from Thermos Scientific Company. The fiber morphology was observed under scanning electron microscope (SEM). The average fiber diameter was calculated from the SEM images using image analysis software. More than 100 fibers were counted from at least 3 SEM images which were taken from different places of a sample.

The main part of electrospinning is shown in figure 1. The polymer solution was contained in a cartridge which has a slit on it; a steel wire acted as an electrode passed the cartridge through the slit and contacted with polymer solution; the cartridge was able to have reciprocating motions onto the steel wire, and left solution droplets on it. Therefore, the solution droplets were exposed to a strong electrical field which provided by high voltage power supply through two electrodes. Once the applied voltage exceeded a critical value, jetting of polymer solution was initiated, and collected by the supporting material. As the charged jets travel toward the supporting material, the solvent evaporates, resulting in solid fibers.

Table 1. Process parameters of electrospinning

Parameters	Values					
Concentration (%)	8	10	12	14		
Cartridge speed (mm/s)	200	300	400	500		
Voltage (kV)	-10/+40	-10/+45	-10/+50	-10/+55	-10/+60	
Electrodes distance (mm)	160	170	180	190	200	
Rotating wire speed (mm/min)	40	80	125	165	215	
Winding speed (mm/min)	100	200	300	400	500	600
Slit diameter (mm)	0.5	0.6	0.7	0.8	1.0	

Note: all the combinations of these process parameters were carried out.

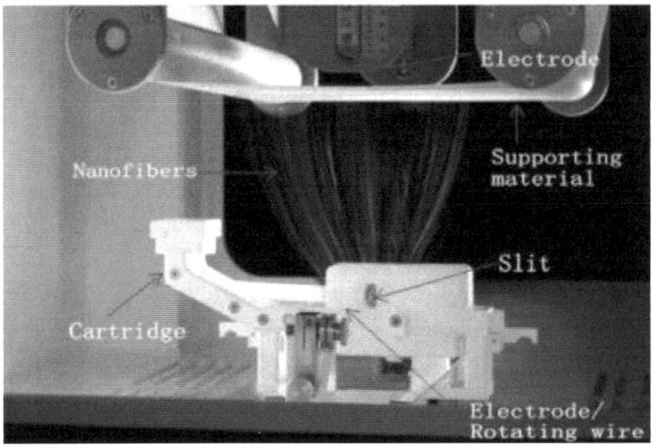

Figure 1. Picture of electrospinning[29]

3. Results and discussions

Both viscosity and conductivity are important factors for electrospinning, which are given in table 2 responding to different solution concentrations. The conductivity of solutions decreased with the increase of solution concentration; however, all these values are higher enough for electrospinning[30]. The increase of viscosity of solutions occurred responding to the solution concentration.

Table 2. Viscosity and conductivity of PVDF solutions

Solution Concentration (%)	Viscosity (mPa·s at 23°C)	Conductivity (μS/cm at 23°C)
8	304.2	103.4
10	553.6	96.1
12	1317.1	90.4
14	1845.3	83.8

3.1. Effect of solution concentration on fiber diameter

Solution concentration was taken as variable while other process parameters were kept constant as given in table 3.

Table 3. Process parameters during electrospinning

Temperature (°C)	Humidity (%)	Electrode distance (mm)	Voltage (kv)	Cartridge speed (mm/s)	Rotation wire speed (mm/s)	Slit diameter (mm)	Winding speed (mm/min)
23	18	180	-10/50	300	40	0.5	200

T: temperature, H: humidity, ED: electrode distance, V: voltage, CS: cartridge speed, RWS: rotating wire speed, SD: slit diameter, WS: winding speed

The images of nanofibers from different solution concentrations are given in figure 2. Almost all the fibers had the beads in their structure when the concentration was 8 wt% (as shown in Fig. 2A). The beads disappeared when concentration increased to 10 wt%; however, many fibers were attaching to each other and extended like webbed hands (as shown in Fig. 2B). When the concentration was over 10 wt%, the more obvious solid fiber-structure fibers appeared (as shown in Fig. 2 C&D). These images revealed that the solution concentration had a very important influence on morphology of fibers.

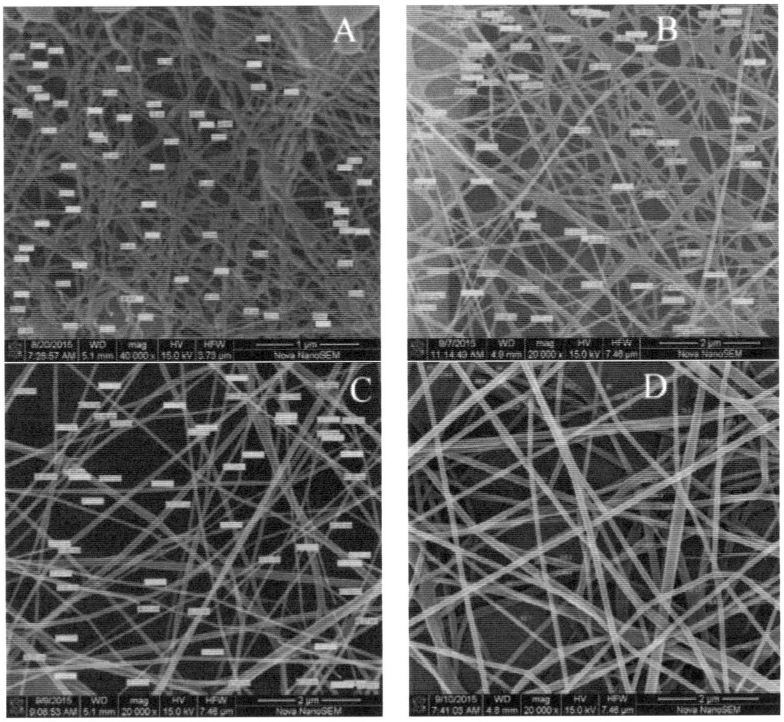

Figure 2. The SEM images of fibers from solutions with different concentrations (A) 8%; (B) 10%; (C) 12%; (D) 14%.

The fiber diameter increased significantly with the increase of concentration; however, the standard deviation indicated that the diameters of fibers were distributed in a large range (as shown in figure 3). The phenomenon of fiber diameter increase with concentration was observed by many researchers [20-22]. This phenomenon could be easily explained by the polymer concentration in solution. However, it's worth noting that the mass of droplet separated from solution induced for electrospinning by strong electric field must be different due to the different surface tensions of solutions with different concentrations. This could also be helpful for explaining the relationship (linear or non-linear) between fiber diameter and solution concentration.

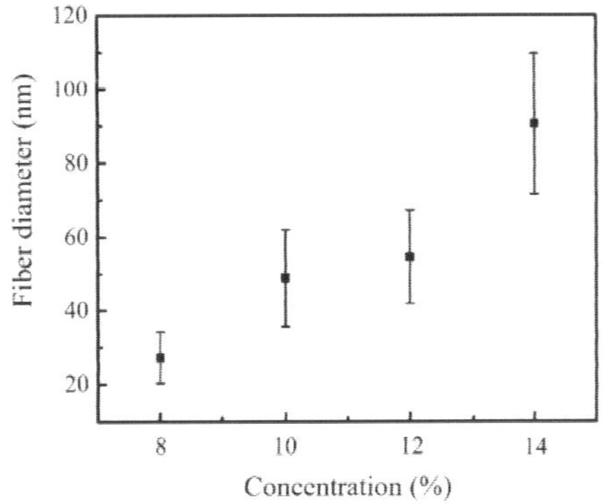

Figure 3. Relationship between solution concentration and fiber diameter

3.2. Effect of cartridge speed on fiber diameter

The solution was evenly deposited onto the wire electrode as the back-and-forth motion of cartridge. The cartridge speed (or frequency) is adjustable; more solution will be deposited onto the wire electrode when the cartridge speed is higher, vice versa. This process is equivalent to the feed rate of needle electrospinning. The SEM images of fibers and the relationship between cartridge speed and fiber diameter are given in figure 4 and figure 5.

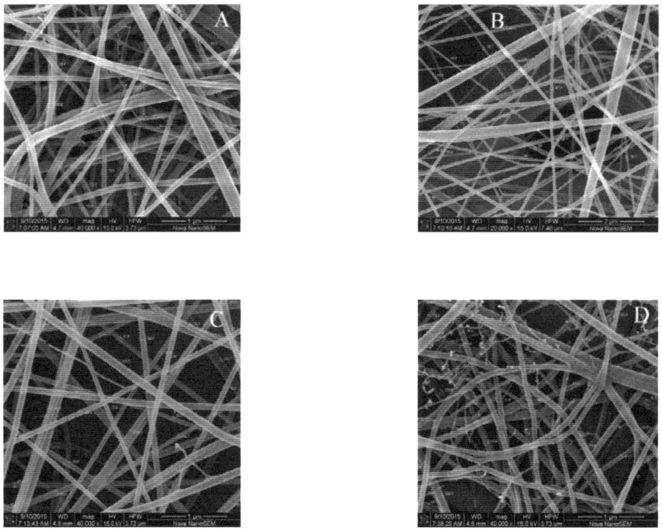

Figure 4. Images of fibers from 14 % solution at different cartridge speed, (A) 200 mm/min, (B) 300 mm/min, (C)400 mm/min , (D) 500 mm/min

There is no significant difference in fiber diameter by changing the cartridge speed (as shown in Fig. 5). Once the feed rate is sufficient for forming fibers, and the electric force is able to overcome the capillary force, then the mass of droplet separated from solution by electric force should be the same. Therefore, the higher cartridge speed only provided more polymer solution and then more fibers. In another case, if the cartridge speed is too high, and the feed solution is over the capacity of fiber formation by electric field, then the over-feed solution will drop down from the wire electrode (this phenomenon was observed during electrospinning process).

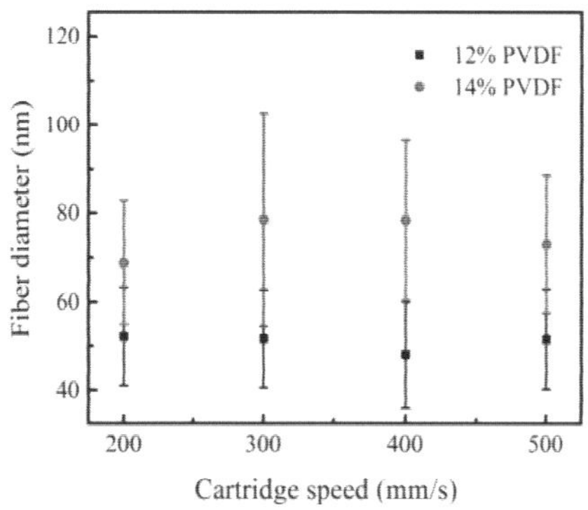

Figure 5. Relationship between cartridge speed and fiber diameter

3.3. Effect of voltage on fiber diameter

The SEM images of fibers under different voltages are given in figure 6, all the fibers had very good morphology, and a few fibers had much larger diameter when the applied voltage was over 60 kV.

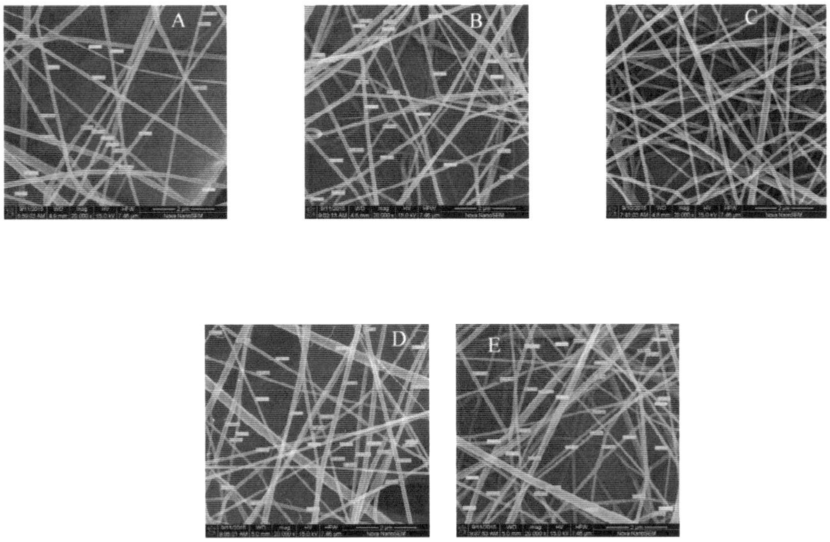

Figure 6. Images of fibers under different voltage (A) 50 kV, (B) 55 kV, (C) 60 kV, (D) 65 kV,(E) 70 kV

The fiber diameter had a very slight increase with the increase of applied voltage, but the difference is insignificant owing to the large range of standard deviation (as seen in figure 7). Vince Beachley and Dhananjay reported the decrease of fiber diameter with the increase of applied voltage for needle electrospinning [21, 31]. In contrary, Huan stated that the fiber diameter increased with the increase of applied voltage for needle electrospinning, and also pointed out the existence of optimal applied voltage for narrowing the fiber diameter distribution [23]. Wang investigated the fiber diameter by free surface electrospinning, revealed the slight finer fibers from higher applied voltage; however, the productivity of fibers dramatically increased under higher voltage[22]. However, the author tends to believe that the higher voltage (stronger electric field intensity) is able to draw bigger droplets from solution; therefore, the fiber diameter could be larger (figure 6 D&E).

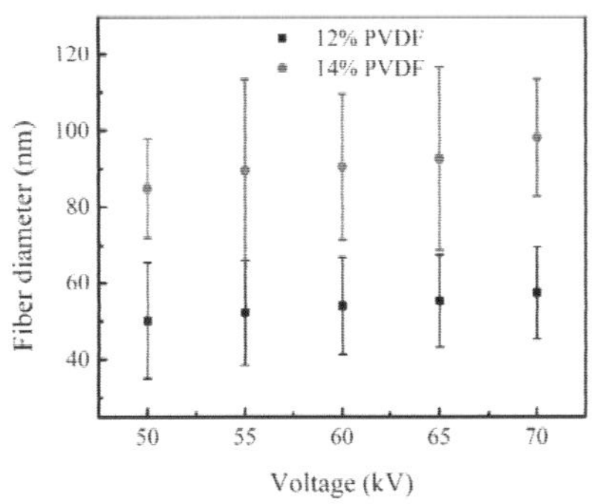

Figure 7. Relationship between voltage and fiber diameter

3.4. Effect of electrode distance on fiber diameter

The images of fibers at different electrode distances are given in figure 8.

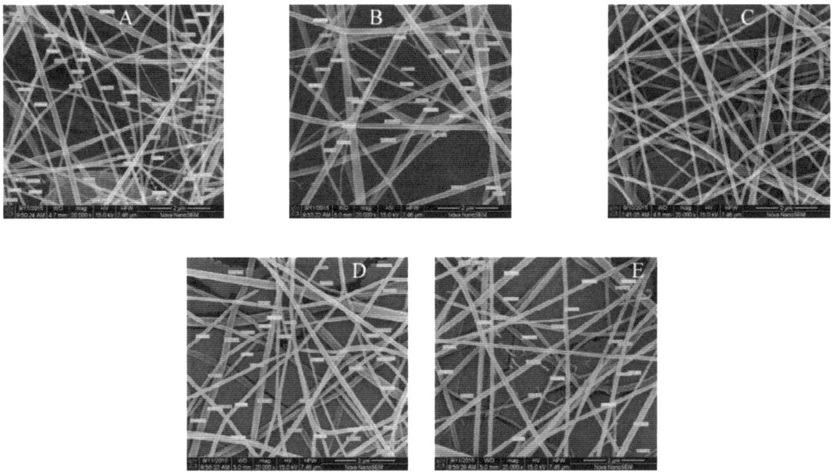

Figure 8. Images of fibers at different electrode distances (A) 160 mm, (B)170 mm, (C)180 mm, (D) 190mm, (E) 200mm

As shown in figure 9, the fiber diameter from 12% solution had a slight decrease when the electrode distance increased; but there is no regular trend to describe the change of fiber diameter from 14% solution. The increase of distance between two electrodes will lead to weaker electric field intensity, and longer travelling path of fibers, which could be the reasons for decrease of fiber diameter at larger distance between electrodes. However, there was an uncertain factor would influence the fiber diameter, i.e. the airflow exchange between electrospinning chamber and external environment. The motion of airflow would change the travelling path of fibers.

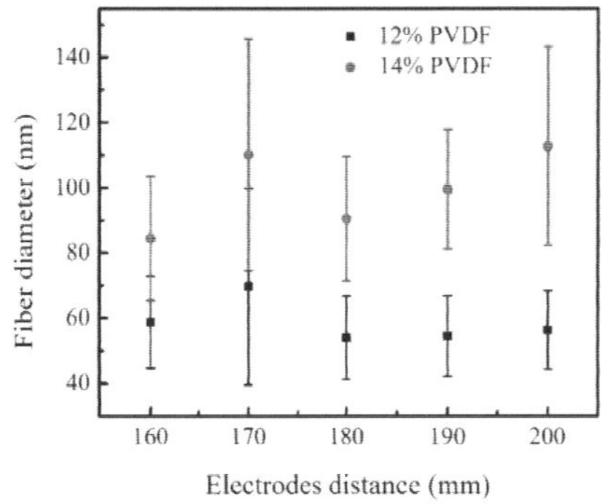

Figure 9. Relationship between electrode distance and fiber diameter

3.5. *Effect of rotating wire speed on fiber diameter*

The bottom wire electrode is dynamic, and transferred from one glide wheel to another in order to make sure that the electrode is always clean keeping the electrospinning process in a good condition. The images of fibers at different rotating wire speed are given in figure 10, and the relationship between rotating wire speed and fiber diameter is given in figure 11. The influence in fiber diameter was small when rotating wire at low speed, only one obvious drop in fiber diameter occurred when rotating wire speed was 215 mm/min.

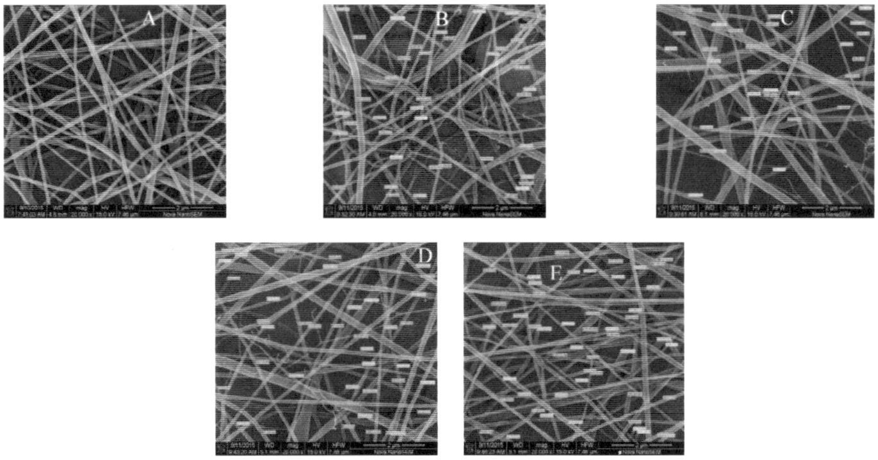

Figure 10. Images of fibers at different rotating wire speed (A) 40 mm/min, (B) 80mm/min, (C) 125mm/min, (D) 165mm/min, (E) 200mm/min

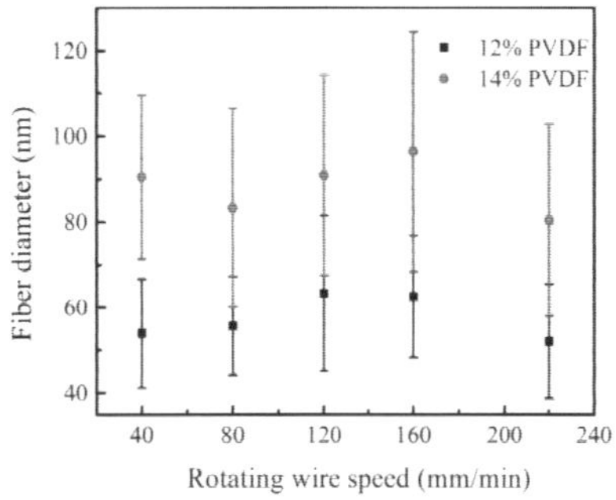

Figure 11. Relationship between rotating wire speed and fiber diameter

3.6. Effect of winding speed on fiber diameter

The winding speed had significant effect on the areal density of nanofibrous mat (figure 12). However, it did not show obvious effect on fiber diameter as shown in figure 13. The higher winding speed provided larger area of supporting material; and in the same period of time, the amount of fibers was the same. Thereby, less fibers and thinner layer of nanofibrous mat were appeared on the supporting material.

Figure 12. Images of fibers at different winding speed (A) 00 mm/min, (B) 200mm/min, (C) 300 mm/min, (D) 400mm/min

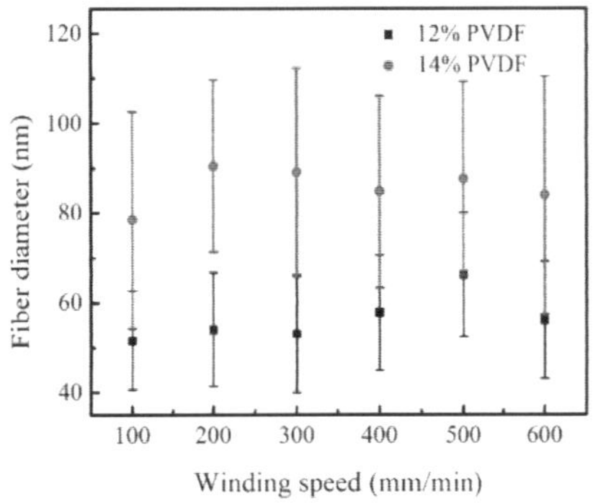

Figure 13. Relationship between winding speed and fiber diameter

3.7. Effect of slit diameter on fiber diameter

The larger the slit diameter is, the more solution would leak out from the cartridge, and attaches onto the wire electrode. The images of fibers are given in figure 14.

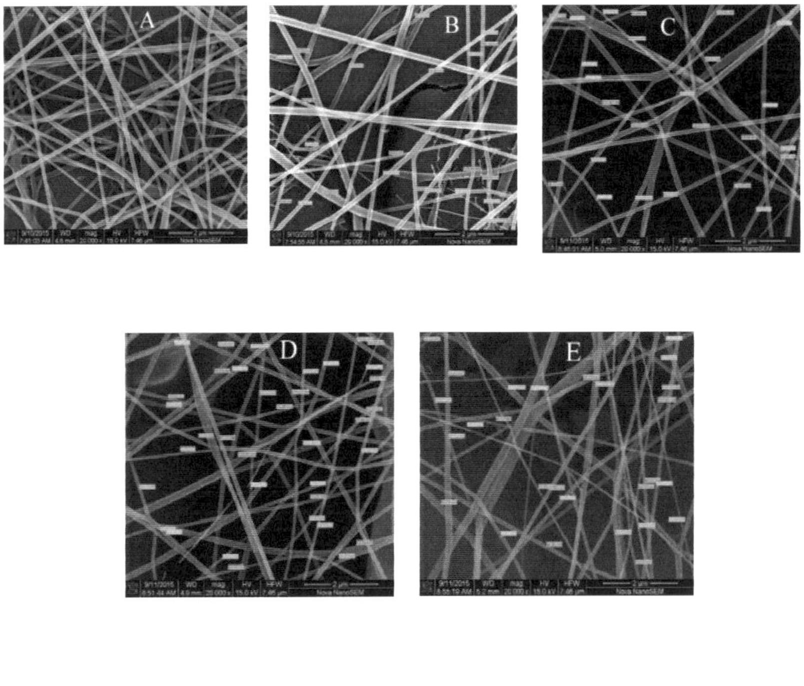

Figure 14. Images of fibers from different slit size (A) 0.5 mm, (B) 0.6 mm, (C) 0.7 mm, (D) 0.9 mm, (E)1 mm

The fiber diameter increased when the slit diameter increased for solution of 12% concentration. But there was a slight decrease in fiber diameter for solution of 14% concentration when slit diameter changed from 0.5 mm to 0.9 mm. An obvious fiber diameter increase for both solutions when 1 mm slit diameter was applied (as shown in figure 15).

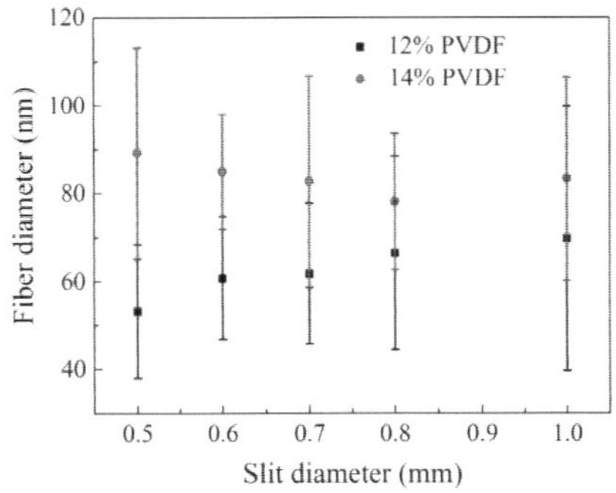

Figure 15. Relationship between slit diameter and fiber diameter

4. Conclusion

Nanofibers were produced by electrospinning equipment with wire electrodes, and the process parameters were adjusted according to the experimental design in order to investigate the influence of process parameters on fiber diameter. After evaluating the fiber diameter, some conclusions can be drawn. The morphology and fiber diameter were significantly influenced by solution concentration; the cartridge speed, voltage, electrode distance and slit diameter had slight effect on fiber diameter, meanwhile, the standard deviation were large; the rotating wire speed and the winding speed hardly had influence on fiber diameter.

5. References

[1] Seeram Ramakrishna, K.F., Wee-Eong Teo.Teik-Cheng Lim and Zuwei Ma, *An introduction to electrospinning and nanofibers* 1st ed. 2005, World Scientific Publishing Co. Pte. Ltd.: New Jersey, london, singapore, beijing, shanghai, hongkong,taipei, chennai.

[2] Frank K. KO and Y. Wan, *Introdiction to nanofiber materials*. 1st ed. 2014, Cambridge university press: New York. 7.

[3] E. Gallo, A. Anwar and B. Nabet 2004, Contact-induced properties of semiconducting nanowires and their local gating., *Nanoengineered Nanofibrous Materials*, 313.

[4] A. El-Aufy, B. Nabet and F. Ko 2003, Carbon nanotube reinforced (PEDT/PAN) nanocomposite for wearable electronics, *Polymer Preprints*, 44, 134-135.

[5] W. H. Otto 1955, Relationship of tensile strength of glass fibers to diameter *Journal of the American Ceramic Society*, 38, 122-125.

[6] B. Yakobson, et al., *"Mechanical properties of carbon nanotubes" in Carbon Nanotubes*. 2001, Springer: Berlin Heidelberg. 287-327.

[7] Formhals, A.,1934, *US Patent 1975504*.

[8] Doshi, J. and D.H. Reneker 1995, Electrospinning process and applications of electrospun fibers, *J. Electrostatics*, 35, 151-160.

[9] D. H. Reneker and I. Chun 1996, Nanometre diameter fibres of polymer, produced by electrospinning, *Nanotechnology* 7, 216-223.

[10] Reneker, D.H., et al. 2000, Bending instability of electrically charged liquid jets of polymer solutions in electospinning, *J. Appl. Phys.*, 87, 4531-4547.
[11] C. J. Luo, et al. 2012, Electrospinning versus fibre production methods: from specifics to technological convergence, *Chem. Soc. Rev.*, 41, 4708-4735.
[12] Tong Lin and X. Wang, *Needleless electrospinning of nanofibers: Technology and Applications*. 2013, Taylor and Francis Group, LLC.
[13] Taylor, G.I. *Electrically driven jets*. in *Proc. Roy. Soc.* 1969. London.
[14] Thandavamoorthy Subbiah, et al. 2005, Electrospinning of nanofibers, *J Appl Polym Sci*, 96, 557-569.
[15] GeunHyung Kim, Young Sam Cho, and W.D. Kim 2006, Stability analysis for multi-jets electrospinning process modified with a cylindrical electrode, *European Polymer Journal*, 42, 2031-2038.
[16] J.S. Varabhas, G.G. Chase, and D.H. Reneker 2008, Electrospun nanofibers from a porous hollow tube, *Polymer*, 49, 4226-4229.
[17] Jirsak, O., et al., *A method of Nanofibres Production from a Polymer Solution Using Electrostatic Spinning and a Device for Carrying Out the Method*. 2004: European Patent.
[18] Liu, Y. and J. He 2007, Bubble electrospinning for mass production of nanofibers, *International Journal of Nonlinear Sciences and Numerical Simulation*, 8, 393-396.
[19] S.R.O, E., *Method for spinning the liquid matrix, device for production of nanofibres through electrostatic spinning of liquid matrix and spinning electrode for such device*. 2009.
[20] Yener, F. and O. Jirsak 2012, Comparison between the Needle and Roller Electrospinning of Polyvinylbutyral, *Journal of Nanomaterials*.
[21] Beachley, V. and X.J. Wen 2009, Effect of electrospinning parameters on the nanofiber diameter and length, *Materials Science & Engineering C-Biomimetic And Supramolecular Systems*, 29, 663-668.
[22] XinWang, et al. 2012, Needleless Electrospinning of UniformNanofibers Using Spiral Coil Spinnerets, *Journal of Nanomaterials*, 1875-1890.
[23] Siqi Huan, et al. 2015, Effect of Experimental Parameters on Morphological, Mechanical and Hydrophobic Properties of Electrospun Polystyrene Fibers, *Materials* 2718-2734.
[24] M.L. Yeow, Y.T. Liu, and K. Li 2004, Morphological study of poly(vinylidene fluoride) asymmetric membranes: effects of the solvent, additive, and dope temperature, *J. Appl. Polym. Sci.* , 1782-1789.
[25] X.Y. Tan, et al. 2006, Polyvinylidene fluoride (PVDF) hollow fibre membranes for ammonia removal from water *J. Membr. Sci.* , 271, 59-68.
[26] D.L. Wang, K. Li, and W.K. Teo 1999, Preparation and characterization of polyvinylidene fluoride (PVDF) hollow fiber membranes, *J. Membr. Sci.*, 163, 211-220.
[27] F. Liu, et al. 2011, Progress in the production and modification of PVDF membranes, *J. Membr. Sci.* , 375, 1-27.
[28] Z.L. Cui, E. Drioli, and Y.M. Lee 2014, Recent progress in fluoropolymers for membranes, *Prog. Polym. Sci.* , 39, 164-198.
[29] *http://www.elmarco.com/photogallery/technology/*.
[30] Uyar, T. and F. Besenbacher 2008, Electrospinning of uniform polystyrene fibers: The effect of solvent conductivity, *Polymer*, 49, 5336-5343.
[31] Dhananjay D. Pise, B. B. Ahuja and S.M. Shendokar 2015, Study of Process Parameters Affecting the Diameter and Morphology of Electrospun Polyvinylidene Fluoride (PVDF) Nanofibers, *International Journal of Science and Research*, 4, 155-160.

Acknowledgment
This work was supported by Zhejiang Sci-Tech University (Project Number: 16012168-Y), The Young Researchers Foundation of Key Laboratory of Advanced Textile Materials and Manufacturing Technology, Ministry of Education, Zhejiang Sci-Tech University (Project Number: 2016QN05)， and

Hubei Key Laboratory of Advanced Textile Materials & Application, Wuhan Textile University (Project Number: Fzxcl2017006).

IOP Conf. Series: Materials Science and Engineering **230** (2017) 012044 doi:10.1088/1757-899X/230/1/012044

Optimized Production of Coal Fly Ash Derived Synthetic Zeolites for Mercury Removal from Wastewater

Z Tauanov[1,3], D Shah[1], G Itskos[2] and V Inglezakis[1,3]
[1]School of Engineering, Chemical Engineering Department, Environmental Science and Technology Group (ESTg), Nazarbayev University, Astana, 010000, Kazakhstan
[2]School of Materials Engineering, Purdue University, West Lafayette, IN 47907-2045, USA.
[3]Corresponding authors: zhtauanov@nu.edu.kz; vasileios.inglezakis@nu.edu.kz

Abstract. Coal fly ash (CFA) derived synthetic zeolites have become popular with recent advances and its ever-expanding range of applications, particularly as an adsorbent for water and gas purification and as a binder or additive in the construction industry and agriculture. Among these applications, perpetual interest has been in utilization of CFA derived synthetic zeolites for removal of heavy metals from wastewater. We herein focus on utilization of locally available CFA for efficient adsorption of mercury from wastewater. To this end, experimental conditions were investigated so that to produce synthetic zeolites from Kazakhstani CFAs with conversion into zeolite up to 78%, which has remarkably high magnetite content. In particular, the effect of synthesis reaction temperature, reaction time, and loading of adsorbent were systematically investigated and optimized. All produced synthetic zeolites and the respective CFAs were characterized using XRD, XRF, PSA and porosimetric instruments to obtain microstructural and mineralogical data. Furthermore, the synthesized zeolites were studied for the removal of mercury from aqueous solutions. A comparison of removal eficiency and its relationship to the physical and chemical properties of the synthetic zeolites were analyzed and interpreted.

1. Introduction

Coal fly ash (CFA) is a by-product from combustion of coal in electric power stations all over the world. The amount of annually discharged CFA is enormous and thus it requires further investigation on the reprocessing of CFA as it causes serious environmental and health problems [1-3]. One efficient way of converting CFA into a value-added product is synthetizing zeolites [4-7], which have a wide range of application fields including agriculture, environment and industry [8]. A number of successful studies on pilot-scale or industrial production of synthetic zeolites from coal fly ash have been conducted already [9-11]. According to these studies it is practical and economically feasible to manufacture zeolitic adsorbents from coal fly ash that allows promising utilization solution for this by-product of coal combustion [11]. Thus, all these sttudies imply the necessity for production of synthetic zeolites in large quantities to effectively utilize value-added raw material.

Synthetic zeolites could be produced using various raw materials as clay minerals [12-14], siliceous minerals [15] and waste materials as coal fly ash [2, 4, 7, 8, 16]. Among the available raw materials coal fly ash is considered as one of the suitable due to low cost, abundance and possibility to produce various zeolites depending on reaction conditions [11], in contrast to natural zeolites that require costly modifications in order to be suitable for industrial application.

The chemical composition of zeolite and CFA is almost the same, but the major difference between these two is the structure: CFA is mainly composed of amorphous structure, while zeolite has a well-defined crystalline structure [17]. Moreover, zeolite has higher cation exchange capacity (CEC), larger surface area and demonstrates superior thermal stability [18] that further enhances the utilization sectors. Over the past decades there have been numerous studies on the synthesis of zeolites from coal fly-ash and one of the widely used methods for production is the hydrothermal conversion using alkaline solutions [8, 19, 20]. There are several other emerging synthetic methods in laboratory scale production as microwave-assisted [2] and ultrasound-assisted production [4, 21, 22], fusion followed by hydrothermal synthesis [23-26] and salt-thermal production [27]. However, hydrothermal synthesis, which is multiphase crystallization process including both amorphous and crystalline solid phases and at least one liquid phase [28], is still the most promising and widely applied method, which is positioned in the core of all of the above mentioned production methods. According to this method the raw coal fly ash is dissolved in alkaline solution such as NaOH or KOH to extract aluminate and silicate constituents and then undergoes heat treatment to produce zeolite crystals [10, 17]. There are several parameters that substantially affect the zeolitization process as reaction temperature and time, Si/Al ratio in the raw fly ash and alkaline solution pH [8, 19, 20, 29]. Depending on these reaction conditions it is possible to obtain various synthetic zeolites, such as zeolite A [6], zeolite Na-P [4], sodalite [30] and many other types of synthetic zeolite.

Despite the fact that there are numerous studies that have already been published the underlying mechanism of zeolite formation and production process are still not well-studied. The questions as how the above-listed main parameters affect the mechanism of zeolite formation and in which way it should be controlled in order to obtain the highest possible conversion from fly ash to synthetic zeolite is not sufficiently explained in literature. Moreover, to the best of our knowledge, there has been no optimization studies conducted based on experimental results that will inherently assist in better understanding of zeolitization mechanism and in finding optimum reaction parameters, which correspondingly emphasizes the necessity for this type of research. In addition, the authors aim to study the removal one of the less investigated heavy metals, namely mercury (II) ions from aqueous solutions.

In this paper, we aim to optimize the zeolitzation process based on experimental results from several different recent references and experimental work carried out in this study by following factorial analysis approach. All produced and raw materials were characterized by means of advanced microstructural and mineralogical characterization instruments. The most successful synthetic zeolites based on conversion yield from raw CFA and purity of the product were tested in batch reactors for the removal of mercury (II) ions from aqueous solutions.

2. Materials and methods

2.1. Materials

A representative coal fly ash (CFA) sample was collected from the electrostatic precipitators of coal-fired power plants of Astana (Ekibastuz CFA or E-CFA) and Oskemen (Karazhyra or K-CFA and Maikuben CFA or M-CFA). Both cities are under maximum electricity load. All CFA samples were used as received without preliminary washing and sieving. Prior to experiment CFA samples were homogenized and dried in oven at 70 °C for 12 h. The sodium hydroxide of analytical grade (Fischer Scientific, pearls, >97%) was used to prepare an alkaline solution with various concentrations.

2.2. Methods

2.2.1. Zeolite synthesis

The E-CFA of Astana city power plant was selected to study the effect of reaction parameters and to optimize production of synthetic zeolites (ZFAs), while after setting optimum parameters for production the other two CFAs were processed and used in the adsorption experiments. To produce synthetic zeolite (ZFA), E-CFA underwent a conventional alkaline hydrothermal treatment at 90-110°C, using NaOH with concentrations of 1M or 3 M as an activation solution, in a 1L heavy-walled glass reactor. The incubation period was set at 24 or 48 hours and a mixing rate was set constant at 126 rpm for all experiments. After set period, the mixture was filtered and washed several times until no NaOH was detected. To study the factorial effect of 3 parameters, namely the reaction temperature, time and loading ($C*V/m$, mole/g) on conversion rate of CFA into ZFAs 8 experiments were conducted in duplicate (table 1).

Table 1. Parameters applied to convert E-CFA into E-ZFA – Factorial analysis.

Experiment	Zeolite acronym	Temperature, °C	Time, h	Loading, mole/g
1	ZFA_1M_50g_90C_126rpm_24h	90	24	0.02
2	ZFA_1M_50g_110C_126rpm_24h	110	24	0.02
3	ZFA_1M_50g_90C_126rpm_48h	90	48	0.02
4	ZFA_3M_50g_90C_126rpm_24h	90	24	0.06
5	ZFA_1M_50g_110C_126rpm_48h	110	48	0.02
6	ZFA_3M_50g_110C_126rpm_24h	110	24	0.06
7	ZFA_3M_50g_90C_126rpm_48h	90	48	0.06
8	ZFA_3M_50g_110C_126rpm_48h	110	48	0.06

2.2.2. Mineralogical characterization

X-Ray Diffraction (XRD) instrument was used to obtain information on mineralogical phases present in CFA and ZFA. The samples analyzed include synthetic zeolites produced from E-CFA of Astana power plant, K-CFA and M-CFA of Oskemen power plant. The CFA samples were analyzed as received without prior purification and sieving, while synthetic zeolites were washed and dried before analysis. The XRD pattern was recorded on a SmartLab X-ray diffraction instrument (Cu, K-β filter, 40 kV and 30 mA) with a diffraction angle of 2θ and a scanning range of 5–100◦ (Rigaku, Japan). Phase identification was made by utilizing the Auto-search option of powder diffraction file library.

X-Ray Fluorescence (XRF, PANalytical Axios) instrument was applied to get the elemental composition of CFAs and ZFAs. The CFA sample was analyzed as received without prior purification and sieving whereas all produced ZFAs were washed and dried beforehand. The ratio of sample with binder was set at 10 g to 3 g (total mass of 13 g) with a diameter of pellet at 5 cm and thickness of pellet at 0.5 cm. The analyses were duplicated and conducted under inert atmosphere.

2.2.3. Microstructural characterization

Particle size analysis (PSA, Malvern Mastersizer 3000) instrument was used to obtain the information on volumetric percentage of each particle size and their distribution across the range of 0.01 μm to 10 000 μm. Distilled water was used as a dispersant for all samples in Hydro-MV mode.

A Nitrogen Porosimeter Autosorb-1 was applied to obtain data on specific surface area (SSA), average pore size and total specific volume. Three widely applied approaches of calculation namely BET, BJH and DFT were used for analysis of raw data. The samples were first degassed for 2-3 hours prior to analysis at stepwise heating from 50°C to 200°C. A 9 mm glass cell without rod was used for all porosimetric studies.

2.2.4. Batch adsorption studies

Mercury (II) solution with 10 ppm concentration was used throughout the study unless otherwise reported. Mercury analyzer (Lumex RA-915M) was used to quantify the amount of mercury adsorbed. All adsorption experiments were carried out in duplication without any stirring and at room temperature. The initial volume of mercury (II) solution was 40 mL and adsorbent amount 0.3 g. After a certain period of time, 0.15-0.2 mL of samples were sampled from the batch reactors for analysis. In all of adsorption experiments, to examine the possible effect of container walls on adsorption a glass and plastic (polypropylene) containers were used. A reference control solution with the same initial concentration of 10 ppm and reference-deionized water with the same pH and adsorbents were used to compare and eliminate any deviation. The total volume of solutions withdrawn was kept to a minimum (1.8 mL) in order to not disturb the initial m/V ratio.

3. Results and discussion

3.1. Zeolite synthesis

Table 2 summarizes the zeolite synthesis results obtained under different reaction conditions. In general, it is noticeable that the effect of reaction temperature and loading is significant with parameter sets No. 6 and 8 demonstrating the highest yield.

Table 2. Zeolite synthesis optimization: Factorial analysis

Experiment No.	Temperature, °C	Time, h	Loading, mole/g	Zeolite yield (%, average of duplication)	Factors effect, %
1	90	24	0.02	19.80	0
2	110	24	0.02	45.30	25.50
3	90	48	0.02	48.25	28.45
4	90	24	0.06	39.60	19.80
5	110	48	0.02	47.00	27.20
6	110	24	0.06	55.90	36.10
7	90	48	0.06	35.50	15.70
8	110	48	0.06	78.00	58.20

According to results, the optimum parameters to produce the highest yield of zeolite (on average 78%) are reaction temperature of 110°C and reaction time of 48 hours with loading of reactants of 0.06. The factors effect under the specified parameters is 58.2% higher than in reference parameters. It should be noted that the effect of reaction duration is also essential, as parameter set No. 3 reveals the third highest yield of zeolite; thus it could be expected that zeolite formation directly proportional to reaction time. In this case, the factors effect is 28.45%, which is 7.65% lower than in No. 6. The conditions (factors) applied in experiment No. 8 was used for all other synthesis of zeolites.

3.2. Mineralogical characterization of zeolites

The phase content of synthetic zeolites produced from CFA varies depending on the origin of raw material and reaction conditions. figures 1-2 show the XRD spectrum of synthetic zeolites produced from two different Kazakhstani CFAs.

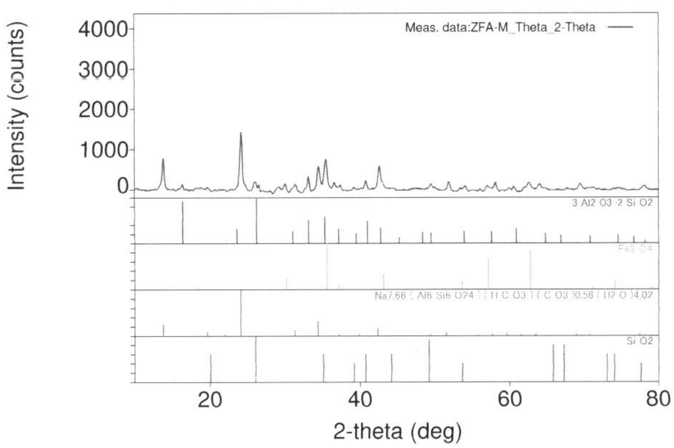

Figure 1. XRD spectrum of M-ZFA.

In general, all zeolites contain phases as mullite, quartz and magnetite/hematite (variations of iron oxides), which are the main phases of parent CFAs. The semi-quantitative tool of software allows identification of the crystalline phases in percentage. The calculation reveals that the content of crystalline mullite and quartz in both ZFAs is negligible. It should be noted that the amount of magnetite/hematite present in CFAs was not detected on this instrument because it is mostly amorphous and only crystalline particles of this phase was used in percentile calculation of software.

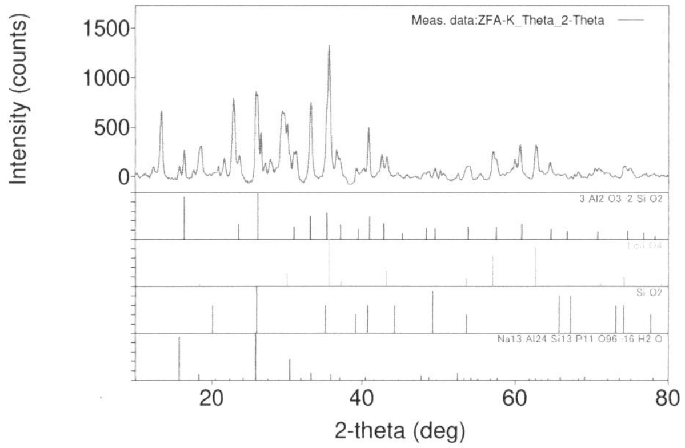

Figure 2. XRD spectrum of K-ZFA.

The detailed chemical composition of K-ZFA and M-ZFA is shown in table 3. According to data the main chemical elements in ZFAs are Al, Si, Fe and Ca, as expected.

Table 3. Elemental composition of Kazakhstani ZFAs.

Compound	K-ZFA	M-ZFA
Na_2O	4.447	5.858
MgO	0.708	1.145
Al_2O_3	30.888	24.859
SiO_2	32.488	32.343
SO_3	0.086	0.874
Cl	0.030	0.023
K_2O	0.150	0.260
CaO	2.390	4.696
TiO_2	1.258	0.959
Cr_2O_3	3.104	3.387
MnO	0.411	0.608
Fe_2O_3	23.066	23.578
CuO	0.028	0.034
ZnO	0.040	0.036
Co_3O_4	0.061	0.070
SrO	0.206	0.309
Y_2O_3	0.013	0.009
ZrO_2	0.074	0.051
BaO	0.119	0.273

However, it is apparent that the content of iron oxide is comparatively lower in K-ZFA (23.066%) in contrast with M-ZFA (23.578%) and noticeably higher than in ZFAs produced from CFAs of other regions of the world [4-10]. It is interesting to note that K-ZFA contain 2 times lower amount of CaO (2.390 and 2.512 wt.%) than in M-ZFA (4.696 wt.%). The total amount of alkaline oxides are relatively higher in M-ZFA (11.959 wt.%) that in turn results in more basic properties. Both ZFAs contain trace amount of heavy metals with concentrations below 0.1wt.% (Co, Cu, Zn, Ga, Sr, Y and Zr). The Si/Al ratio of K-ZFA and M-ZFA are 0.93 and 1.15, respectively. The results are comparable with data published elsewhere [20, 31].

3.3. Microstructural characterization of zeolites

The most characteristic property of all zeolitic materials is their porosity. The general porosimetric analysis using inert gas (argon or nitrogen) could give the porosity type, the SSA, total volume of pores, average pore size.

Table 4 summarizes the experimental results obtained for both ZFAs samples by applying three different adsorption models (isotherms), namely BET, BJH and DFT. It should be noted that the BET method is used for comparison with literature values, as this approach is commonly applied for porosimetric analysis of porous materials.

According to the results obtained by BET method, K-ZFA is the most porous zeolite, where it reveals the average pore diameter of 14.642 nm with the SSA value of 67.734 m^2/g. The total pore volume for pores less than 376.4 nm is 0.248 cm^3/g. This is followed by M-ZFA that demonstrate a wider pore size of 20.104 nm and SSA value of almost 10 m^2/g lower than in K-ZFA. The total pore volume, on the other hand, is about the same as in K-ZFA and reveals the value of 0.291 cm^3/g. In comparison with zeolites, the SSAs of K-CFA and M-ZFA are 32.873 m^2/g and 12.734 m^2/g, respectively.

Table 4. Porosimetric analysis of Kazakhstani ZFAs.

ZFA type	Average pore diameter [nm]	Specific surface area [m^2/g]	Total pore volume [cm^3/g]
	14.642 (BET)	67.734 (BET)	0.248 (BET)
K-ZFA	4.367 (DFT)	46.608 (DFT)	0.138 (DFT)
	3.706 (BJH)	69.253 (BJH)	0.183 (BJH)
	20.104 (BET)	58.191 (BET)	0.291 (BET)
M-ZFA	4.520 (DFT)	62.562 (DFT)	0.179 (DFT)
	3.918 (BJH)	98.052 (BJH)	0.234 (BJH)

According to the results of particle size analysis shown in table 5, K-ZFA and M-ZFA demonstrate similar distribution in terms of average particle size distribution at 26.30 μm and 29.30 μm, respectively. As for particles less than 10% they reveal 6.01 μm (K-ZFA) and 9.26 μm (M-ZFA); while 90% of particle size for these two ZFAs ranges between 66-191 μm.

Table 5. Particle size analysis of ZFAs.

CFA type	Uniformity	Dv(10), μm	Dv(50), μm	Dv(90), μm
K-ZFA	0.711	6.01	26.30	66.60
M-ZFA	1.807	9.26	29.30	191.00

3.4. Batch adsorption studies

The produced zeolites, namely K-ZFA and M-ZFA and their parent CFAs, were tested for adsorption capacity of mercury (II) ions from aqueous solutions with initial Hg(II) ions concentration of 10 ppm under acidic conditions (pH = 2) to avoid precipitation. In total the equilibrium for ZFAs reached after 336 hours or 14 days, while for CFA in reached after 696 hours or 29 days. The results of the adsorption studies is presented in figure 3.

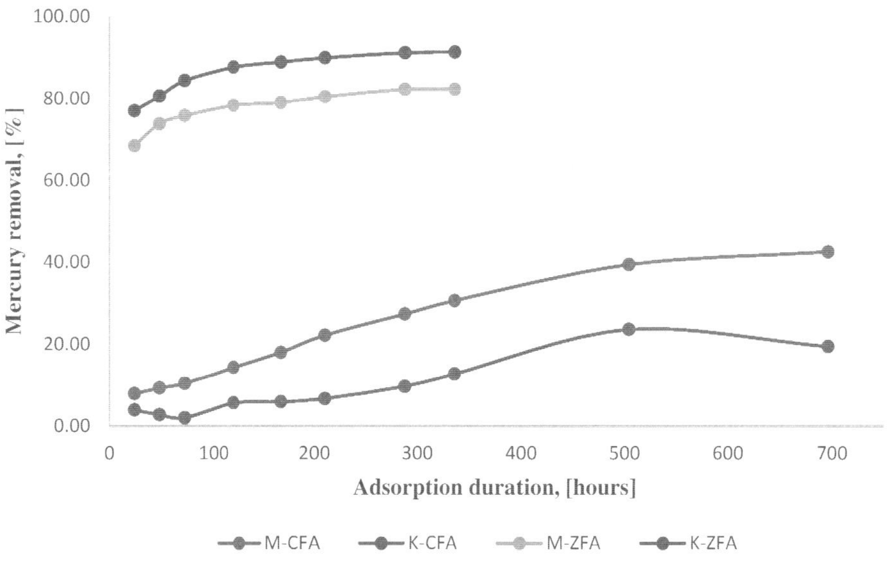

Figure 3. Mercury removal studies of ZFAs.

As it is seen from figure 3, both ZFAs demonstrate significantly higher removal than their respective parent CFAs: K-ZFA removed 91.27% of Hg(II) ions, whereas K-CFA removed 19.43% of Hg (II) ions. On the other hand, M-ZFA demonstrated slightly lower removal of 82.23% Hg (II) ions, while parent M-CFA showed 42.53% that is almost twice as lower as in zeolite. A minor adsorption of Hg (II) ions on container walls and/or evaporation resulted from reduction to Hg (0) was observed during experimental studies. These values were subtracted from obtained values to provide with comparable calculated data.

4. Conclusion

ZFAs produced under optimum reaction conditions were characterized for mineralogy and microstructure and tested for removal efficiency for mercury (II) ions from aqueous phase. The advanced methods, such as XRD, XRF, PSA and porosimetric analysis were successfully applied to obtain valuable characteristic data on these Kazakhstani ZFAs. It was identified that both ZFAs contain mullite, quartz and magnetite/hematite as a residual from parent CFAs, while the major phase being sodalite for M-ZFA and analcime for K-ZFA. The main elements that present in zeolites are Si, Al, Fe, Mn and alkaline metals. Porosimetric analysis revealed that the BET surface area of K-ZFA and M-ZFA are 67.734 m^2/g and 58.191 m^2/g, respectively. The average pore diameter reveals that both ZFAs have mesoporous structure. The results of particle size analysis showed that 50% of particles are less than 26.30-29.30 μm.

5. References

[1] Blissett RS and Rowson NA 2012 *Fuel* 97 1-23.
[2] Bukhari SS, Behin J, Kazemian H, and Rohani S 2015 *Fuel* 140 250-66.
[3] Delkash M, Ebrazi B and Kazemian H 2015 *Micropor.. Mesopor. Mat.* 214 224-41.
[4] Aldahri T, Behin J, Kazemian H and Rohani S 2016 *Fuel* 182 494-501.
[5] Behin J, Bukhari SS, Kazemian H and Rohani S 2016 *Fuel* 171 195-202.
[6] Ojumu TV, Du Plessis PW and Petrik LF 2016 *Ultrason. Sonochem.* 31 342-49.
[7] Visa M 2016 *Powder Technol.* 294 338-47.
[8] Wang J, Li D, Ju F, Han L, Chang L and Bao W 2015 *Fuel Process. Technol.* 136 96-105.
[9] Moriyama R, Takeda S, Onozaki M, Katayama Y, Shiota K, Fukuda T, Sugihara H and Tani Y 2005 *Fuel* 84 1455-61.

[10] Querol X, Umaña JC, Plana F, Alastuey A, Lopez-Soler A, Medinaceli A, Valero A, Domingo MJ and Garcia-Rojo E 2001 *Fuel* 80 857-65.
[11] Wdowin M, Franus M, Panek R, Badura L and Franus W 2014 *Clean Technol. Envir.* 16 1217-23.
[12] El-Mekkawi DM and Selim MM 2012 *Mater. Charact.* 69 37-44.
[13] Ma Y, Yan C, Alshameri A, Qiu X, Zhou C and Li D 2014 *Adv. Powder Technol.* 25 495-99.
[14] Maia AÁB, Neves RF, Angélica RS and Pöllmann H 2015 *Appl. Clay Sci.* 108 55-60.
[15] Liu H, Shen T, Wang W, Li T, Yue Y and Bao X 2015 *Appl.Clay Sci.* 115 201-11.
[16] Mainganye D, Ojumu VT and Petrik L 2013 *Materials* 6.
[17] Querol X, Moreno N, Umaña JC, Alastuey A, Hernández E, López-Soler A and Plana F 2002 *Int. J.* Coal *Geol.* 50 413-23.
[18] Yao ZT, Ji XS, Sarker PK, Tang JH, Ge LQ, Xia MS and Xi YQ 2015 *Earth-Sci. Rev.* 141 105-21.
[19] Belviso C, Cavalcante F, Di Gennaro S, Lettino A, Palma A, Ragone P and Fiore S 2014 *J. Environ. Manage.* 137 16-22.
[20] Wang C, Li J, Sun X, Wang L and Sun X 2009 *J. Environ. Sci.* 21 127-36.
[21] Bukhari SS, Rohani S and Kazemian H 2016 *Ultrason. Sonochem.* 28 47-53.
[22] Musyoka NM, Petrik LF, Hums E, Baser H and Schwieger W 2012 *Catal. Today* 190 38-46.
[23] Molina A and Poole C 2004 *Miner. Eng.* 17 167-73.
[24] Shigemoto N, Hayashi H and Miyaura K 1993 *J. Mater. Sci.* 28 4781-86.
[25] Deng H and Ge Y 2015 *RSC Adv.* 5 9180-88.
[26] Xiao M, Hu X, Gong Y, Gao D, Zhang P, Liu Q, Liu Y and Wang M 2015 *RSC Adv.* 5 100743-749.
[27] Choi CL, Park M, Lee DH, Kim JE, Park BY and Choi J 2001 *Environ. Sci. Technol.* 35 2812-16.
[28] Liu H, Peng S, Shu L, Chen T and Bao T annd Frost RL 2013 *Chemosphere* 91 1539-46.
[29] Cardoso AM, Paprocki A, Ferret LS, Azevedo CMN and Pires M 2015 *Fuel* 139 59-67.
[30] Luo J, Zhang H and Yang J 2016 *Procedia Environ. Sci.* 31 605-14.
[31] Xie J, Wang Z, Wu D and Kong H 2014 *Fuel* 116 71-6.

Acknowledgments
This research work was supported by the Ministry of Education and Science of Kazakhstan. The authors would also like to thank the power stations of Oskemen (East Kazakhstan) and Astana (Central Kazakhstan) cities for generously providing with CFA samples to carry out the studies.

IOP Conf. Series: Materials Science and Engineering **230** (2017) 012045 doi:10.1088/1757-899X/230/1/012045

Electrochemical Study of Unmodified and Inhibitor Doped Silane Films for Corrosion Protection of AA2024-T3

Nauman Mubarak, Jin Hu and Shawei Tang
Department of Materials Science and Engineering, Harbin Institute of Technology, Harbin, China
E-mail: naumanmubarak@yahoo.com

Abstract. Aluminum alloy was coated with unmodified and rare-earth inhibitor doped silane films. The role of number of hydrolysable groups, functional group and cerium ions towards film protective quality was investigated. The anti-corrosion performance was evaluated using electrochemical impedance spectroscopy (EIS), d.c. potentiodynamic polarization and energy dispersive x-ray spectroscopy (EDS). The morphology was studied using scanning electron microscopy (SEM). Results indicate improved corrosion protection performance especially for cerium modified silane films with higher number of hydrolysable groups. Inhibitor doped silanes present a facile method for pre-treatment of aluminium alloys prior to deposition of top coat.

1. Introduction

Chromate conversions coatings have been extensively used for corrosion protections of metals such as aluminum and steels [1]. However, hexavalent chromate species are highly carcinogenic and mutagenic in nature and thus, a great amount of recent research has been focused of discovering feasible alternatives [2]. AA2024 is the most widely used aluminum alloy for aerospace applications such as skin for fuselage. This alloy has an inhomogeneous Cu distribution which lends it high strength and improved toughness, but at the same time make it more susceptible to pitting initiation due to increased chances of a galvanic couple being set up between anodic S phase particles like Al2CuMg and the Al matrix. This pitting can then progress due to increased local alkalization resulting from coupled cathodic reactions [3]. The intermetallic particles are also prone to dissolution as the temperature rises and thus degrade the mechanical properties of the alloy. Hence, treatments that try to improve the corrosion resistance of AA2024 alloys should be carried out at temperatures below 120° C [4].

Organofunctional silanes are capable of functionalizing metallic substrates for subsequent polymeric top coat because of the presence of hydrolysable groups that can form a stable bond with substrate and an organofunctional group that can bond with polymer [5]. Silanes can be broadly classified into two categories as mono-silanes and bis-silanes. Mono-silanes have only three hydrolysable groups present at one end of the molecule while the bis-silanes have a total of six hydrolysable groups with three attached to each silicon atom. The presence of a functional group such as amine, marcapto group can also affect the deposition orientation as well as water uptake properties of the silane coatings. Silanes can only provide passive barrier protection, to provide long term enhanced corrosion protection an active system that can repassivate an initial corrosion attack site is required [6]. Rare-earth ions [7], [8] and nanoparticles [9], [10], [11], [12] have been reported as favorable source for providing such an active protection. However, such addition of inhibitors can deteriorate the stability of the sol-gel and lead to an enhanced rate of corrosion [13], [14].

Two different silanes which will incorporate an inhibitor to provide an active corrosion protection were studied. Gamma-amino propyl triethoxysilane (APS) represents a mono-silane with an amine functional group, while bis-trimethoxy silane (BTMS) is a non-functional bis-silane. Inorganic cerium nitrate salt was used as inhibitor ion source.

2. Experiment

2.1. Materials and reagents

Aluminum alloy 2024-T3 was used as the substrate. Strips having dimensions of 4 cm x 8 cm x 0.1 cm were received from Xionglv Metal Material, Dongguan, China. Gamma-Aminopropyl triethoxysilane (99%) and Bis-trimethoxysilane (≥97%) were obtained from Aladdin, Shanghai, China. The silanes were used as received without any further purification. Cerium nitrate hexa-hydrate salt was procured from Guangfu, Tianjin, China. Ethanol, acetone and distilled water were used to clean the substrate samples and to prepare solutions. The pH, where required, was adjusted by using acetic acid.

2.1.1. Coating fabrication

Prior to use, the silanes were functionalized in an alcohol and water solution. The Gamma-Aminopropyl triethoxysilane (APS) solution was prepared by dissolving 5 vol. % of the silane in 90 vol. % ethanol and 5 vol. % distilled water. The Bis-trimethoxysilane (BTMS) solution was prepared by using the same volume ratios. But in this case the pH of the final solution was adjusted to 4 – 4.5 by using acetic acid. The prepared solutions were stirred for 2 hours and then hydrolyzed further for at least 24 hours before use.

The aluminum alloy 2024-T3 strips were ultrasonically cleaned in acetone and ethanol for 15 minutes each. Then rinsed with ample amount of distilled water and blow dried.

2.2. Inhibitor addition and silane treatment

Cerium nitrate hexa-hydrate ($Ce(NO_3)_3.6H_2O$) was selected as a source for the introduction of inorganic ions into the silane films. In order to incorporate these ions a 100 ppm solution of the salt was prepared by dissolving in distilled water. The prepared solution was stirred and heated at 65o C to allow complete solubility of the inhibitor in the water. A 5 vol. % silane solution, was prepared by adding 5 ml of silane and 5 ml of the cerium salt solution to 90 ml of ethanol. The resulting silane inhibitor solution was hydrolyzed by stirring for 2 hours and then allowed to hydrolyze for 48 hours.

The silane films were applied immediately after the alloys strips had been cleaned and dried. The immersion time was between 100-200 seconds at room temperature. The applied films were then cured at 100o C for an hour. This allows to produce adherent films which are extensively cross-liked. This inhibitor loaded coating will be hereafter referred to as Ce-BTMS coating.

2.3. Characterization

The morphology and performance of the silane films was studied using various optical and electrochemical methods. A Hitachi S-3000N machine was used to carry out scanning electron microscopy (SEM) analysis to determine the uniformity and to detect the presence of pin holes in the cured coatings. The operating voltage was 20 kV.

Accelerated corrosion assessment was carried out by performing a DC polarization tests using Princeton Applied Research (PAR) 273-A potentiostat. A three electrode cell was used. The sample was the working electrode. A graphite rod and saturated calomel electrode (SCE) were used as the counter electrode and the reference electrode respectively. A 3.5 wt. % NaCl solution was used as the electrolyte. The samples were immersed for 2 hours in the electrolyte to achieve steady-state conditions. A range of -250mV and +500mV around the Ecorr was used during the experiment. The scan rate was 0.5mV/s.

Thin coatings on substrates have been widely studied by using the electrochemical impedance spectroscopy (EIS) technique. An Autolab 302N potentiostat equipped with Nova software was employed. The electrode setup was the same as for DC polarization tests. The electrolyte used was a 3.5 wt. % NaCl solution. The exposed area of the sample was $1cm^2$. EIS determination was started after 30 minutes of immersion of any sample in the electrolyte solution. Equivalent circuits were used to study the various electrochemical aspects of the coatings. Seven data points per decade and a frequency range from 10^5 - 10^{-2} Hz, with an alternating current (AC) voltage amplitude of ±10mV was used.

Energy dispersive x-ray spectroscopy (EDS) analysis was performed for the Ce-BTMS coating to provide an insight into the corrosion protection methodology of cerium ions. An FEI Helios NanoLab 600i machine with an accelerating voltage of 20kV was used.

3. Results

3.1. SEM analysis

The coating coverage and uniformity was determined by SEM analysis. The SEM micrographs for the three different coatings are presented in figure1.

Uniform, defect-free coatings are observed for all the three cases. Darker areas are regions with greater silane deposition [15]. Only very few randomly scattered pin holes are observed. For the APS coating dark islands surrounded by less darker regions can be observed. This can be attributed to a greater degree of condensation reaction occurring within the APS silane as well as greater degree of adsorption on Mg-containing particles [15]. The BTMS coating shows a uniform coating devoid of dark island-like formations seen for APS coating representing a greater degree of hydrolysis reaction leading to formation of metallosiloxane bonds and a lesser degree of self-condensation reaction. Even for the Ce-BTMS coating the surface shows overall uniformity indicating that the concentration of cerium ions used is blow the critical limit that can cause agglomeration and excess porosity.

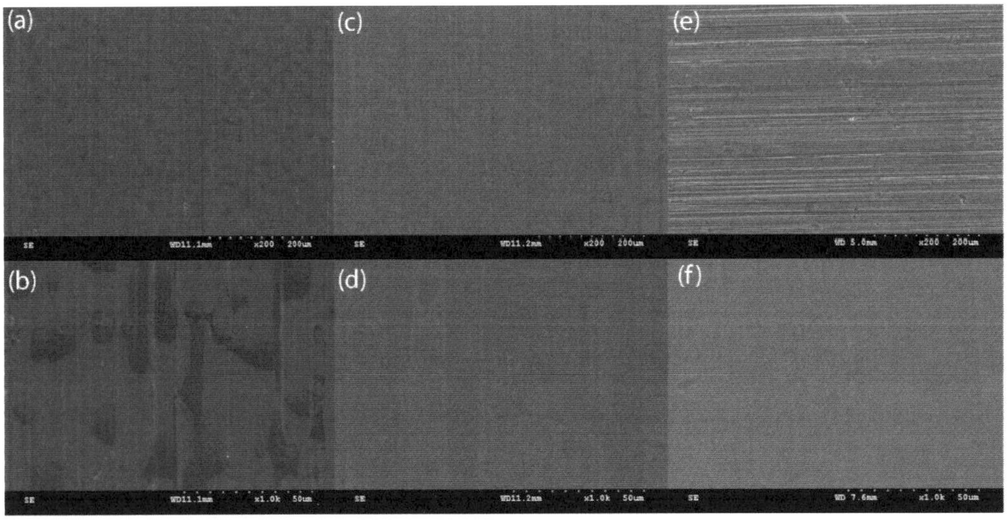

Figure 1. SEM micrographs of (a-b) APS (c-d) BTMS and (e-f) Ce-BTMS silane films

3.2. DC polarization

DC polarization tests were performed in 3.5 wt. % NaCl solution and the results are presented in figure 2.

The pure BTMS coating provides a much enhanced corrosion protection as compared to the APS coating which is evidenced by substantially reduced current densities for both anodic and cathodic parts of polarization curve as compared with the bare alloy. For the BTMS coating the current density is more than one order of magnitude lower than the uncoated sample. This shows the enhanced physical barrier properties of the BTMS as compared to the APS coating. The reduction in cathodic current densities can be explained as the blockage of some of the active cathodic sites by the silanes. The decrease in the anodic current density can be attributed to a decreased rate at which the corrosion products can migrate away from the initial pitting sites due to the formation of dense Al-O-Si interface [16]. For silane coated samples the corrosion products mainly intrude into the pores in the interfacial

IOP Conf. Series: Materials Science and Engineering **230** (2017) 012045 doi:10.1088/1757-899X/230/1/012045

layer leading to pore blockage. This effect leads to an increase in the coating resistance and has been further detailed by EIS analysis.

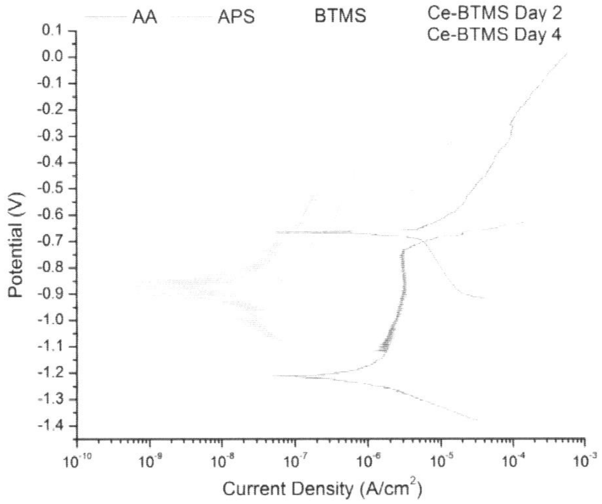

Figure 2. DC polarization curves of silanes at different immersion times

For the Ce-BTMS coating it can be seen that the decrease in anodic and cathodic current densities is virtually the same as un-doped BTMS even after two days of immersion. This shows that the chosen concentration of cerium ions is low enough as to not damage the physical barrier properties of the coating system by leading to an increased number of pin hole defects in the coating. It can also be seen that a further decrease in current densities is observed in the polarization curve conducted after four days of immersion. This decrease in current densities could be due to the deposition of cerium oxide/hydroxides at the sites of corrosion initiation. This phenomenon can be referred to as "active" corrosion protection, where the cerium ions from the coating migrate to corrosion pit initiation sites and deposit as oxide or hydroxides and retard the corrosion progress. This is shown by the shifting of Ecor towards more cathodic values as well as the emergence of a secondary peak around -1.0 V on the cathodic branch of polarization curve. Such a peak is generally observed for cerium coatings [17], which act as cathodic inhibitors. The emergence of this peak also evidences a chemical barrier rather than a simple physical barrier that is provided by the silanes.

The presence of such hydrated oxides of cerium along with the native aluminum oxide layer can lead to the formation of a barrier towards the flow of electrons. This can inhibit corrosion by reducing the rate of oxygen reduction. This process was further explored by EIS and EDS analysis.

3.3. EIS

Efficacy of the silane films was evaluated over a range of immersion period in 3.5 wt. % NaCl solution. figure 3-5 shows the Bode plots after various intervals of immersion. Symbols represent the measured data while the solid lines are the equivalent circuit fitting results.

EIS has been widely used to evaluate silane coatings, both doped and undoped, deposited on aluminum alloys. The Bode plots present a characteristic curve. At high frequencies between 10^5 and 10^2 Hz the impedance curve shows the resistive and capacitive contribution from the silane coatings. While at frequencies slightly lower than 102 Hz the intermediate oxide layer effects are observed. At lowest frequency the information from the corrosion processes characterized by polarization resistance and double layer capacitance can be obtained.

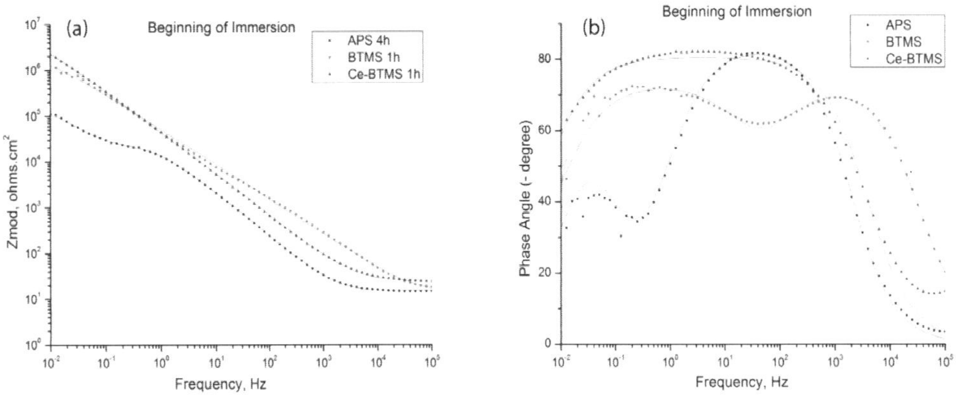

Figure 3. Bode plots at beginning of immersion

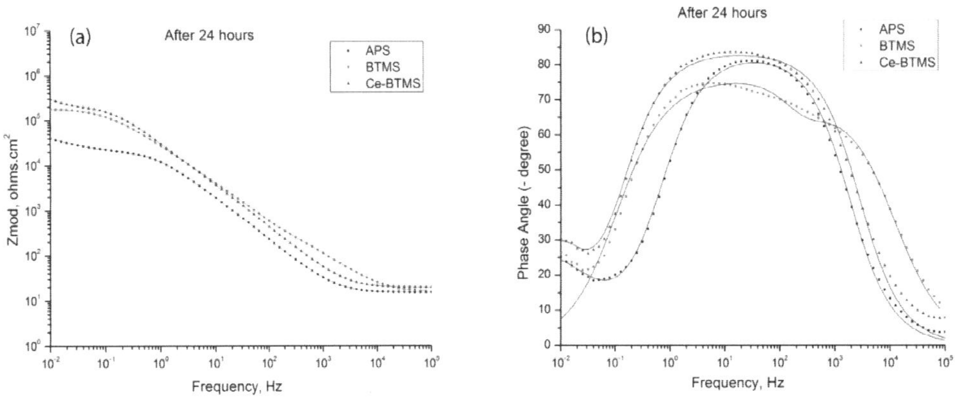

Figure 4. Bode plots after 24 hours of immersion

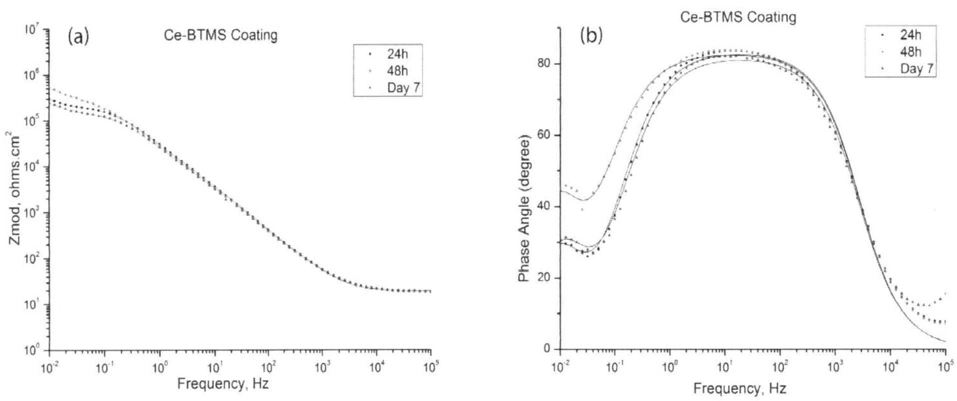

Figure 5. Bode plots for Ce-BTMS coating for prolonged immersion

Figure 3 shows the Bode plots of the silane coatings at the beginning of immersion period. APS shows lowest impedance value at the beginning of immersion along with lowest phase angles in the lower frequency region. Whereas the impedance and phase angle values obtained for un-doped BTMS in the high frequency region 10^3 to 10^5 are the highest indicating good barrier properties of the coating. These values decrease slightly in the lower frequency region. The highest impedance and phase angle values are shown by doped-BTMS sol-gel coating indicating excellent barrier resistance. The results are in keeping with those obtained from DC polarization study.

The Bode plots in figure 3 were obtained just after the start of immersion in the saline solution and already some signs of water permeation into the APS and BTMS coating are emerging. The Ce-BTMS coating seems to be offering the best comparative retardation of water permeation in early stages of immersion.

After 24 hours of immersion it appears that the coatings are saturated with water resulting in decreased impedance values for all the coatings, as in figure 4. The barrier effectiveness of the BTMS coating decreases with values in the high frequency now close to those of Ce-BTMS coating. Even then BTMS coating seems to be faring better than the APS which could be due to the hydrophilic nature of amino functional group in the APS coating. For APS coating signs for initial corrosion are also emerging as evidenced by the appearance of a well-defined time constant in the lowest frequency range. For BTMS coatings although the time constant is not as well-defined as that for APS coating but initiation of pitting corrosion seems likely.

After 48 hours multiple barely visible pits can be observed for the APS coated sample and the values of impedance modulus fall to a low value of around $3x10^4$ ohms. This deterioration of coating could be due to the relatively weak interface formed by the APS coating due to its fewer hydrolysable groups and compounded by the aggressive uptake of water due to the hydrophilic functional group in its molecule. For the BTMS coating the impedance modulus values are nearly one decade higher than those for APS coating after 48 hours and the coating seems to hold its impedance values from 24 hours which could be due to the much slower permeation rate of water into the coating thus allowing the initial corrosion products to retard the pit from growing quickly. In contrast to both other coatings the Ce-BTMS coating shows nearly a four time increase in the impedance values from 24 hours to 48 hours. The values of impedance remain upwards that after 24 hours for more than 7 days of immersion, shown in figure 5. This self-healing property of the coating could be due to the formation of a passive cerium hydroxide layer where cerium ions leach out from the coating in the vicinity of the pitting site and deposit over the defect. The probable mechanism involved has been studied by various authors [14], [18]. Succinctly, in a chloride rich environment S-phase intermetallics having composition Al2CuMg are main culprits for corrosion of 2024 aluminum alloy. These intermetallics can act as effective cathodes where oxygen reduction can occur via four-electron step or two-electron step resulting in hydroxide ions and peroxide formation:

$$O_2 + H_2O + 4e^- \rightarrow 4OH^- \tag{1}$$

$$O_2 + 2H_2O + 2e^- \rightarrow H_2O_2 + 2OH^- \tag{2}$$

A cathodic reaction can take place resulting in the formation of highly insoluble hydroxides as:

$$Ce^{3+} + 3OH^- \rightarrow Ce(OH)_3 \tag{3}$$

Alternatively Ce^{3+} ions can get oxidized by peroxide to Ce^{4+} state as:

$$Ce^{3+} + \frac{1}{2}H_2O_2 + 3OH^- \rightarrow Ce(OH)_4 \tag{4}$$

$$Ce(OH)_4 \rightarrow CeO_2.2H_2O \tag{5}$$

Thus, if a sufficient concentration of cerium ions is available the cathodic process can be greatly retarded by formation of these insoluble hydroxide deposits.

For continued immersion beyond 48 hours for the APS coated sample it can be seen that the impedance values remain nearly constant and pitting progresses at a steady rate. Beyond 96 hours the pits become more easily visible and a very slight increase in impedance value is observed which could be due to the presence of excess corrosion products in and around the pits which can lead to some barrier protection against the corrosion progression.

For the BTMS coating the impedance modulus values fall progressively over the first 7 days of immersion at which point a couple of pits become visible. Upon continued immersion beyond this time a similar phenomenon of pit blockage by excess corrosion product is observed resulting in a slight increase in impedance values to day 11 when the experiment was stopped. For the Ce-BTMS coating the pits only become visible after 10 days of immersion. But the rate of pit progression remains very slow as impedance values after 16 days which are nearly identical to those observed after 10 days.

Equivalent circuits can be used to gather more insight about the influence of various parameters on the sol-gel coated samples. The equivalent circuits which were used to represent various stages during EIS immersion study are shown in figure 6.

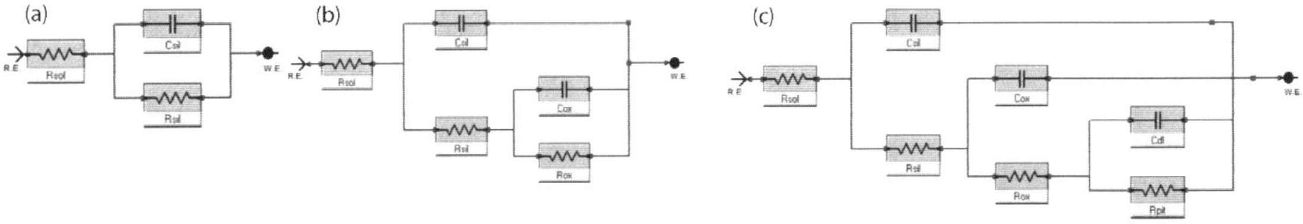

Figure 6. Equivalent circuits (a) beginning of immersion (b) oxide layer resistance (c) on-set of corrosion

In the above circuits Rsol is the solution resistance, Rsil is resistance of the silane layer, Rox is the oxide resistance and Rpit is the pitting or polarization resistance. Csil, Cox and Cdl are silane, oxide and double layer capacitances respectively. Capacitors were replaced by constant phase elements (CPE) as the phase angles were different from -90°, which is a necessary condition for ideal capacitor. The impedance of CPE can be given by [4]:

$$Z_{CPE} = \left(\frac{1}{Y} \right) / (jw)^a \tag{6}$$

The parameter 'Y' corresponds to pseudo-capacitance, while 'a' relates the systems homogeneity. If the values of a =1 and Y = C, this condition would represent a capacitor, whereas the value of a will be less than 1 for a CPE.

The equivalent circuits were used to model different stages of immersion. Circuit (a) is used at the beginning of immersion and gives only the capacitive and resistive response of the coating along with solution resistance. The circuit (b) incorporates the resistive and capacitive responses for the native oxide layer on aluminum alloy as well as the Al-O-Si layer that can form between the silane and the substrate. In circuit (c) double layer capacitance and pitting resistance are incorporated to model the influences of pitting phenomenon. The chi-squared error values were less than 10^{-3} for all fittings.

The low concentration on cerium was chosen to ensure that the coating uniformity is not disrupted by resulting in pin holes or agglomeration of cerium ions. The SEM results show a uniform coating free of agglomerates and a very few scattered pin holes. At the beginning of immersion, figure 7, the coating resistances of BTMS and Ce-BTMS coatings are very close to one another. But, after nearly 24 hours of immersion Ce-BTMS coating shows higher values of resistance and after 48 hours the difference is much more pronounced. The higher R(sil) values for Ce-BTMS coating could be due to pore blocking action of cerium ions or the passivation of pit initiation sites by cerium ions which is more evident around 24 to 48 hours of immersion. This could presumably lead to a slower degradation of coating as the rate at which corrosion products are generated is much slower. BTMS coating loses most of its protective action around 168 hours of immersion and values fall to the range of 10^1 whereas for Ce-BTMS coating the coating only reaches this stage after 216 hours of immersion. For the APS coating the barrier properties are not as effective with coating resistance falling sharply during 24 hours to 48 hours. This can be attributed to the hydrophilic nature of its functional group or upside down deposition [15] of its molecules which can adversely affect its protective nature. The values fall to around 10^2-10^1 range after only 48 hours and maintain these to around 96 hour mark, after which the values rise to around 10^3 which can be due to a large build-up of corrosive products. This phenomenon of increase in resistance due to corrosion products can also be seen for BTMS coating but is less pronounced probably due to a smaller quantity of corrosion products. Whereas for the Ce-BTMS this phenomenon is almost negligible from 240 to 384 hours indicating only a very slow corrosion rate where the corrosion products might have enough time to diffuse a little further away for the corrosion site. But we can presume that if the immersion is continued a similar increase in resistance can be observed due to eventual build up corrosion products.

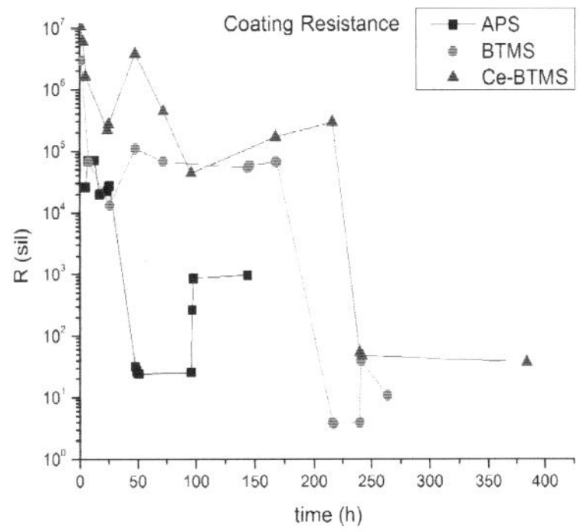

Figure 7. Silane film resistance with immersion time

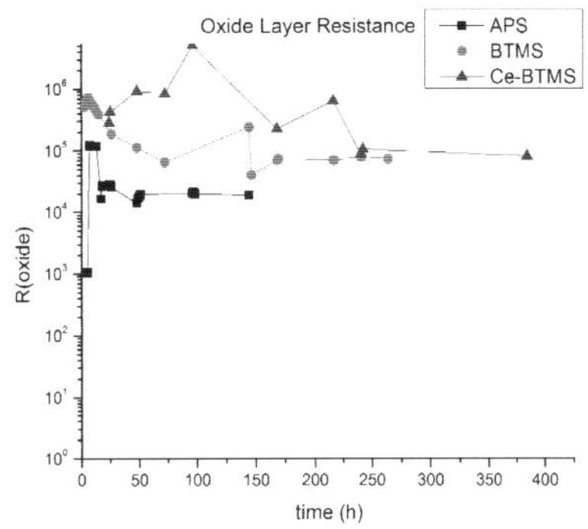

Figure 8. Oxide layer resistance with immersion time

The oxide layer resistance values shown in figure 8, for APS coating are nearly one decade and two decade lower than those for BTMS and Ce-BTMS coating. This most likely occurs due to the low number of interfacial bonds that APS can form resulting is a weaker interface. For the BTMS coating a denser interface can formed due to the presence of more hydrolysable groups present in its structure. The oxide layer only registers a very slow decrease in its resistive values. For the Ce-BTMS coating the oxide layer resistance values show an initial increase in contrast to the other two coatings. This can be explained by considering the corrosion initial phenomenon for aluminum alloys. In aluminum alloys like 2024 the oxide layer is much thinner over the intermetallic phases and due to the electronegativity difference between these phases and the surrounding aluminum phase corrosion will initiate at these sites [3], [15]. For the Ce-BTMS coating not only is a denser interface formed due to more hydrolysable groups, as soon as pit initiation occurs a non-soluble cerium hydroxide layer is formed at these points. Thus, in a matter of speaking, repairing the oxide layer. This phenomenon is easily observable between 24 to 96 hours where resistance values for Ce-BTMS coating are more than a decade higher than those observed for un-doped BTMS coating. Eventually after 240 hours both the BTMS coatings approach similar values which points that the positive influence of cerium ions has disappeared. This points towards one of the key areas which needs to be optimized in future studies, where the release of dopant ions has to be prolonged or a mechanism for preparing a re-loadable coating have to be investigated.

3.4. Corrosion Process

Polarization resistance (Rpit) can be used to evaluate the rate of corrosion processes. The polarization resistance for the coatings is given in Table I.

Table I. Polarization resistance with immersion time.

APS Coating		BTMS Coating		Ce-BTMS Coating	
Time (h)	R(pit) ohms cm^2	Time (h)	R(pit) ohms cm^2	Time (h)	R(pit) ohms cm^2
48	4.28E+04	216	7.41E+04	240	1.18E+05
96	2.51E+04	240	8.81E+04	384	9.37E+04
144	1.79E+03	264	9.39E+04	-	-

The APS coating only provides a fair protection with polarization resistance values in the range of 10^4 after 48 hours of immersion. These values fall to 10^3 after 144 hours of immersion and large corrosion can be observed on the coated sample. The low polarization resistance values are due to the poor barrier properties of the coating. The un-doped BTMS coating offers a much improved polarization resistance with values around $7x10^4$ after 216 hours of immersion and remain in this range even when the immersion time increases to 264 hours. The Ce-BTMS offers the highest polarization resistance values with the third time constant related to pitting appearing only around 240 hours of immersion and values of about $1x10^5$. Even with continued immersion these values are maintained and only after 384 hours the values show a small decrease and falling to around $9x10^4$. The improved polarization resistance values for the BTMS coating are due to much better barrier properties as well as denser interface formation. For the Ce-BTMS coating these values are further enhanced and show decrease only when the cerium ions are released from the coating and the inhibition by cerium ions is discontinued.

3.5. EDS Analysis

In order to ascertain that cerium is present inside the pits an EDS analysis, shown in figure 9, was performed after 4-7 days of immersion in 3.5 wt. % NaCl solution. EDS analysis was also carried out at multiple points for the coating that has not been immersed in NaCl solution.

For the coating that has not been immersed in NaCl solution a few of the tested regions did not show a signal for cerium which could be either due to the non-uniform distribution of the cerium in the coating or due to the very low concentration of the cerium in the final coating. However, a signal for cerium was detected in most tested regions with a maximum of around 0.34 wt. % (0.06 atomic %). But in case of the coating that was immersed for 7 days in NaCl solution, a signal for cerium is always detected in the area of pit. For a relatively smaller pit the signal strength of cerium is higher at 0.65 wt. % (0.1 atomic %) whereas for a larger pit the detected value is lower at 0.22 wt. % (0.03 atomic %). The growth of the pit with time could probably lead to removal of cerium hydroxide deposits from the pit or it possible that the region around the pit initially had a lower concentration of cerium due to non-uniform distribution that the cerium concentration in the pit was not sufficient to actively passivate the pit and thus the pit grew more than the other pit as well as has a lower final cerium concentration. These assumptions also fit the smaller pit where the smaller pit growth could be either due to a larger concentration of cerium and thus a better active protection or the fact that this pit initiated at a later stage and thus has smaller size and higher concentration of cerium. In either case, EDS analysis supports the claim that cerium ions deposit at pit initiation sites and provide an active corrosion protection at least for a moderate time of immersion.

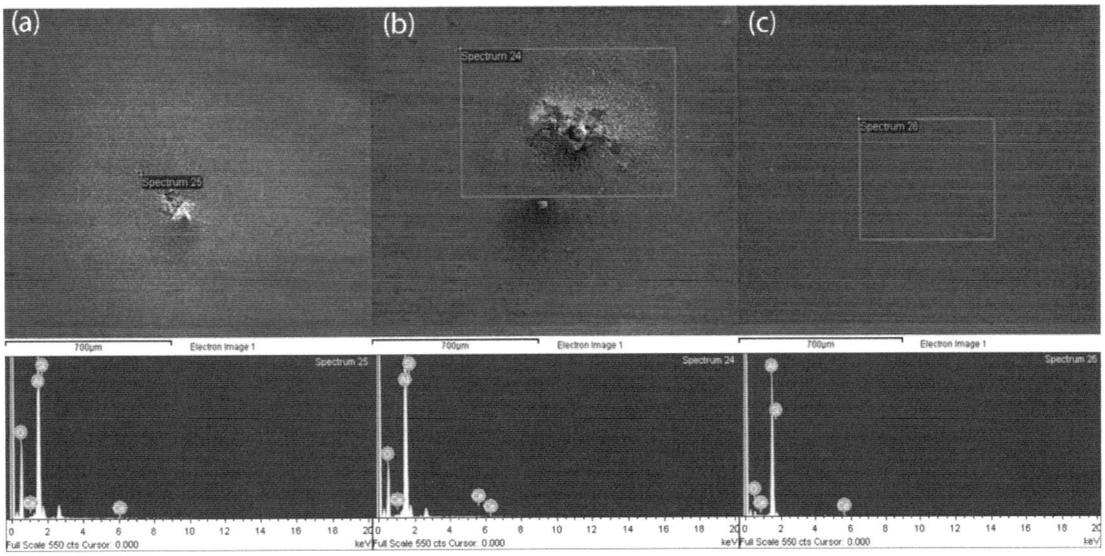

Figure 9. EDS micrographs (a) small pit (b) larger pit (c) un-immersed coating area

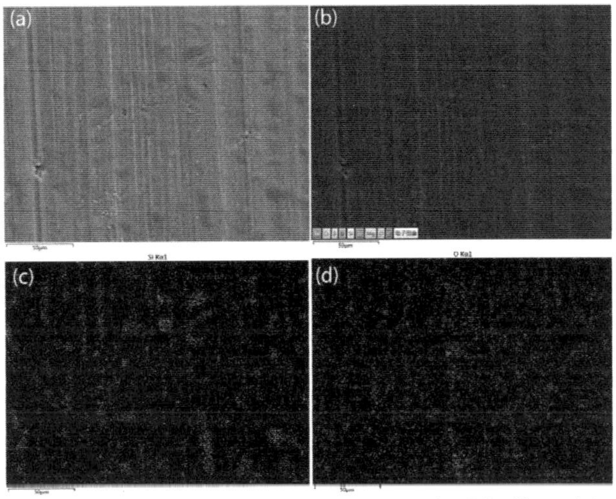

Figure 10. EDS mapping representing self-condensation of APS silane (a) optical (b) all major elements (c) Si (d) O mapping

Figure 10 is an EDS mapping of APS silane film. The mapping again shown island like morphology for the APS. Figure 10(c) provides evidence that these islands are agglomerates of silicon, which result from the excessive self-condensation of the silane. This non-uniformity exhibited by APS is one of the causes of its poorer performance.

4. Conclusion

Coating and metal/interface properties of sol-gel pre-treatments can be studied by using EIS method for aluminum alloys. Three different types of coatings were evaluated for their corrosion protection in 3.5 wt. % NaCl solution.

For the APS coating, a slightly hydrophilic functional group together with fewer number of hydrolysable groups leads a less cross-linked Si-O-Si barrier layer and a less dense interface between the silane and alloy. These lead to only a fair improvement in the corrosion resistance of the substrate. The use of BTMS provides much improved protection due to higher number of hydrolysable groups resulting in a highly cross-linked barrier layer and a dense Al-O-Si interface.

The BTMS coating alone can however only provide a physical barrier against the corrosive medium and thus only passive protection. A small amount of cerium ions were introduced into the BTMS

coating to investigate their influence on the pre-treatment. SEM analysis reveals that the cerium salt concentration is low enough to not disturb the uniformity and homogeneity of the coating. Linear polarization tests reveal the enhanced corrosion protection provided by BTMS and Ce-BTMS coatings as compared with APS coating. Evidence for chemical protection provided by Ce-BTMS coating is provided by a secondary peak in the cathodic branch of tafel curve. EIS testing along with EDS analysis reveals that the cerium ions can deposits at sites of pit initiation as insoluble cerium hydroxide deposits and thus provide an enhanced and active corrosion protection to the AA2024 substrate.
This active protection effect however diminishes with continued immersion and methods to incorporate greater concentration of cerium ions without disturbing the morphology of the coating and providing a prolonged release of inhibitor ions needs to be investigated.

5. References

[1] M. Kendig, S. Jeanjaquet and R. Addison et al., "Role of hexavalent chromium in the inhibition of corrosion of aluminum alloys," Surface and Coatings Technology, vol. 140, no. 1, pp. 58-66, May 2001.
[2] J. Zhao, L. Xia and A. Sehgal et al., "Effects of chromate and chromate conversion coatings on corrosion of aluminum alloy 2024-T3," Surface and Coatings Technology, vol. 140, no. 1, pp. 51-57, May 2001.
[3] D. Zhu and W. J. van Ooij, "Corrosion protection of AA 2024-T3 by bis-[3-(triethoxysilyl)propyl]tetrasulfide in neutral sodium chloride solution. Part 1: corrosion of AA 2024-T3," Corrosion Science, vol. 45, no. 10, pp. 2163-2175, Oct. 2003.
[4] N. C. Rosero-Navarro and S. A. Pellice, A. Durán et al., "Effects of Ce-containing sol–gel coatings reinforced with SiO2 nanoparticles on the protection of AA2024," Corrosion Science, vol. 50, no. 5, pp. 1283-1291, May 2008.
[5] A. M. Cabral, R. G. Duarte and M. F. Montemor et al., "A comparative study on the corrosion resistance of AA2024-T3 substrates pre-treated with different silane solutions: Composition of the films formed," Progress in Organic Coatings, vol. 54, no. 4, pp. 322-331, Dec. 2005.
[6] A. P. More and S. T. Mhaske, "Chemical Modification of Silane-Based Coating with Inhibitor for Anticorrosive Application," Arabian Journal for Science and Engineering, vol. 41, no. 6, pp. 2239-2248, June 2016.
[7] J. Mosa, N. C. Rosero-Navarro and M. Aparicio, "Active corrosion inhibition of mild steel by environmentally-friendly Ce-doped organic-inorganic sol-gel coatings," RSC Advances, vol. 6, no. 46, pp. 39577-39586, April 2016.
[8] M. L. Zheludkevich, R. Serra and M. F. Montemor et al., "Nanostructured sol–gel coatings doped with cerium nitrate as pre-treatments for AA2024-T3: Corrosion protection performance," Electrochimica Acta, vol. 51, no. 2, pp. 208-217, Oct. 2005.
[9] A. Ahmadi, B. Ramezanzadeh and M. Mahdavian, "Hybrid silane coating reinforced with silanized graphene oxide nanosheets with improved corrosion protective performance," RSC Advances, vol. 6, no. 59, pp. 54102-54112, May 2016.
[10] L. Calabrese, L. Bonaccorsi and A. Caprì et al., "Assessment of hydrophobic and anticorrosion properties of composite silane–zeolite coatings on aluminum substrate," Journal of Coatings Technology and Research, vol. 13, no. 2, pp. 287-297, March 2016.
[11] M. Gharagozlou, R. Naderi and Z. Baradaran, "Effect of synthesized NiFe2O4-silica nanocomposite on the performance of an ecofriendly silane sol–gel coating," Progress in Organic Coatings, vol. 90, pp. 407-413, Jan. 2016.
[12] L. Lei, J. Shi and X. Wang et al., "Microstructure and electrochemical behavior of cerium conversion coating modified with silane agent on magnesium substrates," Applied Surface Science, vol. 376, pp. 161-171, July 2016.
[13] C. F. Malfatti, T. L. Menezes and C. Radtke et al., "The influence of cerium ion concentrations on the characteristics of hybrid films obtained on AA2024-T3 aluminum alloy," Materials and Corrosion, vol. 63, no. 9, pp. 819-827, Sep. 2012.
[14] R. Zandi Zand, V. Flexer and M. De Keersmaecker et al., "Self-healing silane coatings of cerium salt activated nanoparticles," Materials and Corrosion, vol. 67, no. 7, pp. 693-701, July 2016.

[15] D. Susac, X. Sun and K. A. R. Mitchell, “Adsorption of BTSE and gamma-APS organosilanes on different microstructural regions of 2024-T3 aluminum alloy,” Applied Surface Science, vol. 207, no. 1-4, pp. 40-50, Feb. 2003.
[16] D. Zhu and W. J. van Ooij, “Corrosion protection of AA 2024-T3 by bis-[3-(triethoxysilyl)propyl]tetrasulfide in sodium chloride solution.: Part 2: mechanism for corrosion protection,” Corrosion Science, vol. 45, no. 10, pp. 2177-2197, Oct. 2003.
[17] H. D. Johansen, C. M. A. Brett and A. J. Motheo, “Corrosion protection of aluminium alloy by cerium conversion and conducting polymer duplex coatings,” Corrosion Science, vol. 63, pp. 342-350, Oct. 2012.
[18] K. A. Yasakau, M. L. Zheludkevich and O. V. Karavai et al., “Influence of inhibitor addition on the corrosion protection performance of sol–gel coatings on AA2024,” Progress in Organic Coatings, vol. 63, no. 3, pp. 352-361, Oct. 2008.

Acknowledgments
The authors acknowledge financial support by the National Natural Science Foundation (Grant No. 51272055 and 51501050).

Mineralogical, Microstructural and Thermal Characterization of Coal Fly Ash Produced from Kazakhstani Power Plants

Z Tauanov[1,4], L Abylgazina[1], C Spitas[2], G Itskos[3] and V Inglezakis[1,4]

[1]School of Engineering, Chemical Engineering Department, Environmental Science and Technology Group (ESTg), Nazarbayev University, Astana, 010000, Kazakhstan
[2]School of Engineering, Mechanical Engineering Department, Nazarbayev University, Astana, 010000, Kazakhstan
[3]School of Materials Engineering, Purdue University, West Lafayette, IN 47907-2045, USA.
[4]Corresponding authors: zhtauanov@nu.edu.kz; vasileios.inglezakis@nu.edu.kz

Abstract. Coal fly ash (CFA) is a waste by-product of coal combustion. Kazakhstan has vast coal deposits and is major consumer of coal and hence produces huge amounts of CFA annually. The government aims to recycle and effectively utilize this waste by-product. Thus, a detailed study of the physical and chemical properties of material is required as the data available in literature is either outdated or not applicable for recently produced CFA samples. The full mineralogical, microstructural and thermal characterization of three types of coal fly ash (CFA) produced in two large Kazakhstani power plants is reported in this work. The properties of CFAs were compared between samples as well as with published values.

1. Introduction

Coal fly ash (CFA) is a waste by-product from coal combustion in power plants all over the world. Kazakhstan produces around 19 million tons of CFA annually and has accumulated more than 300 million tons of CFA in ash disposal areas across the country [1-2]. According to statistics, the utilization rate of CFA in Kazakhstan is under 10%, which is considerably lower as compared to EU states (over 90%), India (60%), China (67%) and the USA (nearly 50%) [1, 3]. Characterization of this material is of utmost importance as it determines how and in which sectors of industry it could be utilized.

To this aim, in this work a full-scale mineralogical, microstructural, thermal characterization is provided and analytical methods of spectroscopy/chromatography are utilized. Three types of Kazakhstani CFAs, namely Ekibastuz (E-CFA), Karazhyra (K-CFA) and Maikuben (M-CFA) from Astana and Oskemen city power plants (Central and East Kazakhstan, respectively) were analysed. In order to obtain a full data on chemical and physical properties all samples underwent characterization using advanced instruments as X-Ray diffraction (XRD) analysis to retrieve mineralogical phases present in CFAs, X-Ray Fluorescence (XRF) analysis to obtain elemental composition of CFAs and FTIR for specific functional groups identification. In addition to this, thermogravimetric and differential calorimetric analysis (TGA/DSC) were applied to acquire information regarding thermal properties of CFAs, while Total Carbon (TC) analysis to quantify the residual carbon. The Particle Size Analyzer (PSA) and Porosimetric analysis were used to estimate the particle size groups present in crude CFAs, their respective specific surface area, average pore size and total pore volume.

2. Materials and methods

2.1. Materials

A representative CFA sample with a composition typical of the average CFA output was collected under maximum electricity load from the electrostatic precipitators of Astana city coal-fired power plant (E-CFA) and coal-fired power plant of Oskemen city (K-CFA and M-CFA). All CFA samples were used as received without preliminary washing and sieving. Prior to experiment CFA samples were homogenized and dried in oven at 70 °C for 12 h.

2.2. Methods

2.2.1. Mineralogical characterization of fly ash

XRD was used to obtain information on mineralogical phases present in CFA. The samples were analyzed as received without prior purification and sieving. The XRD pattern was recorded on a SmartLab X-ray diffraction instrument (Cu, K-β filter, 40 kV and 30 mA) with a diffraction angle of 2θ and a scanning range of 5–100° (Rigaku). Phase identification was made by searching the Auto-search option of powder diffraction file library. All analysis are duplicated and the average values are presented in this study.

XRF (PANalytical Axios) was used to get the elemental composition of each CFAs. The CFA sample was analyzed as received without prior purification and sieving. The ratio of CFA sample with commercial borate flux was set at 10 g to 3 g (total mass of 13 g) with a diameter of pellet at 5 cm and thickness of pellet at 0.5 cm. The analyses were conducted under inert atmosphere.

2.2.2. Microstructural characterization of fly ash

PSA (Malvern Mastersizer 3000) instrument was used to obtain the information on volumetric percentage of each particle size and their distribution across the range of 0.01 μm to 10 000 μm. Distilled water was used as a dispersant for all samples of CFA in Hydro-MV mode.

A Nitrogen Porosimeter (Quantachrome, Autosorb-1) was applied to obtain data on specific surface area (SSA) and porosity of the particles of fly ash. The dry sample of fly ash was first degassed for 2-3 hours prior to analysis at stepwise heating from 50°C to 200°C. A 9 mm glass cell without rod was used for all porosimetric analysis.

2.2.3. Thermal properties characterization of fly ash

TGA-DSC was used to determine the thermal properties of CFA. The temperature range applied was 50°C to 950°C with heating rate of 10°C/min under inert atmosphere. The isothermal heating was hold at 50°C for 1 minute with minimum initial mass of 10 mg.

2.2.4. FTIR characterization of fly ash

In order to determine specific functional groups present on the surface of CFA samples, they were analyzed on FTIR spectroscopy (ThermoFisher Scientific, Nicolet iS5) with attenuated total reflection (ATR) unit. This analysis required approximately 0.1 g of sample with each measurement being duplicated.

2.2.5. Chemical analysis of fly ash

Total carbon analysis of each CFA sample was conducted applying Multianalyzer N/C 3100 instrument (Analytik Jena) using 100 mg of CFA sample for analysis at furnace temperature of 950°C under pure oxygen atmosphere. All measurements were triplicated and the average values are presented in this study.

3. Results and discussion

3.1. Mineralogical characterization of fly ash

Mineralogical analysis of E-CFA, K-CFA and M-CFA on XRD give information on phase content with estimated percentage of each containing mineral. According to figures 1-3, the main minerals containing in all CFA samples are quartz, mullite and magnetite/hematite.

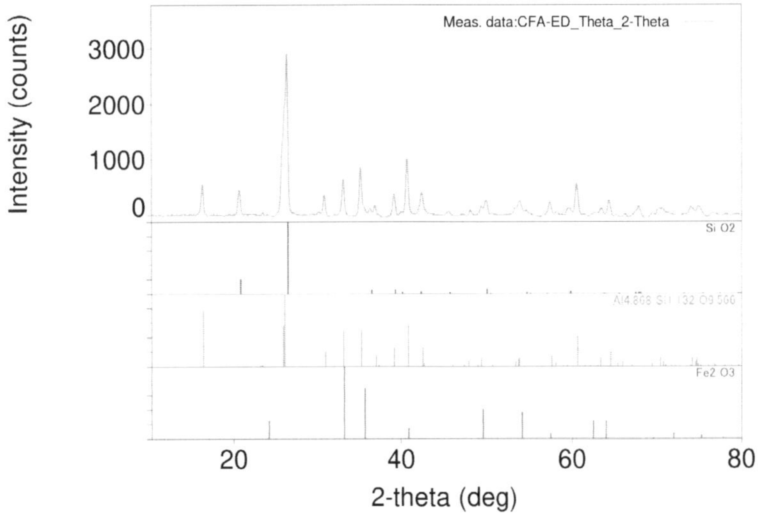

Figure 1. XRD spectrum of E-CFA.

The additional semi-quantitative tool of software allows assuming the crystalline phases in percentage. According to that calculation, the content of crystalline mullite and quartz in E-CFA are 67% and 33%, respectively. K-CFA demonstrated more even distribution of mullite and quartz: 56% and 44%, accordingly; whereas M-CFA showed 11% iron oxide (magnetite), 56% quartz and 33% mullite. It should be noted that the amount of magnetite/hematite present in CFAs was not detected on this instrument because it is mostly amorphous and only crystalline particles of this phase was used in percentile calculation of software.

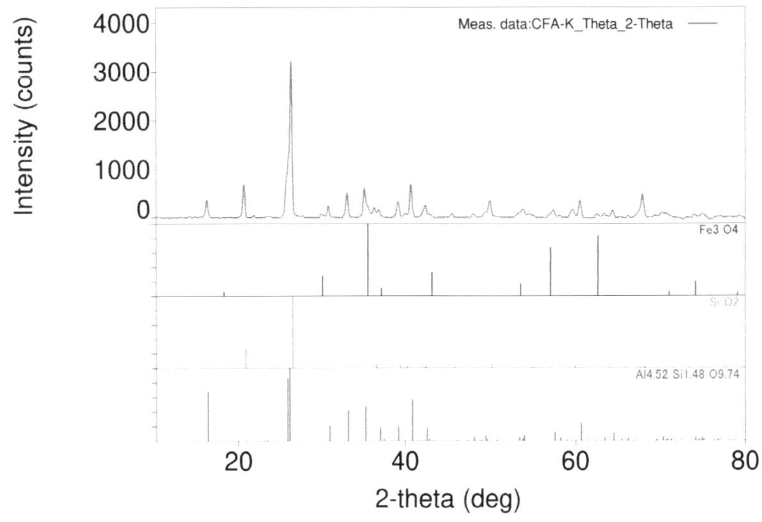

Figure 2. XRD spectrum of K-CFA.

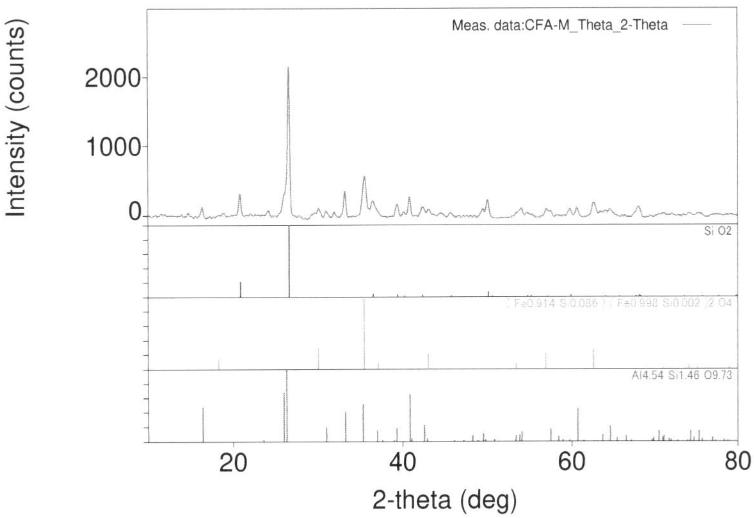

Figure 3. XRD spectrum of M-CFA.

The detailed elemental composition of E-CFA, K-CFA and M-CFA obtained on XRF are shown in table 1. According to data the main elements containing in raw CFA are Al, Si, Fe and Ca, which was expected. However, it is apparent that the content of iron oxide is comparatively higher in E-CFA in contrast with K-CFA (16.076%) and M-CFA (18.138%) samples and significantly higher than in CFA from other regions in the world [4-10], which could be of benefit for practical applications.

Table 1. Elemental composition of Kazakhstani CFAs in wt.%.

Compound	E-CFA	K-CFA	M-CFA
Na_2O	0.187	0.678	0.763
MgO	0.628	0.716	1.423
Al_2O_3	18.071	25.761	21.750
SiO_2	44.128	49.802	44.425
SO_3	0.204	0.237	2.248
Cl	0.138	0.026	0.021
K_2O	0.514	1.324	1.857
CaO	6.077	2.798	6.775
TiO_2	1.353	1.636	1.169
Cr_2O_3	0.035	0.033	0.024
MnO	0.608	0.216	0.432
Fe_2O_3	27.786	16.076	18.138
CuO	0.032	0.042	0.035
ZnO	0.013	0.052	0.024
Ga_2O_3	0.008	-	0.008
SrO	0.059	0.195	0.342
Y_2O_3	0.008	0.012	0.008
ZrO_2	0.048	0.060	-
BaO	0.103	0.143	0.369

It is interesting to note that M-CFA and E-CFA contain similar amount of Ca (6.775 and 6.077wt.%) that is more than 2 times higher as in Karazhyra CFA (2.798 wt.%). The total amount of alkaline metals are relatively higher M-CFA (11.187wt.%). All CFA samples contain some trace amount of heavy metals with concentrations below 1wt.% (Cr, Cu, Zn, Ga, Sr, Y and Zr). The Si/Al ratio of E-

CFA, K-CFA and M-CFA are 2.15, 1.70 and 1.80, respectively. These results are comparable with data obtained elsewhere [6-7,11].

3.2. Microstructural characterization of fly ash

The characteristic property of all coal derived materials are their porosity, which is important to investigate because they demonstrate a high adsorption or separation potential. The general porosimetric analysis using inert gas (argon or nitrogen) could give the following valuable data as porosity type, the SSA, total volume of pores, average pore size.

Table 2 summarizes the experimental results obtained for all CFAs samples by applying three different methods, namely BET, BJH and DFT. It should be noted that the BET method is mainly used for comparison with literature values, since this approach is generally applied for porosimetric analysis of porous materials.

It is clearly seen from the results obtained that the most porous is K-CFA, where it shows the average pore diameter of 8.457 nm with the SSA value at 32.873 m^2/g. The total pore volume for pores less than 3091 nm is 0.069 cm^3/g. This is followed by M-CFA that demonstrate very close pore size at 7.573 nm and SSA at 29.319 m^2/g. The total pore volume is lower for 0.011 cm^3/g than in K-CFA. On the other hand, E-CFA show the lowest values in all characteristics. The literature values of average micro-pores size using BET, DFT and BJH methods show that it could be as small as 0.7 nm [10] and reach up to 15.3- 27.2 nm [12, 13] and 570-780 nm for macro-pores [11, 14]. The SSA values by BET method could range from 1.1 to 15.6 m2/g depending on phase content and porosity [12-16]; while total volume could vary from 0.004 to 0.022 cm3/g [13,15].

Table 2. Porosimetric analysis of Kazakhstani CFAs.

CFA type	Average pore diameter [nm]	Specific surface area [m^2/g]	Total pore volume [cm^3/g]
	8.457 (BET)	32.873 (BET)	0.069 (BET)
K-CFA	7.573 (DFT)	19.203 (DFT)	0.044 (DFT)
	3.362 (BJH)	31.440 (BJH)	0.053 (BJH)
	7.938 (BET)	29.319 (BET)	0.058 (BET)
M-CFA	4.520 (DFT)	21.392 (DFT)	0.046 (DFT)
	4.223 (BJH)	34.073 (BJH)	0.056 (BJH)
	5.525 (BET)	23.018 (BET)	0.032 (BET)
E-CFA	6.160 (DFT)	12.935 (DFT)	0.027 (DFT)
	3.413 (BJH)	17.734 (BJH)	0.028 (BJH)

According to the results of particle size analysis shown in table 3, K-CFA and E-CFA demonstrate similar distribution with 90% of particles being less than 229 μm and 260 μm, respectively. In terms of average particle size for these two CFAs it ranges between 94-109 μm. As for M-CFA, it reveals much lower particle size being of 8.6 μm (10%) and at the same time wider distribution with 90% of particle size being under 419 μm.

Table 3. Particle size analysis of CFAs.

CFA type	Uniformity	Dv(10), μm	Dv(50), μm	Dv(90), μm
E-CFA	0.702	15.2	109	260
K-CFA	0.675	29.4	94.1	229
M-CFA	1.921	8.6	60.6	419

3.3. Thermal properties characterization of fly ash

Thermal behavior of CFAs was determined using TGA-DSC. According to results shown in figure 4, E-CFA lost 4.58% of its weight in total upon heating to 950°C, which means that the main content of CFA is inorganic compounds. The derivative weight demonstrates the highest peak of loss at 641.67°C

(-0.012 mg), and two more noticeable losses at 73.51°C (-0.006 mg) and 337.96°C (-0.003 mg). The rest of the curve shows a steady loss throughout the temperature range. A heat flow change, on the other hand, reveals a minor endothermic peak at around 790°C that might be due to reactions (decomposition and oxidation) or other types of transformations occurring while heating.

On the other hand, a distinct two-step mass loss curse for M-CFA could be observed in figure 5. According to analysis data, the first batch of loss was at 123.1°C (-0.199 mg), then the mass of sample decreased smoothly until it reached the region of temperature between at 565.8-667.25°C (-0.071 mg). The results reveal that in total the sample loss was 0.85 mg or 2.4wt.% that is almost 2 times lower than in E-CFA, which in turn means that M-CFA coal absorbed less water phase and contains less volatile organic matters in its structure. In figure 6 we can see the results for K-CFA, where the total mass loss showed the lowest value at approximately 0.28 mg or 0.57wt.%. The shape of the curve is similar to E-CFA with steady decrease by time.

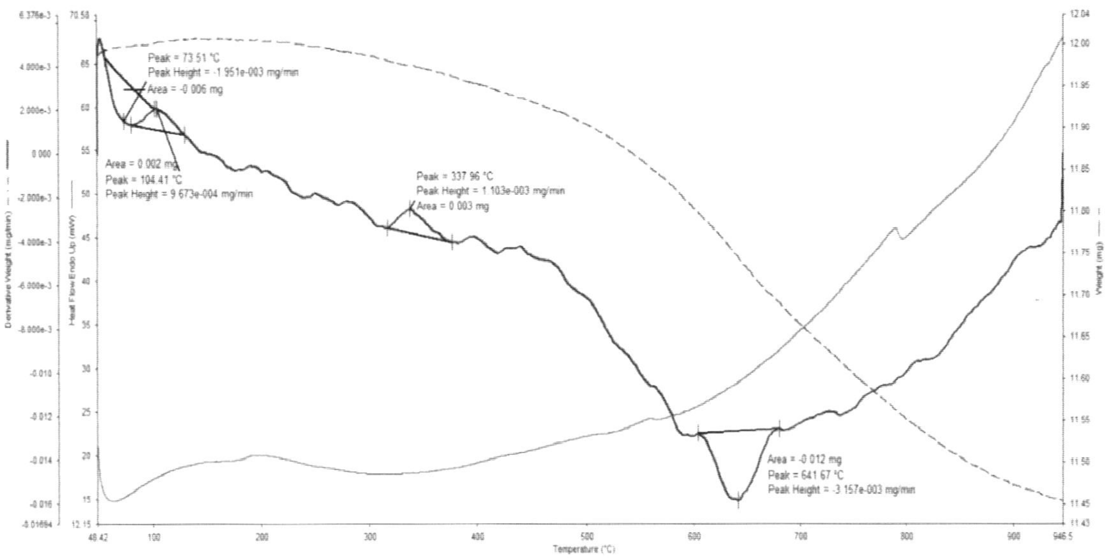

Figure 4. Thermal properties of E-CFA from Astana Power Plant.

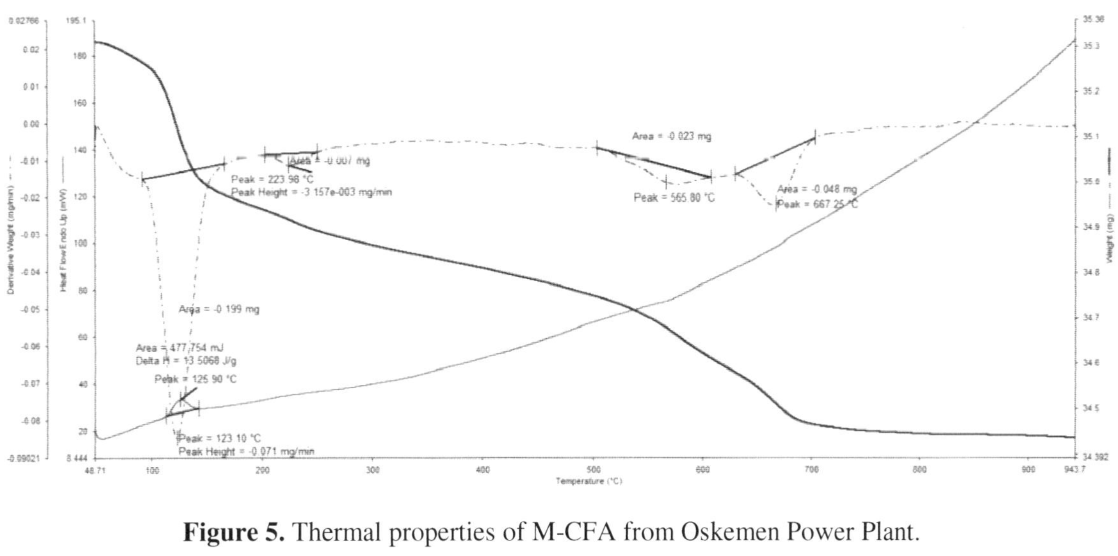

Figure 5. Thermal properties of M-CFA from Oskemen Power Plant.

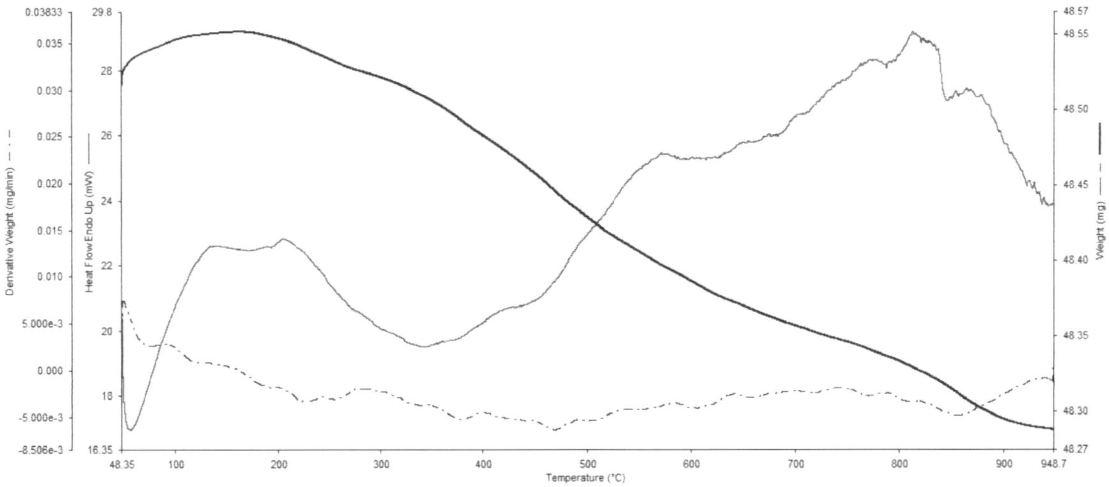

Figure 6. Thermal properties of K-CFA from Oskemen Power Plant.

3.4. FTIR characterization of fly ash

All CFA samples were characterized on FTIR spectroscopy to determine specific functional groups. Figure 7 shows the comparative spectrum of E-CFA, K-CFA and M-CFA. According to spectrums all of them have similar functional groups. For example, a characteristic peak at 1020-1050 cm-1 demonstrate the presence of Si-O- and Al-O-, while peaks at 795-775 cm^{-1} are the symmetric stretches. Similarly, there is a minor peak at 660-670 cm^{-1}, which might correspond to Al-O-, Si-O-, Fe-O-. There are also two peaks at 2160 cm^{-1} and 2025 cm^{-1} that also correspond to aluminosilicate functional groups.

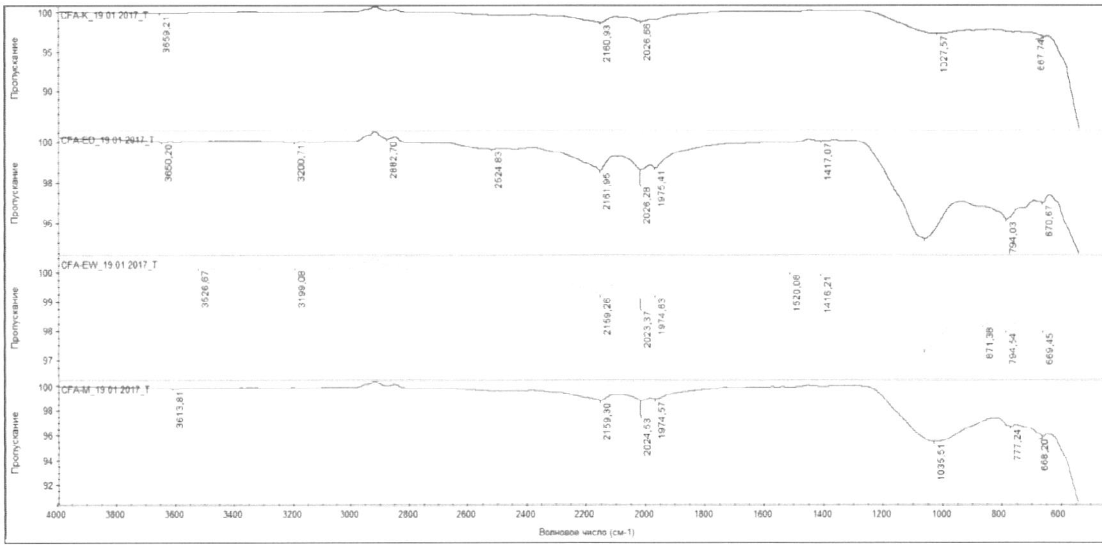

Figure 7. FTIR spectrum of all CFAs (note CFA ED and CFA EW are wet and dry E-CFAs).

3.5. Chemical analysis of fly ash

Total carbon content of all 3 CFAs was determined using HT-300 instrument. According to the results that amount of total carbon in Ekibastuz, Karazhyra and Maikube CFAs are 2.15wt.%, 0.58wt.% and 0.82wt.%. These values correspond to mass loss results in TGA/DSC analysis, where the highest and lowest loss were recoded for Ekibastuz and Karazhyra CFAs, respectively.

4. Conclusion

The E-CFA, K-CFA and M-CFA were fully characterized for mineralogy, microstructure and thermal properties. The advanced instruments such as XRD, XRF, Nitrogen Porosimeter, PSA, TGA-DSC, FTIR and TC were successfully applied to obtain valuable data on these Kazakhstani CFA, which helps to update the database. It was identified that the main phases in all three CFAs are mullite, quartz and magnetite/hematite, while the major elements containing in raw fly ash being Si, Al, Fe and alkaline metals. Porosimetric analysis revealed that the BET surface area of CFAs range between 23-32 m^2/g with mesoporous structure. The results of particle size analysis showed that 90% of particles are less than 419 μm. In addition to that the thermal analysis showed a curve of steady decrease upon heating till 950C with TC amount ranging between 0.82-2.15wt.%.

5. References

[1] http://www.kap.kz/upload/files/27144_989850_15.pdf (accessed on 14 May 2017)
[2] http://www.kap.kz/upload/files/25204_134179_15.pdf (accessed on 14 May 2017)
[3] Yao ZT , Ji XS, Sarker PK, Tang JH, Ge LQ, Xia MS and Xi YQ 2015 *Earth-Sci Rev.* **141** 105-21.
[4] Querol X, Moreno N, Umaña JC, Alastuey A, Hernández E, López-Soler A and Plana F 2002 *Int. J Coal Geol.* **50** 413-23.
[5] Mainganye D, Ojumu VT and Petrik L 2013 *Materials* 6 2074-89
[6] Wdowin M, Franus M, Panek R, Badura L and Franus W 2014 *Clean Technol. Envir.* **16** 1217-23.
[7] Cardoso AM, Paprocki A, Ferret LS, Azevedo CMN and Pires M 2015 *Fuel* **139** 59-67.
[8] Aldahri T, Behin J, Kazemian H and Rohani S 2016 *Fuel* **182** 494-501.
[9] Luo J, Zhang H and Yang J 2016 *Procedia Environ.* Sci. **31** 605-14.
[10] Ojumu TV, Du Plessis PW and Petrik LF 2016 *Ultrason. Sonochem.* **31** 342-49.
[11] Derkowski A, Franus W, Beran E and Czímerová A 2006 *Powder Technol.* **166** 47-54.
[12] Wang J, Li D, Ju F, Han L, Chang L, Bao W *Fuel Process. Technol.* **136** 96-105.
[13] Duta A and Visa M 2015 *J. Photoch. Photobio. A* **306** 21-30.
[14] Visa M and Chelaru AM 2014 *Appl. Surf. Sci.* **303** 14-22.
[15] Jha B and Singh DN 2014 *Appl. Clay Sci.* **90** 122-29.
[16] Xie J, Wang Z, Wu D and Kong H 2014 *Fuel* **116** 71-6.

Acknowledgments

This research work was supported by the internal fund for research of Nazarbayev University (grant №SOE2016002) and the Ministry of Education and Science of Kazakhstan. The authors would also like to thank the thermal power plants of Oskemen and Astana cities for generously providing with CFA samples to carry out the studies.

IOP Conf. Series: Materials Science and Engineering **230** (2017) 012047 doi:10.1088/1757-899X/230/1/012047

A Kinematic Analysis of Flat Leverage Mechanism of the Fourth Class for Manipulators

B Tultayev[1], G Balbayev[1,2] and A Zhauyt[3*]

[1]Institute of Mechanics and Engineering named after academician U. Dzholdasbekov, Almaty, Kazakhstan
[2]Almaty University of Power Engineering and Telecommunication, Almaty 050013, Kazakhstan
[3]Kazakh National Research Technical University, Almaty 050013, Kazakhstan
[*]E-mail: ali84jauit@mail.ru

Abstract. Linkwork, in its early applications, consisted mainly of revolute-connected members and was widely used for converting the continuous rotation of a water wheel into a reciprocating motion suited to piston pumps. The piston-cylinder combination at the end of the line represents a prismatic pair, of course, but ahead of this there are only the revolute connections generally associated with linkwork. Agricola's arrangements show wheel and pump-power source and point of work-fairly close together. Such compactness did not always prevail; link-works of magnificent proportions were also part of the past. A linkwork is a means of power transmission as well as being a motion transformer. Before the introduction of rope transmissions and the now universal electric wire, linkwork was employed for long-distance transmission of power. Gigantic linkages, principally for mine pumping operations, connected water wheels at the riverbank to pumps high up on the hillside.

1. Introduction

The theory of synthesis of mechanisms and cars are closely connected with a task of the kinematic analysis. The unity of tasks of the analysis and synthesis is caused by coincidence of the equations defining interdependence of variables and fixed parameters of mechanisms and used in each case in their purposes [1]. Moreover, at the solution of a problem of synthesis it is necessary to use managements of kinematic analysis as restrictions. For example [2], at the solution of a problem of synthesis of directed and transmission gears of high classes it is necessary to use the kinematics equations for definition of a rotation angle φ of the plane if Q_2 it isn't set in a case when the considered problem of synthesis of the opened OABCDEF six-bar chain [3-5]. Therefore, the vector method of the kinematic analysis is offered in this chapter which will be coordinated well with the method of synthesis of IKC stated in the previous chapter, and also it can independently be used for the solution of tasks on provisions, speeds and accelerations of the synthesizable mechanism [6-8].

2. Materials and methods

At the solution of synthesis problems of transmission mechanisms with difficult transfer function, for example, providing stops of an output link, in most cases we come to the leverage mechanism of the fourth class. This mechanism can be considered formally as a six-bar linkage with a variable length of connecting rod, than, it is obviously reached with implementation of difficult laws of the movement of an output link [1]. In this subsection the vector method of the kinematic analysis of the mechanism fourth class is offered. We work out four vector equations of isolation of mechanism circuits (see figure. 1):

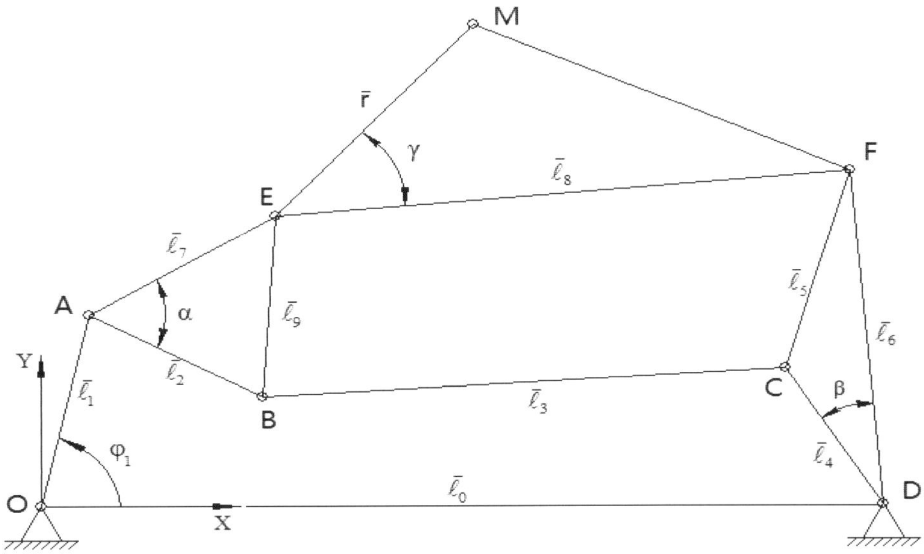

Figure 1. Kinematic scheme of flat lever mechanism of the fourth class.

$$\left.\begin{array}{l}\bar{l}_1+\bar{l}_2+\bar{l}_3+\bar{l}_4=\bar{l}_0\\ \bar{l}_1+\bar{l}_7+\bar{l}_8-\bar{l}_6=\bar{l}_0\\ \bar{l}_2+\bar{l}_9=\bar{l}_7\\ \bar{l}_4+\bar{l}_6=\bar{l}_5\end{array}\right\} \quad (1)$$

We rewrite the system Eq. (1) of rather unknown vectors in the following type:

$$\left.\begin{array}{l}\bar{l}_2+\bar{l}_3+\bar{l}_4=\bar{L}\\ \bar{l}_7+\bar{l}_8-\bar{l}_6=\bar{L}\\ \bar{l}_2+\bar{l}_9-\bar{l}_7=0\\ \bar{l}_4+\bar{l}_6-\bar{l}_5=0\end{array}\right\} \quad (2)$$

where $l_0=\overline{OO_1}$ is the vector connecting mechanism racks; $\bar{l}=l_0-l_1$ is the known vector-function of the generalized coordinate φ_1. We will choose any two from required vectors in system Eq. (2) and we spread out them on orthogonal vectors $\bar{L}$ and ($\bar{K}\times\bar{L}$). Let it be a vector $\bar{l}_2$ and $\bar{l}_4$:

$$\begin{array}{l}\bar{\ell}_2=P_2\bar{L}+q_2\left(\bar{k}\times\bar{L}\right)\\ \bar{\ell}_4=P_4\bar{L}+q_4\left(\bar{k}\times\bar{L}\right)\end{array} \quad (3)$$

where P_2, q_2, P_4, q_4 are unknown coordinates of decomposition.
The others are unknown vectors of the system Eq. (1) it is possible to express through chosen as follows. Vectors $\bar{l}_7$ and $\bar{l}_5$ are also defined by turning movement on the corresponding corners α

IOP Conf. Series: Materials Science and Engineering **230** (2017) 012047 doi:10.1088/1757-899X/230/1/012047

and β vectors $\bar{l}_2$ and $\bar{l}_4$ where α and β are also known corners between vectors $\bar{l}_2$, $\bar{l}_7$ and $\bar{l}_4, \bar{l}_7$ accordingly:

$$\begin{aligned} \bar{\ell}_7 &= P_7 \bar{L} + q_7 \left(\bar{k} \times \bar{L}\right) \\ \bar{\ell}_5 &= P_5 \bar{L} + q_5 \left(\bar{k} \times \bar{L}\right) \end{aligned} \tag{4}$$

where:

$$\left.\begin{aligned} p_7 &= \frac{l_7}{l_2}(p_2 \cos\alpha - q_2 \sin\alpha) \\ p_7 &= \frac{l_7}{l_2}(p_2 \sin\alpha - q_2 \cos\alpha) \\ p_5 &= \frac{l_5}{l_4}(p_4 \cos\beta - q_4 \sin\beta) \\ p_5 &= \frac{l_5}{l_4}(p_4 \sin\beta - q_4 \cos\beta) \end{aligned}\right\}$$

Vectors $\bar{\ell}_3$, $\bar{\ell}_6$, $\bar{\ell}_8$, $\bar{\ell}_9$ are defined from the system Eq. (1) taking into account Eq. (3-4) in the following type:

$$\left.\begin{aligned} \bar{l}_3 &= (1 - p_2 - p_4) - (q_2 + q_4)\,(\bar{k} \times \bar{L}) \\ \bar{l}_6 &= (p_5 - p_4)\bar{L} + (q_5 - q_4)\,(\bar{k} \times \bar{L}) \\ \bar{l}_8 &= (1 + p_5 - p_4 - p_7)\bar{L} + (q_5 - q_4 - q_7)\,(\bar{k} \times \bar{L}) \\ \bar{l}_9 &= (p_7 - p_2)\bar{L} + (q_7 - q_2)\,(\bar{k} \times \bar{L}) \end{aligned}\right\} \tag{5}$$

In the systems Eq. (3-5) P_2, P_4, q_2, q_4 are unknown. For the determination of these unknown parameters we square both parts of the equations of the system Eq. (4), and also the first and third equations of the system Eq. (5) as a result we receive system of the equations with four unknown P_2, P_4, q_2, q_4:

$$\left.\begin{aligned} &(p_2^2 + q_2^2)L^2 = l_2^2 \\ &(p_3^2 + q_3^2)L^2 = l_4^2 \\ &\left[(1 - p_4 - p_2)^2 + (q_2 + q_4)^2\right]L^2 = l_6^2 \\ &\left[(1 + p_5 - p_4 - p_7)^2 + (q_5 - q_4 - q_7)^2\right]L^2 = l_8^2 \end{aligned}\right\} \tag{6}$$

Set of roots of the system Eq. (6) define number of assemblies of the mechanism [2]. The system of the equations Eq. (6) can be reduced to two equations relatively q_2 and q_4. For this purpose from the first and second equation of the system Eq. (6) parameters p_2 and p_4 we also express through q_2 and q_4:

$$\left.\begin{aligned} p_2 &= \pm\left[\sqrt{\frac{l_2^2}{L^2} - q_2^2}\,\right]^{\frac{1}{2}} \\ p_4 &= \pm\left[\sqrt{\frac{l_4^2}{L^2} - q_4^2}\,\right]^{\frac{1}{2}} \end{aligned}\right\} \tag{7}$$

The last two equations of the system Eq. (6) taking into account Eq. (4) and Eq. (7) take the form:

$$\left.\begin{aligned} (1 - p_2 - p_4)^2 + (q_2 + q_4)^2 &= \frac{l_3^2}{L^2} \\ [a_{11}p_4 + a_{12}q_4 + a_{13}p_2 + a_{14}q_2 + 1]^2 + [-a_{12}p_4 + a_{11}q_4 - a_{14}p_2 + a_{13}q_2]^2 &= \frac{l_8^2}{L^2} \end{aligned}\right\} \tag{8}$$

where $a_{11} = \frac{\ell_5}{\ell_4} cos\,\beta - 1,\quad a_{13} = -\frac{\ell_7}{\ell_2} cos\,\alpha,\quad a_{12} = -\frac{\ell_5}{\ell_4} sin\beta,\quad a_{14} = \frac{\ell_7}{\ell_2}\sin\alpha.$

The signs plus and minus in the system Eq. (7) correspond to different assemblies of the mechanism. From decisions Eq. (8) taking into account Eq. (7) we determine unknown parameters P_2, P_4, q_2, q_4. As initial approximations for the solution of system Eq. (7) projections of required vectors on orthogonal vector $\vec{L}$ and ($\overline{K} \times \overline{L}$) for the set conditions of synthesis position of the mechanism [3]. We designate cross headings of rectangular system of coordinates $Oxyz$ through $\vec{i}$, $\vec{j}$, $\vec{k}$. Then position functions of output links of the mechanism are also found from scalar products of cross headings of the system of coordinates $Oxyz$ and found vectors:

$$\left.\begin{aligned}
tg\varphi_2 &= \frac{\overline{l}_2\overline{j}}{\overline{l}_2\overline{i}} = \frac{p_2L_y + q_2L_x}{p_2L_x - q_2L_y} \\
tg\varphi_3 &= \frac{\overline{l}_3\overline{j}}{\overline{l}_3\overline{i}} = \frac{(1 - p_2 - p_4)L_y - (q_2 + q_4)L_x}{(1 - p_2 - p_4)\overline{L}_x + (q_2 + q_4)\overline{L}_y} \\
tg\varphi_4 &= \frac{\overline{l}_4\overline{j}}{\overline{l}_4\overline{i}} = \frac{p_4L_y + q_4L_x}{p_4L_x - q_4L_y} \\
tg\varphi_5 &= \frac{\overline{l}_5\overline{j}}{\overline{l}_5\overline{i}} = \frac{p_5L_y + q_5L_x}{p_5L_x - q_5L_y} \\
tg\varphi_6 &= \frac{\overline{l}_6\overline{j}}{\overline{l}_6\overline{i}} = \frac{(p_5 - p_4)L_y + (q_5 - q_4)L_x}{(p_5 - p_4)\overline{L}_x - (q_5 - q_4)\overline{L}_y} \\
tg\varphi_7 &= \frac{\overline{l}_7\overline{j}}{\overline{l}_7\overline{i}} = \frac{p_7L_y + q_7L_x}{p_7L_x - q_7L_y} \\
tg\varphi_8 &= \frac{\overline{l}_8\overline{j}}{\overline{l}_8\overline{i}} = \frac{(1 + p_5 - p_4 - p_7)L_y + (q_5 - q_4 - q_7)L_x}{(1 + p_5 - p_4 - p_7)L_x - (q_5 - q_4 - q_7)L_y} \\
tg\varphi_9 &= \frac{\overline{l}_9\overline{j}}{\overline{l}_9\overline{i}} = \frac{(p_7 - p_2)L_y + (q_7 - q_2)L_x}{(p_7 - p_2)L_x - (q_7 - q_2)L_y}
\end{aligned}\right\} \tag{9}$$

where: $L_x = l_{Ox} - l_1\cos\varphi_1$, $L_y = l_{Oy} - l_1\sin\varphi_1$.

The analogousness of formulas of the system Eq. (9) allows calculating functions of situation on the uniform algorithm [4]. We pass to definition of analogs of speeds. For this purpose we differentiate the first two equations of the system Eq. (1) on coordinate φ_1 taking into account $\varphi_2' = \varphi_7'$, $\varphi_6' = \varphi_4'$ we receive:

$$\begin{aligned} \bar{l}_2\varphi_2' + \bar{l}_3\varphi_3' + \bar{l}_4\varphi_4' &= -l_1' \\ \bar{l}_7\varphi_2' + \bar{l}_8\varphi_8' - \bar{l}_6\varphi_4' &= -l_1' \end{aligned} \tag{10}$$

Let's multiply salary the first equation on $\bar{k} \times \bar{l}_3$ the second on $\bar{k} \times \bar{l}_8$:

$$\begin{aligned} \left[\bar{l}_2\ \bar{k}\ \bar{l}_8\right]\varphi_2' + \left[\bar{l}_4\ \bar{k}\ \bar{l}_3\right]\varphi_4' &= -\left[\bar{l}_1\ \bar{k}\ \bar{l}_3\right] \\ \left[\bar{l}_7\ \bar{k}\ \bar{l}_8\right]\varphi_2' - \left[\bar{l}_6\ \bar{k}\ \bar{l}_8\right]\varphi_4' &= -\left[\bar{l}_1\ \bar{k}\ \bar{l}_8\right] \end{aligned} \tag{11}$$

The solution of the system Eq. (11) can be written down in the form of determinants from the mixed vector products:

$$\begin{aligned} \varphi_2' &= -\frac{1}{\Delta}\begin{vmatrix} \left[\bar{l}_3\ \bar{k}\ \bar{l}_1\right] & \left[\bar{l}_4\ \bar{k}\ \bar{l}_3\right] \\ \left[\bar{l}_1\ \bar{k}\ \bar{l}_8\right] & \left[\bar{l}_6\ \bar{k}\ \bar{l}_8\right] \end{vmatrix} \\ \varphi_4' &= -\frac{1}{\Delta}\begin{vmatrix} \left[\bar{l}_2\ \bar{k}\ \bar{l}_3\right] & \left[\bar{l}_3\ \bar{k}\ \bar{l}_1\right] \\ \left[\bar{l}_7\ \bar{k}\ \bar{l}_8\right] & \left[\bar{l}_8\ \bar{k}\ \bar{l}_6\right] \end{vmatrix} \end{aligned} \tag{12}$$

where:

$$\Delta = \begin{vmatrix} \left[\bar{l}_2\ \bar{k}\ \bar{l}_8\right] & \left[\bar{l}_4\ \bar{k}\ \bar{l}_3\right] \\ \left[\bar{l}_7\ \bar{k}\ \bar{l}_8\right] & \left[\bar{l}_8\ \bar{k}\ \bar{l}_6\right] \end{vmatrix}$$

The other transfer functions we find from the system Eq. (10):

$$\left.\begin{aligned} \varphi_3' &= -\frac{1}{l_3^2}\left[\left(\bar{l}_1\ \bar{l}_3\right) + \left(\bar{l}_2\ \bar{l}_3\right)\varphi_2' + \left(\bar{l}_3\ \bar{l}_4\right)\varphi_4'\right] \\ \varphi_8' &= \frac{1}{l_8^2}\left[\left(\bar{l}_6\ \bar{l}_8\right)\varphi_4' - \left(\bar{l}_7\ \bar{l}_8\right)\varphi_2' - \left(\bar{l}_1\ \bar{l}_8\right)\right] \end{aligned}\right\} \tag{13}$$

For definition of analogs of speedups of links we differentiate Eq. (10) on φ_1:

$$\left.\begin{aligned} \bar{l}_2\varphi_2'' + \bar{l}_3\varphi_3'' + \bar{l}_4\varphi_4'' &= -\bar{L}_1 \\ \bar{l}_7\varphi_2'' + \bar{l}_8\varphi_8'' - \bar{l}_6\varphi_4'' &= -\bar{L}_2 \end{aligned}\right\} \tag{14}$$

Here vectors $\bar{L}_1$ and $\bar{L}_2$ is also determined by the calculated analogs of speeds as follows:

$$\left.\begin{aligned} \bar{L}_1 &= \bar{k}\times\bar{l}_1 + \bar{k}\times\bar{l}_2\varphi_2'^2 + \bar{k}\times\bar{l}_3\varphi_3'^2 + \bar{k}\times l_4'\varphi_4'^2 \\ \bar{L}_2 &= \bar{k}\times\bar{l}_1 + \bar{k}\times\bar{l}_7\varphi_2'^2 + \bar{k}\times\bar{l}_8\varphi_8'^2 - \bar{k}\times l_6'\varphi_4'^2 \end{aligned}\right\} \tag{15}$$

The system Eq. (13) is by the form similar to the system Eq. (5) therefore the solution of the last system can be written down in the following type:

$$\left.\begin{aligned}
\varphi_2'' &= -\frac{1}{\Delta}\left\|\begin{matrix}[\bar{l}_3\ \bar{k}\ \bar{L}_1] & [\bar{l}_4\ \bar{k}\ \bar{l}_3] \\ [\bar{L}_2\ \bar{k}\ \bar{l}_8] & [\bar{l}_6\ \bar{k}\ \bar{l}_8]\end{matrix}\right\| \\
\varphi_3'' &= -\frac{1}{l_3^2}\left[(\bar{L}_1\ \bar{l}_3)+(\bar{l}_2\ \bar{l}_3)\varphi_2''+(\bar{l}_3\ \bar{l}_4)\varphi_1''\right] \\
\varphi_4'' &= \frac{1}{\Delta}\left\|\begin{matrix}[\bar{l}_2\ \bar{k}\ \bar{l}_3] & [\bar{l}_3\ \bar{k}\ \bar{L}_1] \\ [\bar{l}_7\ \bar{k}\ \bar{l}_8] & [\bar{l}_8\ \bar{k}\ \bar{L}_2]\end{matrix}\right\| \\
\varphi_8'' &= \frac{1}{l_8^2}\left[(\bar{l}_6\ \bar{l}_8)\varphi_4''-(\bar{l}_7\ \bar{l}_8)\varphi_2''-(\bar{L}_2\ \bar{l}_8)\right]
\end{aligned}\right\} \quad (16)$$

On the basis of the derived formulas the algorithm of the task solution of the kinematic analysis of mechanisms of fourth class is made [5]. The replacement of the rotary guide link on the onward doesn't make essential changes to formulas of the kinematic analysis. It is possible to determine by the known vector-functions Eq. (4) trajectories, speeds and speedups of any points of links of the mechanism. For example, for coordinates and speeds of the point M of the link 7 it is easy to write down:

$$\begin{aligned}
X_M &= l_1\cos\varphi_1 + l_7\cos\varphi_7 + r\cos(\gamma+\varphi_8) \\
Y_M &= l_1\sin\varphi_1 + l_7\sin\varphi_7 + r\sin(\gamma+\varphi_8)
\end{aligned} \quad (17)$$

$$\begin{aligned}
X_M' &= -l_1\sin\varphi_1 - l_7\sin\varphi_7\cdot\varphi_7' - r\sin(\gamma+\varphi_8)\cdot\varphi_8' \\
Y_M' &= l_1\cos\varphi_1 + l_7\cos\varphi_7\cdot\varphi_7' + r\cos(\gamma+\varphi_8)\cdot\varphi_8'
\end{aligned} \quad (18)$$

This technique was used at kinematic study of the synthesized transmission mechanisms of the fourth class.

3. Results and discussion

A proper dynamic model has been developed within GIM software to provide information on the feasibility of the proposed design solution. Simulation tests have been carried out and results are discussed for validating the proposed design and characterizing its operation (see figures. 2-3).

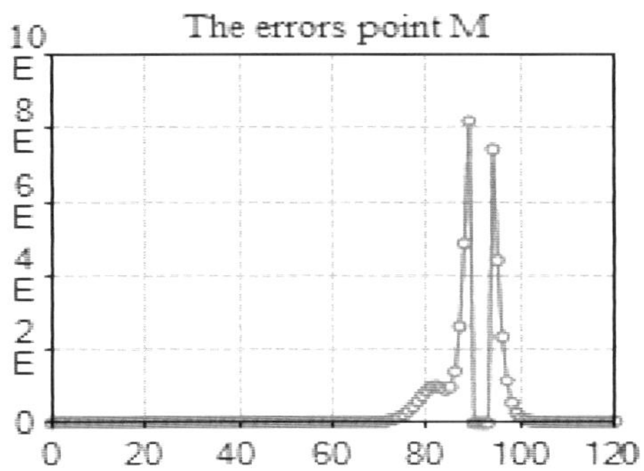

Figure 2. Computed plot of the errors point M.

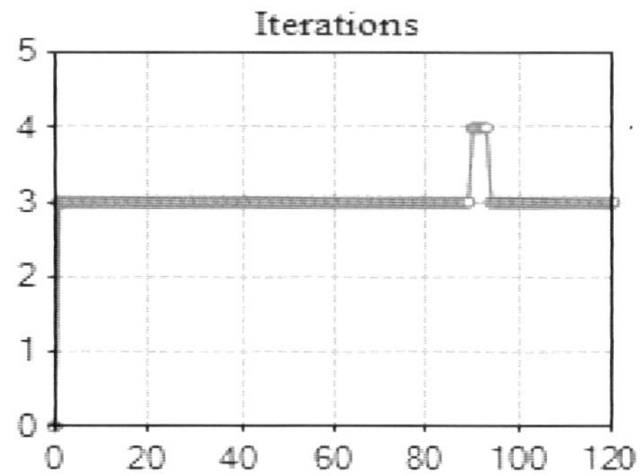

Figure 3. Computed plot of the iterations point M.

4. Conclusions

As practice has shown the researcher can make the procedure of drawing of mechanical system for this purpose it is necessary to calculate coordinates of characteristic points and in them to display the corresponding elements. The set of standard procedures for receiving Karina's animation of mechanisms movement, some of them has been created for this purpose, necessary for the solution of an objective are given above. For engineers and researchers the created program has great practical value by designing directing lever mechanisms i.e. on set at the law movement point within several seconds it is possible to receive the real mechanism with animation.

5. References

[1] Lesser M 1992 A geometrical interpretation of Kane's equations *Mathematical and Physical Sciences* **436** 69–87

[2] Liu X 2008 A Lie group formulation of Kane's equations for multibody systems *Multibody Syst. Dyn* **20** 29–49

[3] Zhauyt A, Zhahanova I, Smailova G, Murzakhmetova U, Kurmangalieva L and Kurenbay A 2017 The kinematic analysis of the third class mechanism. *Vibroengineering PROCEDIA* **12** 208-212

[4] Zhauyt A, Alipov K, Sakenova A, Zhankeldi A, Abdirova R and Abilkaiyr Zh 2016 The synthesis of four-bar mechanism *Vibroengineering PROCEDIA* **10** 486-491

[5] Musayev J and Zhauyt A 2015 Analysis of disturbing influence of traffic load on soil body *Advances in Materials Science and Engineering* **2015** 1-7

[6] Askarov E, Zhauyt A, Abilkaiyr Zh, Zhankeldi A and Naurushev B 2017 A new type cam-screw mechanical press *Engineering for rural development* **16** 36-41

[7] Zhauyt A, Mamatova G, Alipov K, Sakenova A and Abdirova R 2016 The kinematic analysis of flat lever mechanisms with application of vector calculation *Vibroengineering PROCEDIA* **8** 1-5

[8] Zhauyt A 2015 The substantiating of the dynamic parameters of the shaking conveyor mechanism *Vibroengineering PROCEDIA* **5** 15-20

Effect of Cooling Rate on Microstructure of Two Kinds of High Nb Containing Tial Alloys

L H Chai, Z Y Feng, Z L Xiang, Y S Cui, F Zhou and Z Y Chen

Beijing University of Technology, Beijing, China
E-mail: lh.chai@163.com

Abstract. In this paper, high Nb-TiAl alloys with Cr and W additions were prepared by Vacuum induction melting method, and then were heat treated under three different cooling rates of slow cooling, furnace cooling and air cooling. The phase composition of the alloy was analyzed by X ray diffraction, and the microstructure of the alloy was observed by optical microscope (OM), scanning electron microscope (SEM) and energy dispersive analyzer. The results show that the microstructure of Ti45Al8Nb0.2Cr and Ti45Al8Nb0.2W are fully lamellar structure with the main phase composition of α+γ after 3 different heat treatment conditions. The grain size of the two alloys decreases with decreasing of cooling rate, and the grain size of the alloyed with Cr alloy is smaller than that of the alloyed with W alloy. Most of the original massive β phase at grain boundaries and lamellar interfaces dissolved after heat treatment, and the transformation of β phase is easier for Ti45Al8Nb0.2Cr.

1. Introduction

TiAl alloys are attractive for application on aircraft and automotive for their excellent high temperature specific strength, especially the alloy with high Nb content[1-2]. β phase stabilized elements such as W, Mo, Cr, Ta were used to improve the mechanical properties[3-4]. β phase segregation on the grain boundaries was thought to be unbenefited to the mechanical properties of high Nb containing TiAl alloys due to the brittle characters of β(B2) phase at room temperature and even lower the brittle-to-ductile transition temperature. It has been observed that microcracks at interface of β and matrix occurring after tensile test. And β also reduce the creep resistance at higher temperature[5-8]. It has been reported that heat treatment in the α+γ phase field, such as Ti-45Al-7Nb-0.4W-0.15B (at%) alloy heated in the temperature of 1220-1260°C for 8-36h can eliminate the β phase. Increasing the heat treatment temperature will reduce the holding time, and impede grain coarsening [9]. In the earlier work, it has been found that the transformation of β phase was related with the cooling rate[10]. So it is necessary to study the dynamics of β phase transformation. In the present work, two alloys with compositions of Ti45Al8Nb0.2Cr and Ti45Al8Nb0.2W (at%) were used to investigate the effect of cooling rate on β phase dissolution during heat treatment in α+γ phase field.

2. Experimental

Two alloy ingots with nominal compositions of Ti-45Al-8Nb-0.2Cr and Ti-45Al-8Nb-0.2Cr (at.%) were prepared by vacuum induction skull melting furnace. Then small pieces with dimension of 10mm*10mm*10mm were cut of the ingot for heat treatment. The specimens were heat treated under vacuum at the temperature of 1300°C (about 60°C below the Tα), which is in the α+γ phase field, held for 10min then cooled at different cooling rates. Three cooling rates were selected, such as controlled slow cooling in furnace (about 5-6°C/min), furnace cooling (about 20 °C/min), and air cooling (100-200 °C/min). The X-ray diffraction measurements were performed on a Bruker AXS D/MAX-3C

IOP Conf. Series: Materials Science and Engineering **230** (2017) 012048 doi:10.1088/1757-899X/230/1/012048

using Cu-Kα radiation. The SEM measurement was carried out on FEI QUANTA FEG 650 equipped with Energy Dispersive X-ray Detector(EDX). The specimens for SEM observation were electron-polished in a solution of methanol, 2-Butoxyethanol and perchloric acid at room temperature and 40 kV for 15 s.

3. Results and discussion

3.1. Phase constitution of Ti45Al8Nb0.2Cr alloy and Ti45Al8Nb0.2W alloy

The XRD patterns of Ti45Al8Nb0.2Cr and Ti45Al8Nb0.2W alloys at different cooling rates are shown in figure 1, which indicates that the main phases of the two alloys are γ and α_2 phase, and the content of α_2 phase decreases with decreasing cooling rates. The reason is that α_2 phase transformed to γ phase directly and the eutectoid reaction $\alpha \rightarrow \alpha_2 + \gamma$ was inhibited at higher cooling rate, thus the content of α_2 phase in the alloy at air cooling condition is lower than that of the alloy at furnace cool and the alloy at controlled slow cooling rate shows the highest owing to the sufficient transformation of α_2 phase.

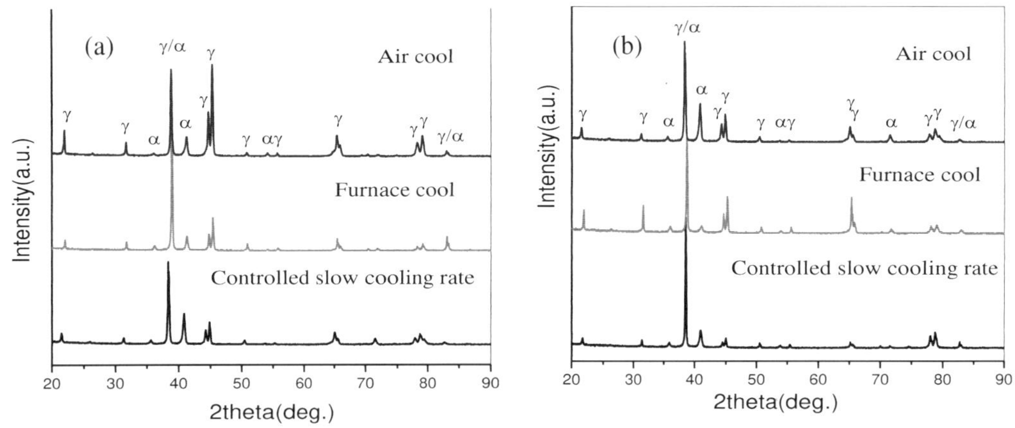

Figure 1. XRD patterns of two alloys at different cooling rates, (a) Ti45Al8Nb0.2Cr, (b) Ti45Al8Nb0.2W

3.2. Microstructure of Ti45Al8Nb0.2Cr alloy

The microstructure of Ti45Al8Nb0.2Cr alloy after different cooling rate is shown in figure 2 and figure 3. The microstructure of casting ingot also shows here for comparison, which presents near lamellar microstructure with massive β phase located at the grain boundaries and the α/γ interphase. After heat treatment the alloys show near fully lamellar microstructure, and the grain size increases with increasing the cooling rate. However the lamellar spacing decreases with increasing the cooling rate. The alloy after air cool shows massive α_2 phase at grain boundaries. EDX analysis (Table 1.) shows that the light gray contrast area contains less Al content and higher Nb content, but the dark area contains higher Al content and less Nb content, which are α_2 phase and γ phase respectively based on the result of XRD. The partition coefficient of Nb between α and γ phases is corresponding to the results of R. Kainuma[11]. After heat treatment the β phase transformed to α_2 and γ phase.

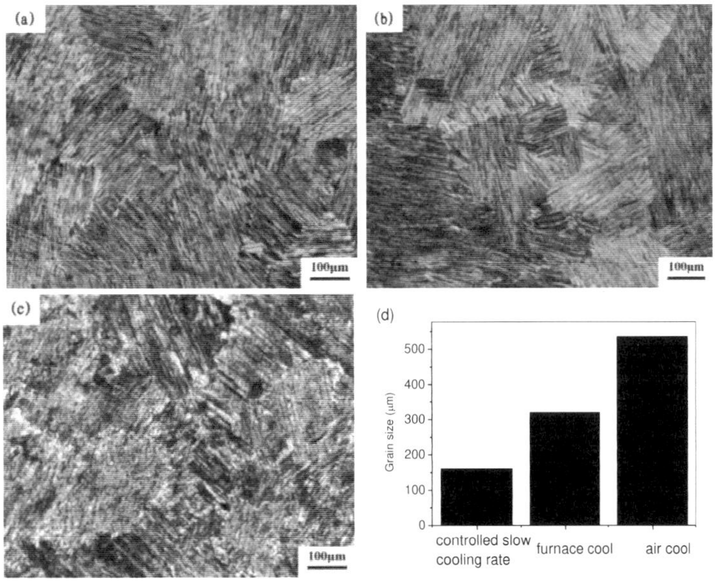

Figure 2. Optical images of Ti45Al8Nb0.2Cr alloy at different cooling rates (a) controlled slow cooling rate, (b) furnace cool, (c) air cool, (d) grain size vs cooling condition.

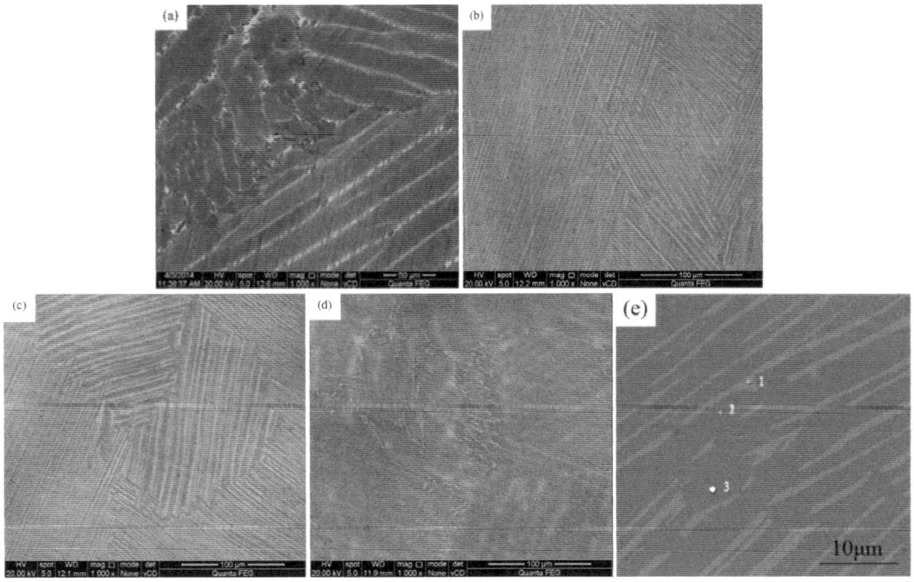

Figure 3. SEM images of Ti45Al8Nb0.2Cr alloy at different cooling rates (a) casting ingot, (b)controlled slow cooling rate, (c)furnace cool, (d) air cool, (e) locations for EDX analysis of the alloy under furnace cool.

Table 1. EDX analysis of Ti45Al8Nb0.2Cr alloy under furnace cool condition.

Test Locations	Element Content (at%)			
	Ti	Al	Nb	Cr
1	52.79	38.50	8.46	0.25
2	45.95	45.61	8.44	
3	45.78	46.02	8.01	

3.3. Microstructure of Ti45Al8Nb0.2W alloy

The microstructure of Ti45Al8Nb0.2W alloy are shown in figure 4, figure 5 and Table 2. Similar to Ti45Al8Nb0.2Cr alloy, the Ti45Al8Nb0.2W alloy also has near lamellar microstructure, with grain size decreasing as decreasing the cooling rate. The main phases are α_2+γ. But some white contrast phases are found at the grain boundaries, especially in the alloy under air cool condition, which contains less Al content and higher Nb and W content by EDX analysis. They are untransformed β phase comparing to the microstructure of casting ingot.

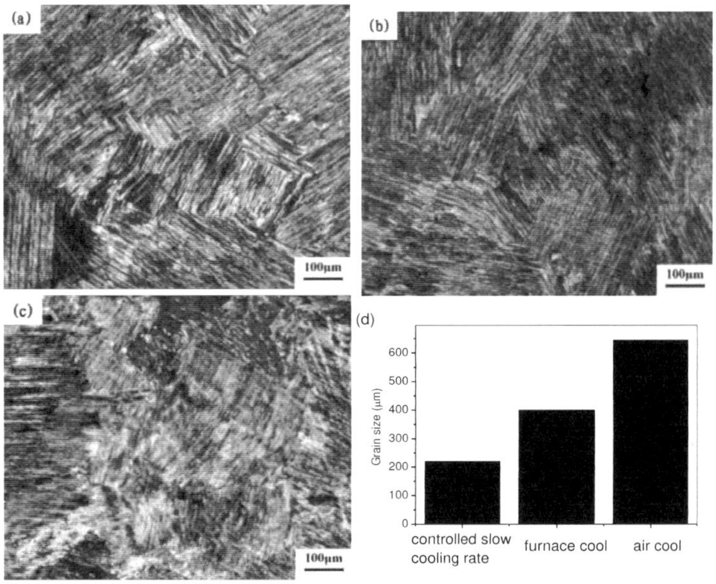

Figure 4. Optical images of Ti45Al8Nb0.2Cr alloy at different cooling rates (a) controlled slow cooling rate, (b) furnace cool, (c) air cool, (d) grain size vs cooling condition.

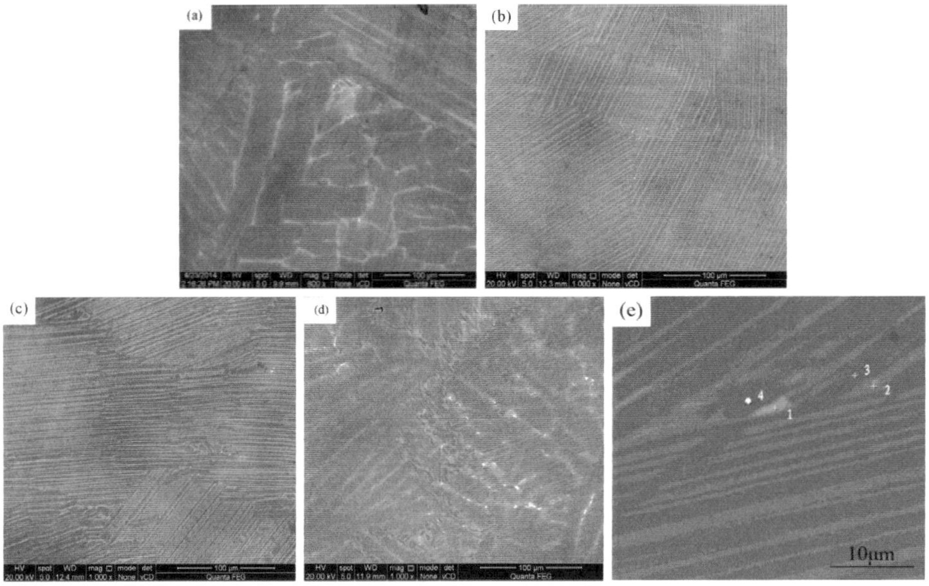

Figure 5. SEM images of Ti45Al8Nb0.2W alloy at different cooling rates (a) casting ingot, (b)controlled slow cooling rate, (c)furnace cool, (d) air cool, (e) locations for EDX analysis of the alloy under furnace cool.

Table 2. EDX analysis of Ti45Al8Nb0.2W alloy undr furnace cool.

Test Locations	Element Content (at%)			
	Ti	Al	Nb	W
1	50.45	38.22	10.47	0.86
2	49.66	41.20	9.13	-
3	45.06	45.84	8.73	0.37
4	44.62	46.15	8.88	0.35

Comparing the two alloys, the grain size of Ti45Al8Nb0.2Cr alloy is smaller and the transformation of β phase is easier than that of Ti45Al8Nb0.2W alloy. The reason is maybe that the diffusion of W is difficult, the eutectoid reaction of $\alpha \rightarrow \alpha_2 + \gamma$ and the reaction of $\beta \rightarrow \alpha + \gamma$ needs more time, so the nucleation and growth of subgrain of $\alpha_2 + \gamma$ is difficult at the same condition.
The above results indicate that heat treatment in the $\alpha + \gamma$ area for short time can dissolve the original massive β phase of the two alloys, but the transformation extent depends on the cooling rate. The dissolution of β phase is a diffusion controlled transformation. Holding at the higher temperature gives the driving force to break through the energy barrier for elements diffusion. The accomplishment of the diffusion is occurred during the cooling process. The element has sufficient time for diffusion at controlled cooling rate, so the β phase transformed completely, but at the higher cooling condition, such as the air cool, only part of β phase transformed to α and γ phase. So heating at higher temperature and slow cooling can effectively dissolve the β phase.

4. Conclusions

The microstructures of Ti45Al8Nb0.2Cr and Ti45Al8Nb0.2W alloy cooling from the $\alpha + \gamma$ phase field at three different cooling rates were investigated. The main conclusions are drawn below.
The main microstructure of Ti45Al8Nb0.2Cr and Ti45Al8Nb0.2W alloys are fully lamellar structure, with grain size decreases as decreasing the cooling rate.
Comparing the two alloys, Ti45Al8Nb0.2Cr shows smaller grain size and than that of Ti45Al8Nb0.2W alloy.
In the two alloys, most of β phases transformed to α and γ phases except the Ti45Al8Nb0.2W alloy at air cool condition. And the dissolution of β phase is easier in Ti45Al8Nb0.2Cr than that of Ti45Al8Nb0.2W alloy.

References

[1] Kim S, Hong J and Na Y 2014 *Mater. Des.* **54** 814
[2] Pflumm R, Friedle S and Schütze M 2015 *Intermetallics.* **56** 1
[3] Liu Z C, Lin P J and Chen G L 2001 *Trans. Mater. Heat. Treat.* **22** 8
[4] Dong L M, Yang R 2003 *J. Mat. Sci. Technol.* **19(4)** 330
[5] Wang J G and Nieh T G 2000 *Intermetallics.* **8** 737
[6] Herrouin F, Hu D, Bowen P and Jones I P 1998 *Acta Mater.* **46**, 4963
[7] Hamzah E, Kanniah M and Harun M 2009 *Mech Adv Mater Struc.* **16** 384
[8] Xu X J et al. 2006 *J. Alloys Compd.* **414** 131
[9] Zhang W, Liu Y and Huang J S 2009 *Rare Metal Mat. Eng.* **38** 1711
[10] Chai L H, Yang L, Zhang J P, Zhang Z Y, Zhang L Q and Lin J P 2013 *Mater. Sci. Forum.* **747-748** 38
[11] Kainuma R, Fujita Y, Mitsui H, Ohnuma I and Ishida K 2000 *Intermetallics.* **8**, 855

Acknowledgments

This work is financially supported by National Natural Science Foundation of China (No. 51301005)

IOP Conf. Series: Materials Science and Engineering **230** (2017) 012049 doi:10.1088/1757-899X/230/1/012049

Tool Wear Feature Extraction Based on Hilbert Marginal Spectrum

Shan Guan[1,a], Weijie Song[1] and Hongyang Pang[1]
[1]School of Mechanical Engineering, Northeast Electric Power University, Jilin 132012, China

Abstract. In the metal cutting process, the signal contains a wealth of tool wear state information. A tool wear signal's analysis and feature extraction method based on Hilbert marginal spectrum is proposed. Firstly, the tool wear signal was decomposed by empirical mode decomposition algorithm and the intrinsic mode functions including the main information were screened out by the correlation coefficient and the variance contribution rate. Secondly, Hilbert transform was performed on the main intrinsic mode functions. Hilbert time-frequency spectrum and Hilbert marginal spectrum were obtained by Hilbert transform. Finally, Amplitude domain indexes were extracted on the basis of the Hilbert marginal spectrum and they structured recognition feature vector of tool wear state. The research results show that the extracted features can effectively characterize the different wear state of the tool, which provides a basis for monitoring tool wear condition.

1. Introduction
In the metal cutting process, the tool wear state has great influence on machining quality of the workpiece. Therefore, it is of great significance to grasp the tool wear condition in time to improve the machining quality of the workpiece and control the machining cost of the parts. Tool wear is a complex phenomenon in the metal cutting process with different variety of the way, such as flank, rake, boundary wear [1]. It is related to the characteristics of the cutting tool, workpiece material, cutting parameters and so on. Acoustic emission (AE) signal is one of the most effective methods to monitor tool wear. AE signal is a kind of instantaneous elastic wave phenomenon, which is generated by the internal *or* local fast release of the material and directly derived from the tool. So it contains a wealth of tool wear state information.

In this paper, the author collected the acoustic emission signals of different wear stages in the cutting process and used the Hilbert Huang transform (HHT) and Hilbert marginal spectrum to analyze the acoustic emission signals. Present method extracted amplitude domain index on the basis of Hilbert marginal spectrum and used statistical method and the two-dimensional scatterplot to screen five indicators including: skewness, kurtosis, crest factor, margin index, pulse index. They form a feature vector which describes the tool wear state and provides some basis for tool wear condition monitoring and identification.

2. The experiment system and experimental method
The experimental program is shown in figure 1. Experiments were carried out on a C616 lathe. The workpiece material is T10 Carbon Tool Steel. The cutting tool is YT15 Carbide Tool. The experimental used R15-ALPHA resonant acoustic emission sensor which center frequency is 150kHz and preamplifier which bandwidth is 20kHz ~ 1.2MHz as well as Gain is 40dB.The Experiment select PXI - 6366 acoustic emission acquisition card. The Sampling frequency is 2MHz.

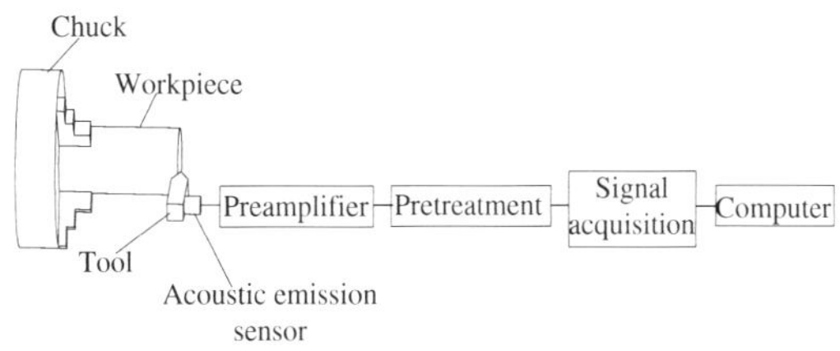

Figure 1. Experimental system design

In the process of tool wear, the tool wear condition is judged according to the wear of tool flank (VB) [2]. Wear loss in 0.15mm to 0 mm is defined as the initial wear. Wear loss in 0.15 mm to 0.3 mm is defined as medium wear. Wear loss greater than 0.3mm is defined as severe wear which is considered to be invalid.

In the cutting process, the cutting parameters which affect the cutting tool wear state are: cutting speed, feed rate and cutting depth. In order to collect a comprehensive and accurate tool wear acoustic emission signal, the orthogonal test method is used to carry out a comprehensive combination of the three main cutting parameters, and a total of 27 different combinations of cutting parameters. The experimental method is shown in figure 2.

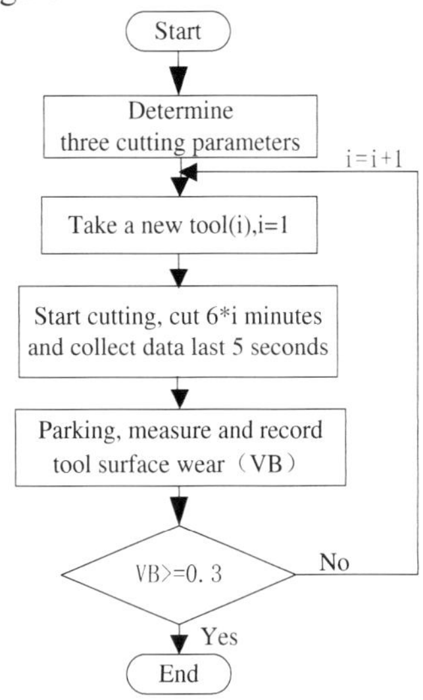

Figure 2. Experimental flow chart

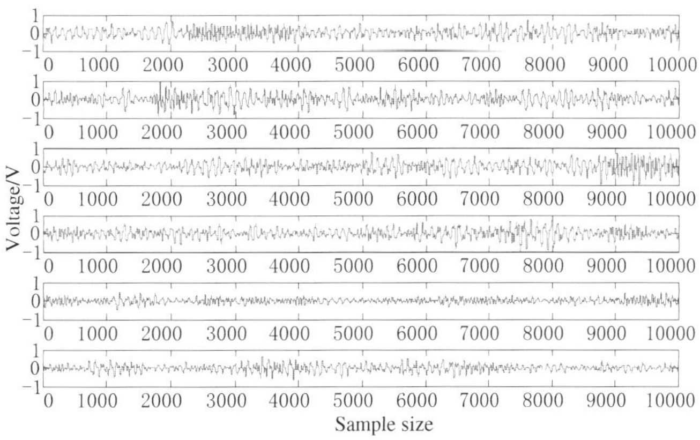

Figure 3. Signal timing diagram of different wear

During the experiment, only the last 5seconds data of each tool is collected. Figure3 is the time sequence diagram of the signal with different tool wear, the cutting condition is the cutting speed which is 520r/min, feed rate which is 0.176mm/r, cutting depth which is 0.4mm. The wear loss from top to bottom is 0.05mm, 0.13mm, 0.16mm, 0.20mm, 0.27mm, 0.35mm.

3. Signal preprocessing

3.1. Empirical mode decomposition principle

Empirical mode decomposition (EMD) is adaptive. The signal is decomposed by empirical mode decomposition and a number of intrinsic mode functions (IMF) are obtained. Each IMF components includes local features of different time scales of the original signal. The IMF components must satisfy two conditions: (1) the number of extreme points and the number of zero crossing is the same or the maximum difference of one. (2) the upper and lower envelope lines are about the local symmetry of the time axis. So that the signal can be decomposed into a finite number of IMF components and a residual function. The original signal ($S(t)$) is decomposed by EMD, which is expressed as:

$$S(t)=\sum_{i=1}^{n}c_{i}(t)+r_{n}(t) \tag{1}$$

where the average trend of the signal is represented by the residual function($r_n(t)$).And each IMF component is $c_1(t)$, $c_2(t) \ldots c_n(t)$, respectively including different frequency segment composition of the signal from high frequency to low frequency .

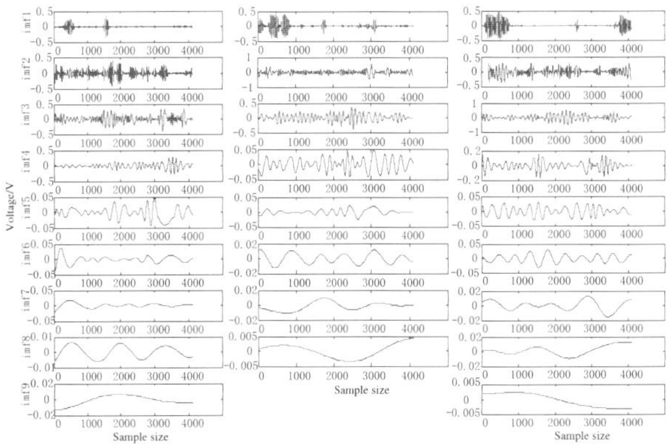

Figure 4. EMD decomposition results of different wear state signals

Figure 4 is the EMD decomposition results of the signal in different wear states, the wear loss from left to right is 0.05mm, 0.20mm, 0.35mm. As can be seen from the graph, the signal is adaptively decomposed into a number of IMF components, and the final IMF component is a residual function. The IMF component obtained by the decomposition of AE signals with different wear loss is different.

3.2. Principal component analysis

The IMF component obtained by the decomposition of AE signals with different wear amounts is different. The IMF components include mechanical noise, electromagnetic noise, spurious components, and so on, which make the Hilbert time-frequency spectrum aliasing and affect the main frequency of the signal. Xiren M employed the variance contribution rate to analyze the principal component of the signal and the magnitude of variance contribution rate of IMF component is used as a measure of the principal component [3]. The IMF component with large variance contribution rate contains the main information of the signal. The formula for calculating the variance contribution rate of IMF component is:

$$M_i = \frac{D_i}{\sum_{i=1}^{n} D_i} \tag{2}$$

$$D_i = \frac{1}{N}\sum_{k=1}^{N}\left|c_i\left(k\Delta t\right)\right|^2 - \left[\frac{1}{N}\sum_{k=1}^{N}c_i\left(k\Delta t\right)\right]^2 \tag{3}$$

where D_i is the variance of the i-th IMF components; Δt is the time interval of the sampling point.

Gang L used the principle of mutual relation to filter the signal, and the main components of the signal were obtained [4]. He measured the true or false of the IMF component by the size of the correlation of the IMF component and the original signal. The correlation coefficient among the false component and the noise component and the original signal is very small. The related formula is:

$$\rho_{XY} = \operatorname{cov}\left(X,Y\right)\Big/\left(\sqrt{D(X)}\sqrt{D(Y)}\right) \tag{4}$$

where $\operatorname{cov}(X,Y)$ is the covariance of the two signal, $D(X)$ and $D(Y)$ is the variance of the signal X and the signal Y.

In this paper, the variance contribution rate and the correlation coefficient method are used to analyze the signal [5]. The combination of the two methods is used to select the main components that are sensitive to tool wear and remove the unwanted components, so that the extracted signal features are more and more accurate. Figure 5 is Principal component analysis of IMF component.

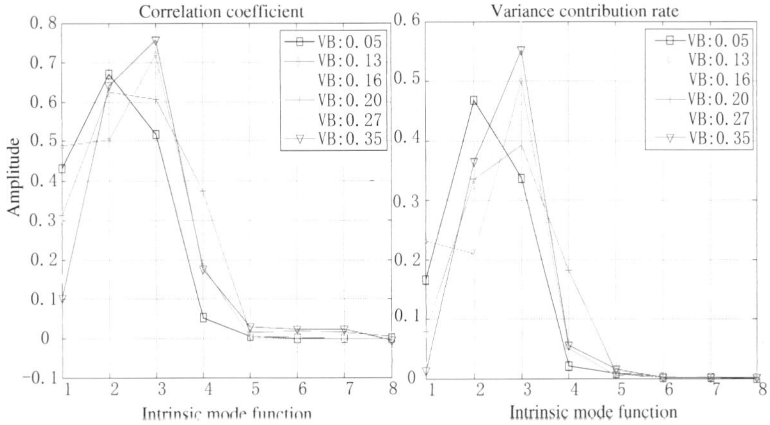

Figure 5. Principal component analysis of IMF component

From Figure 5, we can see that the correlation coefficient of the IMF component and the original signal and the variance contribution rate of each IMF component gradually increased to a certain extent. But after reaching a certain degree, it will decrease dramatically. Through the study, it was found that correlation coefficient is large between IMF1 to 4 and the original signal. After IMF5, the correlation coefficient is almost close to zero, and the variance contribution rate is less than 2%. By combining the correlation coefficient and variance contribution rate, we can know that the main components of the signal are concentrated in the first 4 order IMF components, and the residual IMF component can be eliminated.

4. Signal feature extraction based on Hilbert marginal spectrum

Hilbert marginal spectrum is obtained by Hilbert Huang transform. Hilbert Huang transform is a new nonlinear and non-stationary signal analysis method proposed by N.E.Huang et al. [6,7], which is widely used in the field of fault diagnosis [8]. Hilbert Huang transform (HHT) consists of two parts: Empirical Mode Decomposition and Hilbert transform. First the signal is through the empirical mode decomposition to get a series of intrinsic mode functions; then the intrinsic mode function is used to carry out Hilbert transform to get the Hilbert time spectrum which contains the time, frequency, amplitude (energy). Hilbert marginal spectrum is obtained by integrating the Hilbert time-frequency spectrum.

4.1. Hilbert time spectrum

Every IMF component obtained from the EMD decomposition can be carried out by Hilbert transform, the instantaneous amplitude, instantaneous phase and instantaneous frequency of each IMF component with the time variation can be obtained. The IMF component $c_i(t)$ of the signal $S(t)$ is transformed by the Hilbert transform, then its complex analytic signal is:

$$H\left[c_i(t)\right] = c_i(t) + j\tilde{c_i}(t) = a_i(t)e^{j\phi_i(t)} \quad (5)$$

where $a_i(t)$ is the amplitude function of the *i*-th IMF component, and $\phi_i(t)$ is the phase function of the *i*-th IMF component.

After EMD decomposition and Hilbert transform, the original signal $S(t)$ can be expressed as:

$$S(t) = \mathrm{Re}\sum_{i=1}^{n} a_i(t)e^{j\int \omega_i(t)dt} \quad (6)$$

$$H(\omega,t) = \mathrm{Re}\sum_{i=1}^{n} a_i(t)e^{j\int \omega_i(t)dt} \quad (7)$$

where the $\omega_i(t)$ is instantaneous frequency and the Hilbert time spectrum of the original signal is called the $H(\omega,t)$.

Figure 6 is the Hilbert time-frequency spectrum of the original signal and after principal component analysis. The results show that the Hilbert time-frequency spectrum of the principal component analysis can eliminate the influence of false component and the residual frequency of low frequency part, so that the main frequency curve of the signal is more clear and accurate.

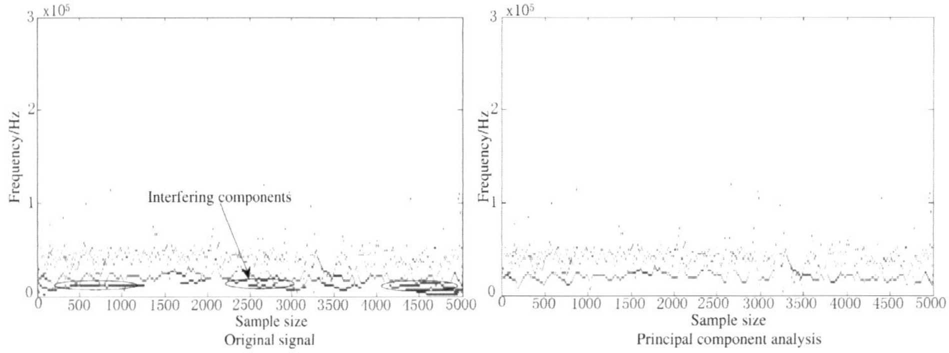

Figure 6. Hilbert time-frequency spectrum of the original signal and the principal component analysis

Figure 7 is the Hilbert time-frequency spectrum of AE signal under different tool wear condition. The Hilbert time-frequency spectrum of different tool wear state is more obvious difference, the change can be clear in time-frequency space. In figure7 (a), tool is in the initial stage of wear. From the time-frequency chart, we can see that the signal frequency is relatively single, the frequency is mainly concentrated in the following 50 kHz, the frequency is evenly distributed over time. In figure 7 (b), tool is in the middle stage of wear, tool in continuous cutting process is accompanied by the rebound of the high temperature and high pressure, material on tool flank extrusion effect, the inner cutter began to trend micro cracks, acoustic emission signal began to become more and more complex,

IOP Conf. Series: Materials Science and Engineering **230** (2017) 012049 doi:10.1088/1757-899X/230/1/012049

frequency began to appear in more than 50 kHz, the high frequency information mainly concentrate between 50 kHz to 200kHz, frequency curve appear mixed phenomenon. In figure 7 (c), tool is in the stage of severe wear, the internal crack of the cutting tool is extended to form a serious wear, and a large number of high frequency information appears in the signal. High frequency information can be clearly seen from the time-frequency spectrum between 50 kHz and 400 kHz. The Hilbert time-frequency spectrum has obvious change in different tool wear condition, therefore the Hilbert time-frequency spectrum can be used as an auxiliary feature to monitor and identify the tool wear condition.

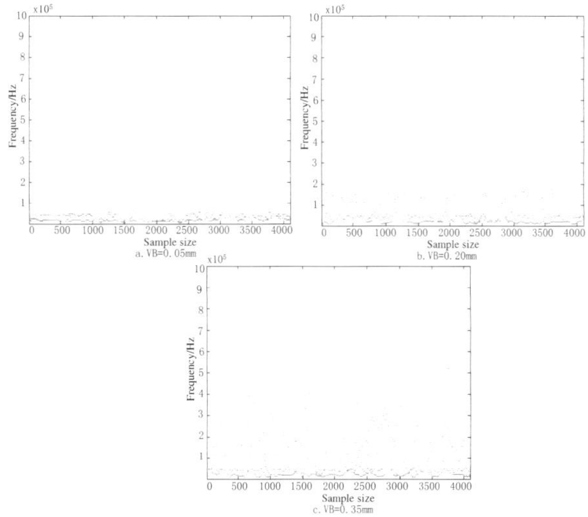

Figure 7. Hilbert time-frequency spectrum of tool under different wear condition

4.2. Hilbert marginal spectrum

Hilbert marginal spectrum is different from the Fourier spectrum, it can handle the non-stationary signal. On the basis of the spectrum of Hilbert, it will be the time for integration. we can get the corresponding Hilbert marginal spectrum($h(\omega)$):

$$h(\omega) = \int_0^T H(\omega,t)dt \tag{8}$$

where T is the signal length in the formula.

Hilbert marginal spectrum can more accurately reflect the signal actual frequency components .With the change of tool wear condition, the frequency components of the AE signal can be changed correspondingly, so the Hilbert marginal spectrum can be used as the basis for identifying the tool wear condition. From figure 8 (a), we can see that the main frequency of the signal concentrate 0 kHz to 100 kHz when the tool in the initial stage of wear. The figure 8 (b) and (c) are the Hilbert marginal spectrum of the signal when the tool is in the stage of medium wear and severe wear. The signal's main frequency is still concentrated in 0 kHz to 100 kHz, but with the tool wear signal, frequency components change gradually. Between 100 kHz to 200 kHz, we can see high frequency information compared to the initial wear stage.

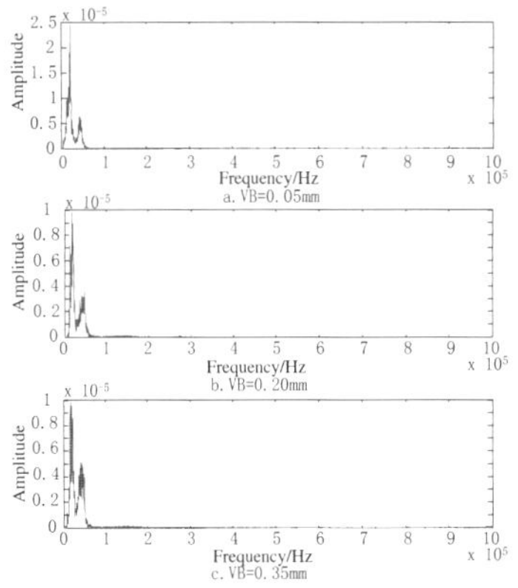

Figure 8. Hilbert marginal spectrum of tool under different wear condition

4.3. Amplitude index

Signal's Hilbert marginal spectrum contains a wealth of tool wear information. But through directing observation signal Hilbert marginal spectrum we are difficult to accurately distinguish out corresponding changes, By extracting amplitude characteristic index of the Hilbert marginal spectrum, the author used the amplitude index to distinguish different tool wear Hilbert marginal spectrum and then identify the different tool wear condition [9].Amplitude domain indexes which include: mean absolute, peak, rms, root mean square amplitude, variance, peak to peak, skewness, kurtosis, crest factor, waveform index, pulse index and margin index is a total of 12 indicators.

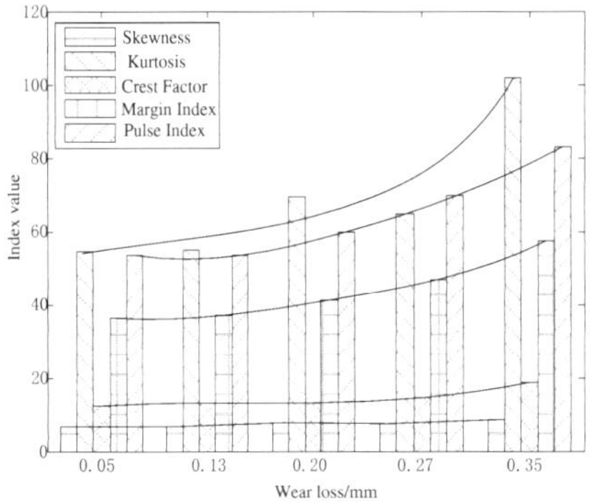

Figure 9. Amplitude domain index chart under different wear state states

The two-dimensional scatterplot help analysis and select feature index [10]. By visual observation, we can know the characteristic distribution of the tool wear state is dense and the degree of dispersion is low, and the degree of dispersion between the different states is large and the distance is far, such features can be more accurate classification of tool wear condition. In this paper, the author used statistical methods to combine the 12 characteristic indexes and observed the clustering of the samples by two-dimensional scattered point diagram and compared different feature combination classification results and selected from the features of good classification effect including: skewness, kurtosis, crest factor, margin index and pulse index. Study found that with the tool wear, this five amplitude domain characteristic index can better distinguish different wear conditions and the size of the eigenvalues has a clear trend with tool wear, as shown in figure 9. Therefore, they can be used as indicators of tool wear condition identification.

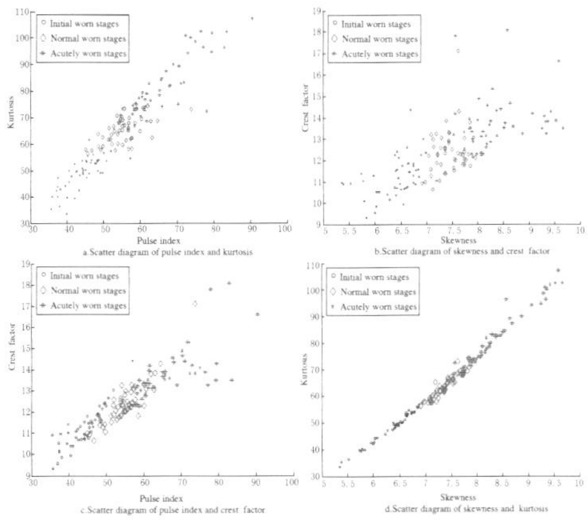

Figure 10. Two dimensional scatter diagram of the combination of different characteristic indexes

Five kinds of extracted amplitude domain index were combined with each other and found a total of 10 kinds of combinations, here only lists the four combinations, figure 10 is 2D scatter plot of 48 samples of different wear stages of different feature combinations. By observing the scattered point diagram we can find different combinations of features classification results, the feature combination classification effect of figure a, b and d is relatively good, and feature combination of figure c did not distinguish the mid wear and severe wear. The feature combined with each other to identify the tool wear state. In different wear state there are varying degrees of mixed phenomenon. Therefore, the five feature together described different tool wear states, this method is more completeness, its information is more comprehensive and can be better for tool wear state recognition and provide a basis for prediction.

5. Conclusions

By using the correlation coefficient and the variance contribution rate, the author analyzes the principal component of the IMF component of EMD decomposition. The results show that this method can effectively remove the interference in the Hilbert time-frequency spectrum. With the change of tool wear, the Hilbert time-frequency spectrum can clearly reflect the changes of the tool wear acoustic emission signal in time-frequency space. Hilbert marginal spectrum can more accurately reflect the signal frequency components. Its five amplitude domain index that include: skewness, kurtosis, crest factor, pulse index and margin index can distinguish Hilbert marginal spectrum of different wear state and different tool wear state. Therefore it can provide a certain basis for tool wear condition monitoring and identification.

6. References

[1] Yousheng L, Jianxin D, Hui Z, et al. J. Tribology, 28(5): 443-447. (2008)
[2] shan G, J. Journal of Northeast Dian li University, 33(3): 5. (2013)
[3] Xiren M, Xiaomei W , Dunyi S , et al. J. Transactions of China Electrometrical Society, 29(11): 154-161. (2014)
[4] Gang L, Chuang L, Xiangyang X, et al. J. Journal of Vibration and Shock, 34(12): 212-218. (2015)
[5] Feng X, Yunfei L. J. Journal of Basic Science and Engineering, 22(6): 1238-1247. (2014)
[6] Huang N E, Shen Z, Long S R, et al. Huang, N.E. et al. The Empirical Mode Decomposition and the Hilbert Spectrum for Nonlinear and Non-stationary Time Series Analysis. Proc. R. Soc. Lond. A 454, 903-995 J. Proceedings of the Royal Society A Mathematical Physical & Engineering Sciences, 454(1971):903-995. (1998)
[7] Huang N E, Shen Z, Long S R. A new view of nonlinear water waves J. Annual review of fluid mechanics, 31(1): 417-457. (1999)
[8] Moufa G, Lilan X, Xiren L, et al. J. Proceedings of the CSEE, 34(28): 4990-4997. (2014)

IOP Conf. Series: Materials Science and Engineering **230** (2017) 012049 doi:10.1088/1757-899X/230/1/012049

[9] Xiaohan C, Aiming W, Ruxiang H, et al. J. Journal of Vibration, Measurement & Diagnosis, 36(2): 351-358+406. (2016)
[10] Nantian H, Dianguo X, Xiaosheng L. J. Transactions of China Electro technical Society, 26(10): 23-30. (2011)